Student Solutions Manual

Intermediate Algebra

NINTH EDITION

Charles P. McKeague
Cuesta College

Prepared by

Ross Rueger
College of the Sequoias

BROOKS/COLE
CENGAGE Learning™

Australia • Brazil • Japan • Korea • Mexico • Singapore • Spain • United Kingdom • United States

For product information and technology assistance, contact us at **Cengage Learning Customer & Sales Support, 1-800-354-9706**

For permission to use material from this text or product, submit all requests online at **www.cengage.com/permissions** Further permissions questions can be emailed to **permissionrequest@cengage.com**

ISBN-13: 978-1-111-57159-7
ISBN-10: 1-111-57159-7

Brooks/Cole
20 Davis Drive
Belmont, CA 94002-3098
USA

Cengage Learning is a leading provider of customized learning solutions with office locations around the globe, including Singapore, the United Kingdom, Australia, Mexico, Brazil, and Japan. Locate your local office at: **www.cengage.com/global**

Cengage Learning products are represented in Canada by Nelson Education, Ltd.

To learn more about Brooks/Cole, visit **www.cengage.com/brookscole**

Purchase any of our products at your local college store or at our preferred online store **www.cengagebrain.com**

Printed in the United States of America
1 2 3 4 5 6 7 14 13 12 11 10

Contents

Preface

This *Student Solutions Manual* contains complete solutions to all odd-numbered exercises and all chapter test exercises of *Intermediate Algebra* by Charles P. McKeague. Every attempt has been made to format solutions for readability and accuracy, and I apologize for any errors that you may encounter. If you have any comments, suggestions, error corrections, or alternative solutions please feel free to drop me a note or send an email (address below).

Please use this manual with some degree of caution. Be sure that you have attempted a solution, and re-attempted it, before you look it up in this manual. Algebra can only be learned by *doing*, and not by observing! As you use this manual, do not just read the solution but work it along with the manual, using my solution to check your work. If you use this manual in that fashion then it should be helpful to you as you do homework and study for tests.

I would like to thank a number of people for their assistance in preparing this manual. Thanks go to Shaun Williams at Cengage Learning for her valuable assistance and support. Also thanks go to Patrick McKeague for keeping me updated to manuscript corrections and prompt response to my emails.

I wish to express my appreciation to Pat McKeague for asking me to be involved with this textbook. This book provides a complete course in intermediate algebra, and you will find the text very easy to read and understand. Good luck!

Ross Rueger
College of the Sequoias
rossrueger@gmail.com

Chapter 1
Basic Properties and Definitions

Getting Ready for Chapter 1

1. Calculating: $5.7 - 1.9 = 3.8$

2. Calculating: $10 - 9.5 = 0.5$

3. Calculating: $0.3(6) = 1.8$

4. Calculating: $39 \div 13 = 3$

5. Calculating: $28 \div 7 = 4$

6. Calculating: $125 - 81 = 44$

7. Calculating: $630 \div 63 = 10$

8. Calculating: $210 \div 10 = 21$

9. Calculating: $2 \cdot 3 \cdot 5 \cdot 7 = 210$

10. Calculating: $3 \cdot 7 \cdot 11 = 231$

11. Calculating: $2 \cdot 3 \cdot 3 \cdot 5 \cdot 7 = 630$

12. Calculating: $18,000 - 4,500 = 13,500$

13. Calculating: $1.6 + 2.0 = 3.6$

14. Calculating: $7(0.05) = 0.35$

15. Calculating: $7,546 \div 35 = 215.6$

16. Calculating: $1.2052 - 1 = 0.2052$

17. Calculating: $250(3.14) = 785$

18. Calculating: $109 - 36 + 14 = 73 + 14 = 87$

19. Calculating: $0.08(4,000) = 320$

20. Calculating: $3.9 \div 1.3 = 3$

1.1 Fundamental Definitions and Notation

1. Written in symbols, the expression is: $x + 5 = 2$

3. Written in symbols, the expression is: $6 - x = y$

5. Written in symbols, the expression is: $2t < y$

7. Written in symbols, the expression is: $x + y < x - y$

9. Written in symbols, the expression is: $3(x - 5) > y$

11. Multiplying: $6^2 = 6 \cdot 6 = 36$

13. Multiplying: $10^2 = 10 \cdot 10 = 100$

15. Multiplying: $2^3 = 2 \cdot 2 \cdot 2 = 8$

17. Multiplying: $2^4 = 2 \cdot 2 \cdot 2 \cdot 2 = 16$

19. Multiplying: $10^4 = 10 \cdot 10 \cdot 10 \cdot 10 = 10,000$

21. Multiplying: $11^2 = 11 \cdot 11 = 121$

23.
    a. Using the rule for order of operations: $3 \cdot 5 + 4 = 15 + 4 = 19$
    b. Using the rule for order of operations: $3(5 + 4) = 3 \cdot 9 = 27$
    c. Using the rule for order of operations: $3 \cdot 5 + 3 \cdot 4 = 15 + 12 = 27$

25.
    a. Using the rule for order of operations: $6 + 3 \cdot 4 - 2 = 6 + 12 - 2 = 16$
    b. Using the rule for order of operations: $6 + 3(4 - 2) = 6 + 3 \cdot 2 = 6 + 6 = 12$
    c. Using the rule for order of operations: $(6 + 3)(4 - 2) = 9 \cdot 2 = 18$

27.
    a. Using the rule for order of operations: $(7 - 4)(7 + 4) = 3 \cdot 11 = 33$
    b. Using the rule for order of operations: $7^2 - 4^2 = 49 - 16 = 33$

1

29.   **a.**   Using the rule for order of operations: $(5+7)^2 = 12^2 = 144$

      b.   Using the rule for order of operations: $5^2 + 7^2 = 25 + 49 = 74$

      c.   Using the rule for order of operations: $5^2 + 2 \cdot 5 \cdot 7 + 7^2 = 25 + 70 + 49 = 144$

31.   **a.**   Using the rule for order of operations: $2 + 3 \cdot 2^2 + 3^2 = 2 + 3 \cdot 4 + 9 = 2 + 12 + 9 = 23$

      b.   Using the rule for order of operations: $2 + 3\left(2^2 + 3^2\right) = 2 + 3(4 + 9) = 2 + 3 \cdot 13 = 2 + 39 = 41$

      c.   Using the rule for order of operations: $(2 + 3)\left(2^2 + 3^2\right) = (5)(4 + 9) = 5 \cdot 13 = 65$

33.   **a.**   Using the rule for order of operations: $40 - 10 \div 5 + 1 = 40 - 2 + 1 = 38 + 1 = 39$

      b.   Using the rule for order of operations: $(40 - 10) \div 5 + 1 = 30 \div 5 + 1 = 6 + 1 = 7$

      c.   Using the rule for order of operations: $(40 - 10) \div (5 + 1) = 30 \div 6 = 5$

35.   **a.**   Using the rule for order of operations: $40 + \left[10 - (4 - 2)\right] = 40 + (10 - 2) = 40 + 8 = 48$

      b.   Using the rule for order of operations: $40 - 10 - 4 - 2 = 30 - 4 - 2 = 26 - 2 = 24$

37.   **a.**   Using the rule for order of operations:

$$3 + 2\left(2 \cdot 3^2 + 1\right) = 3 + 2(2 \cdot 9 + 1) = 3 + 2(18 + 1) = 3 + 2 \cdot 19 = 3 + 38 = 41$$

      b.   Using the rule for order of operations: $(3 + 2)\left(2 \cdot 3^2 + 1\right) = 5(2 \cdot 9 + 1) = 5(18 + 1) = 5 \cdot 19 = 95$

39.   Using the rule for order of operations: $5 \cdot 10^3 + 4 \cdot 10^2 + 3 \cdot 10 + 1 = 5,431$

41.   Using the rule for order of operations:

$$3\left[2 + 4(5 + 2 \cdot 3)\right] = 3\left[2 + 4(5 + 6)\right] = 3(2 + 4 \cdot 11) = 3(2 + 44) = 3 \cdot 46 = 138$$

43.   Using the rule for order of operations:

$$6\left[3 + 2(5 \cdot 3 - 10)\right] = 6\left[3 + 2(15 - 10)\right] = 6(3 + 2 \cdot 5) = 6(3 + 10) = 6 \cdot 13 = 78$$

45.   Using the rule for order of operations:

$$5(7 \cdot 4 - 3 \cdot 4) + 8(5 \cdot 9 - 4 \cdot 9) = 5(28 - 12) + 8(45 - 36) = 5 \cdot 16 + 8 \cdot 9 = 80 + 72 = 152$$

47.   Using the rule for order of operations: $25 - 17 + 3 = 8 + 3 = 11$

49.   Using the rule for order of operations: $109 - 36 + 14 = 73 + 14 = 87$

51.   Using the rule for order of operations: $20 - 13 - 3 = 7 - 3 = 4$

53.   Using the rule for order of operations: $63 - 37 - 4 = 26 - 4 = 22$

55.   Using the rule for order of operations: $36 \div 9 \cdot 4 = 4 \cdot 4 = 16$

57.   Using the rule for order of operations: $75 \div 3 \cdot 25 = 25 \cdot 25 = 625$

59.   Using the rule for order of operations: $64 \div 16 \div 4 = 4 \div 4 = 1$

61.   Using the rule for order of operations: $75 \div 25 \div 5 = 3 \div 5 = 0.6$

63.   Calculating the value: $18,000 - 9,300 = 8,700$

65.   Calculating the value: $3.45 + 2.6 - 1.004 = 6.05 - 1.004 = 5.046$

67.   Calculating the value: $275 \div 55 = 5$

69.   Calculating the value: $4(2)(4)^2 = 4(2)(16) = 8 \cdot 16 = 128$

71.   Calculating the value: $250(5) - 25(5)^2 = 250(5) - 25(25) = 1,250 - 625 = 625$

73.   Calculating the value: $5 \cdot 2^3 - 3 \cdot 2^2 + 4 \cdot 2 - 5 = 5 \cdot 8 - 3 \cdot 4 + 4 \cdot 2 - 5 = 40 - 12 + 8 - 5 = 31$

75.   Calculating the value: $125 \cdot 2^3 = 125 \cdot 8 = 1,000$ **77.**   Calculating the value: $500(1.5) = 750$

79.   Calculating the value: $5(0.10) = 0.5$ **81.**   Calculating the value: $0.20(8) = 1.6$

83.   Calculating the value: $0.08(4,000) = 320$ **85.**   Using a calculator: $0.6931 + 1.0986 = 1.7917$

87.   Using a calculator: $3(0.6931) = 2.0793$ **89.**   Using a calculator: $250(3.14) = 785$

91.   Using a calculator: $4,628 \div 25 = 185.12$ **93.**   Using a calculator: $65,000 \div 5,280 \approx 12.3106$

95.   Using a calculator: $1 - 0.8413 = 0.1587$ **97.**   Using a calculator: $16(3.5)^2 = 16(12.25) = 196$

99. Using a calculator: $11.5(130) - 0.05(130)^2 = 11.5(130) - 0.05(16,900) = 1,495 - 845 = 650$

101. **a.** Evaluating when $x = 5$: $x + 2 = 5 + 2 = 7$ **b.** Evaluating when $x = 5$: $2x = 2(5) = 10$

  c. Evaluating when $x = 5$: $x^2 = 5^2 = 5 \cdot 5 = 25$ **d.** Evaluating when $x = 5$: $2^x = 2^5 = 2 \cdot 2 \cdot 2 \cdot 2 \cdot 2 = 32$

103. **a.** Evaluating when $x = 10$: $x^2 + 2x + 1 = (10)^2 + 2(10) + 1 = 100 + 20 + 1 = 121$

  b. Evaluating when $x = 10$: $(x + 1)^2 = (10 + 1)^2 = 11^2 = 121$

  c. Evaluating when $x = 10$: $x^2 + 1 = 10^2 + 1 = 100 + 1 = 101$

  d. Evaluating when $x = 10$: $(x - 1)^2 = (10 - 1)^2 = 9^2 = 81$

105. **a.** Evaluating when $a = 2, b = 5,$ and $c = 3$: $b^2 - 4ac = (5)^2 - 4(2)(3) = 25 - 24 = 1$

  b. Evaluating when $a = 10, b = 60,$ and $c = 30$: $b^2 - 4ac = (60)^2 - 4(10)(30) = 3,600 - 1,200 = 2,400$

  c. Evaluating when $a = 0.4, b = 1,$ and $c = 0.3$: $b^2 - 4ac = (1)^2 - 4(0.4)(0.3) = 1 - 0.48 = 0.52$

107. **a.** Evaluating when $x = -1$ and $y = 3$: $2x^2 - 3y + 2 = 2(-1)^2 - 3(3) + 2 = 2 - 9 + 2 = -5$

  b. Evaluating when $x = 3$ and $y = -1$: $2x^2 - 3y + 2 = 2(3)^2 - 3(-1) + 2 = 18 + 3 + 2 = 23$

  c. Evaluating when $x = 4$ and $y = 9$: $2x^2 - 3y + 2 = 2(4)^2 - 3(9) + 2 = 32 - 27 + 2 = 7$

109. **a.** Evaluating when $x = -4$: $x^2 + 3x + 4 = (-4)^2 + 3(-4) + 4 = 16 - 12 + 4 = 8$

  b. Evaluating when $x = 2$: $x^2 + 3x + 4 = (2)^2 + 3(2) + 4 = 4 + 6 + 4 = 14$

  c. Evaluating when $x = -5$: $x^2 + 3x + 4 = (-5)^2 + 3(-5) + 4 = 25 - 15 + 4 = 14$

111. **a.** Substituting $n = 1, x = 5, y = 8,$ and $z = 13$: $x^n + y^n = 5^1 + 8^1 = 5 + 8 = 13 = 13^1 = z^n$

  b. Substituting $n = 1, x = 2, y = 3,$ and $z = 5$: $x^n + y^n = 2^1 + 3^1 = 2 + 3 = 5 = 5^1 = z^n$

  c. Substituting $n = 2, x = 3, y = 4,$ and $z = 5$: $x^n + y^n = 3^2 + 4^2 = 9 + 16 = 25 = 5^2 = z^n$

  d. Substituting $n = 2, x = 7, y = 24,$ and $z = 25$: $x^n + y^n = 7^2 + 24^2 = 49 + 576 = 625 = 25^2 = z^n$

1.2 Sets and Venn Diagrams

1. The set is $A \cup B = \{0, 1, 2, 3, 4, 5, 6\}$.

3. The set is $A \cap B = \{2, 4\}$.

5. The set is $B \cap C = \{1, 3, 5\}$.

7. The set is $A \cup (B \cap C) = \{0, 1, 2, 3, 4, 5, 6\}$.

9. The set is $\{0, 2\}$.

11. The set is $\{0, 6\}$.

13. The set is $\{0, 1, 2, 3, 4, 5, 6, 7\}$.

15. The set is $\{1, 2, 4, 5\}$.

17. Drawing a Venn diagram:

19. Drawing a Venn diagram:

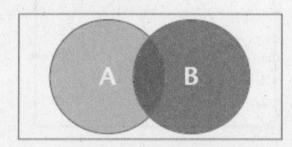

21. Drawing a Venn diagram:

23. Drawing a Venn diagram:

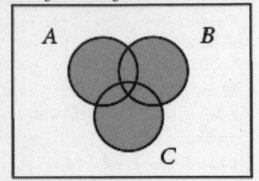

25. Drawing the Venn diagram:

27. Drawing the Venn diagram:

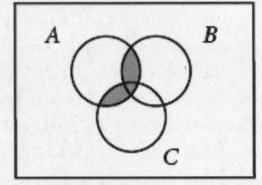

29. Drawing the Venn diagram:

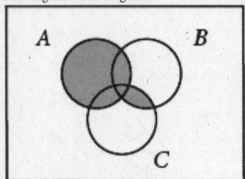

31. Drawing the Venn diagram:

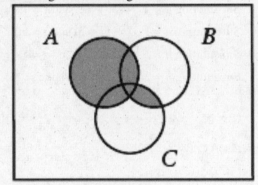

33. Drawing the Venn diagram:

35. Drawing the Venn diagram:

37.   Drawing the Venn diagram:

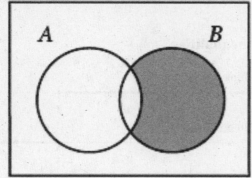

39.   Drawing the Venn diagram:

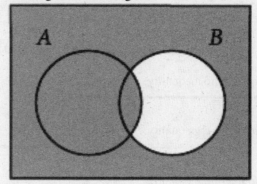

41.   Drawing the Venn diagram:

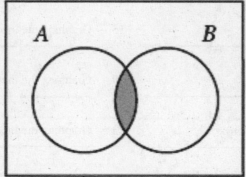

43.   Performing the operation: $175 \div 35 = 5$

45.   Performing the operation: $189 \div 9 = 21$

47.   Performing the operation: $156 \div 4 = 39$

49.   Performing the operation: $3 \cdot 5 \cdot 7 = 15 \cdot 7 = 105$

51.   Performing the operation: $2 \cdot 3^2 \cdot 7 = 2 \cdot 9 \cdot 7 = 18 \cdot 7 = 126$

1.3 The Real Numbers

1.   The counting numbers are: $1, 2$

3.   The rational numbers are: $-6, -5.2, 0, 1, 2, 2.3, \dfrac{9}{2}$

5.   The irrational numbers are: $-\sqrt{7}, -\pi, \sqrt{17}$

7.   The nonnegative integers are: $0, 1, 2$

9.   Factoring: $60 = 4 \cdot 15 = 2^2 \cdot 3 \cdot 5$

11.   Factoring: $266 = 14 \cdot 19 = 2 \cdot 7 \cdot 19$

13.   Factoring: $111 = 3 \cdot 37$

15.   Factoring: $369 = 9 \cdot 41 = 3^2 \cdot 41$

17.   Reducing the fraction: $\dfrac{165}{385} = \dfrac{3 \cdot 55}{7 \cdot 55} = \dfrac{3}{7}$

19.   Reducing the fraction: $\dfrac{385}{735} = \dfrac{35 \cdot 11}{35 \cdot 21} = \dfrac{11}{21}$

21.   Reducing the fraction: $\dfrac{111}{185} = \dfrac{37 \cdot 3}{37 \cdot 5} = \dfrac{3}{5}$

23.   Reducing the fraction: $\dfrac{525}{630} = \dfrac{105 \cdot 5}{105 \cdot 6} = \dfrac{5}{6}$

25.   Reducing the fraction: $\dfrac{75}{135} = \dfrac{15 \cdot 5}{15 \cdot 9} = \dfrac{5}{9}$

27.   Reducing the fraction: $\dfrac{6}{8} = \dfrac{2 \cdot 3}{2 \cdot 4} = \dfrac{3}{4}$

29.   Reducing the fraction: $\dfrac{200}{5} = \dfrac{40 \cdot 5}{1 \cdot 5} = 40$

31.   Reducing the fraction: $\dfrac{10}{22} = \dfrac{2 \cdot 5}{2 \cdot 11} = \dfrac{5}{11}$

33.   **a.** $10 = 3 + 7$ or $10 = 5 + 5$

     b. $16 = 11 + 5$ or $16 = 13 + 3$

     c. $24 = 19 + 5$ or $24 = 17 + 7$ or $24 = 13 + 11$

     d. $36 = 19 + 17$ or $36 = 23 + 13$ or $36 = 29 + 7$ or $36 = 31 + 5$

35.   The equations are $x = 200$ and $x = 100$.

37.   Calculating the value: $5! = 5 \cdot 4 \cdot 3 \cdot 2 \cdot 1 = 120$

39.   Calculating the value: $6! = 6 \cdot 5 \cdot 4 \cdot 3 \cdot 2 \cdot 1 = 6 \cdot 5!$. So the statement is true.

41. **a.** An accident will be approximately 80 times more likely.

  b. Computing the percent: $80(0.02\%) = 1.6\%$

1.4 Simple and Compound Inequalities

1. Graphing the inequality:

3. Graphing the inequality:

5. Graphing the inequality:

7. Graphing the inequality:

9. Graphing the inequality:

11. Graphing the inequality:

13. Graphing the inequality:

15. Graphing the inequality:

17. Graphing the inequality:

19. Graphing the inequality:

21. Graphing the inequality (no solution):

23. Graphing the inequality (no solution):

25. Graphing the inequality:

27. Graphing the inequality:

29. Graphing the inequality:

31. Graphing the inequality:

33. Graphing the inequality:

35. Graphing the inequality:

37. The inequality is: $x \geq 5$

39. The inequality is: $x \leq -3$

41. The inequality is: $x \leq 4$

43. The inequality is: $-4 < x < 4$

45. The inequality is: $-4 \leq x \leq 4$

47. The numbers are either $2 - 5 = -3$ or $2 + 5 = 7$.

49. The numbers are between 3 and 13, which is the inequality $3 < x < 13$.

51. The numbers are either less than 3 or more than 13, which is the inequality $x < 3$ or $x > 13$.

53. The inequality is $50 \leq F \leq 270$.

1.5 Arithmetic with Real Numbers

1-11. Completing the table:

	Number	Opposite	Reciprocal
1.	4	–4	$\frac{1}{4}$
3.	$-\frac{1}{2}$	$\frac{1}{2}$	–2
5.	5	–5	$\frac{1}{5}$
7.	$\frac{3}{8}$	$-\frac{3}{8}$	$\frac{8}{3}$
9.	$-\frac{1}{6}$	$\frac{1}{6}$	–6
11.	3	–3	$\frac{1}{3}$

13. –1 and 1 are their own reciprocals.

15. 0 is its own opposite.

17. Writing without absolute values: $|-2| = 2$

19. Writing without absolute values: $\left|-\frac{3}{4}\right| = \frac{3}{4}$

21. Writing without absolute values: $|\pi| = \pi$

23. Writing without absolute values: $-|4| = -4$

25. Writing without absolute values: $-|-2| = -2$

27. Writing without absolute values: $-\left|-\frac{3}{4}\right| = -\frac{3}{4}$

29. Finding the sum: $6 + (-2) = 6 - 2 = 4$

31. Finding the sum: $-6 + 2 = -4$

33. Finding the sum: $19 + (-7) = 19 - 7 = 12$

35. Finding the sum: $-17 + 4 = -13$

37. Finding the difference: $-7 - 3 = -7 + (-3) = -10$

39. Finding the difference: $-7 - (-3) = -7 + 3 = -4$

41. Finding the difference: $\frac{3}{4} - \left(-\frac{5}{6}\right) = \frac{3}{4} + \frac{5}{6} = \frac{3}{4} \cdot \frac{3}{3} + \frac{5}{6} \cdot \frac{2}{2} = \frac{9}{12} + \frac{10}{12} = \frac{19}{12}$

43. Finding the difference: $\frac{11}{42} - \frac{17}{30} = \frac{11}{42} \cdot \frac{5}{5} - \frac{17}{30} \cdot \frac{7}{7} = \frac{55}{210} - \frac{119}{210} = -\frac{64}{210} = -\frac{32}{105}$

45. Subtracting: $-3 - 5 = -8$

47. Subtracting: $-4 - 8 = -12$

49. Subtracting: $-3x - 4x = -7x$

51. The number is 13, since $5 - 13 = -8$.

53. Computing the value: $-7 + (2 - 9) = -7 + (-7) = -14$

55. Simplifying the value: $(8a + a) - 3a = 9a - 3a = 6a$

57. Finding the product: $3(-5) = -15$

59. Finding the product: $-3(-5) = 15$

61. Finding the product: $2(-3)(4) = -6(4) = -24$

63. Finding the product: $-2(5x) = -10x$

65. Finding the product: $-\frac{1}{3}(-3x) = \frac{3}{3}x = x$

67. Finding the product: $-\frac{2}{3}\left(-\frac{3}{2}y\right) = \frac{6}{6}y = y$

69. Multiplying the fractions: $\frac{3}{5} \cdot \frac{7}{8} = \frac{21}{40}$

71. Multiplying the fractions: $\frac{1}{3} \cdot 6 = \frac{6}{3} = 2$

73. Multiplying the fractions: $\left(\frac{2}{3}\right)^3 = \frac{2}{3} \cdot \frac{2}{3} \cdot \frac{2}{3} = \frac{8}{27}$

75. Multiplying the fractions: $\left(\frac{1}{10}\right)^4 = \frac{1}{10} \cdot \frac{1}{10} \cdot \frac{1}{10} \cdot \frac{1}{10} = \frac{1}{10,000}$

77. Multiplying the fractions: $\frac{3}{5} \cdot \frac{4}{7} \cdot \frac{6}{11} = \frac{72}{385}$

79. Multiplying the fractions: $\frac{4}{3} \cdot \frac{3}{4} = \frac{12}{12} = 1$

81. Dividing: $\frac{8}{-4} = 8 \cdot \left(-\frac{1}{4}\right) = -2$

83. $\frac{4}{0}$ is undefined

85. Dividing: $\frac{0}{-3} = 0 \cdot \left(-\frac{1}{3}\right) = 0$

87. Dividing: $-\frac{3}{4} \div \frac{9}{8} = -\frac{3}{4} \cdot \frac{8}{9} = -\frac{2}{3}$

89. Dividing: $-8 \div \left(-\dfrac{1}{4}\right) = -8 \cdot \left(-\dfrac{4}{1}\right) = 32$ **91.** Dividing: $-40 \div \left(-\dfrac{5}{8}\right) = -40 \cdot \left(-\dfrac{8}{5}\right) = 64$

93. Dividing: $\dfrac{4}{9} \div (-8) = \dfrac{4}{9} \cdot \left(-\dfrac{1}{8}\right) = -\dfrac{1}{18}$

95. Adding the fractions: $\dfrac{2}{5} + \dfrac{1}{15} = \dfrac{2}{5} \cdot \dfrac{3}{3} + \dfrac{1}{15} = \dfrac{6}{15} + \dfrac{1}{15} = \dfrac{7}{15}$

97. Adding the fractions: $\dfrac{17}{30} + \dfrac{11}{42} = \dfrac{17}{30} \cdot \dfrac{7}{7} + \dfrac{11}{42} \cdot \dfrac{5}{5} = \dfrac{119}{210} + \dfrac{55}{210} = \dfrac{174}{210} = \dfrac{29}{35}$

99. Adding the fractions: $\dfrac{9}{48} + \dfrac{3}{54} = \dfrac{9}{48} \cdot \dfrac{9}{9} + \dfrac{3}{54} \cdot \dfrac{8}{8} = \dfrac{81}{432} + \dfrac{24}{432} = \dfrac{105}{432} = \dfrac{35}{144}$

101. Adding the fractions: $\dfrac{25}{84} + \dfrac{41}{90} = \dfrac{25}{84} \cdot \dfrac{15}{15} + \dfrac{41}{90} \cdot \dfrac{14}{14} = \dfrac{375}{1260} + \dfrac{574}{1260} = \dfrac{949}{1260}$

103. Adding the fractions: $\dfrac{5}{72} + \dfrac{11}{84} = \dfrac{5}{72} \cdot \dfrac{7}{7} + \dfrac{11}{84} \cdot \dfrac{6}{6} = \dfrac{35}{504} + \dfrac{66}{504} = \dfrac{101}{504}$

105. Adding the fractions: $\dfrac{4}{39} + \dfrac{7}{65} = \dfrac{4}{39} \cdot \dfrac{5}{5} + \dfrac{7}{65} \cdot \dfrac{3}{3} = \dfrac{20}{195} + \dfrac{21}{195} = \dfrac{41}{195}$

107. Adding the fractions: $\dfrac{7}{54} + \dfrac{7}{60} = \dfrac{7}{54} \cdot \dfrac{10}{10} + \dfrac{7}{60} \cdot \dfrac{9}{9} = \dfrac{70}{540} + \dfrac{63}{540} = \dfrac{133}{540}$

109. Simplifying: $\dfrac{3}{14} + \dfrac{7}{30} = \dfrac{3}{14} \cdot \dfrac{15}{15} + \dfrac{7}{30} \cdot \dfrac{7}{7} = \dfrac{45}{210} + \dfrac{49}{210} = \dfrac{94}{210} = \dfrac{47}{105}$

111. Simplifying: $32\left(\dfrac{3}{4}\right) - 16\left(\dfrac{3}{4}\right)^2 = 32\left(\dfrac{3}{4}\right) - 16\left(\dfrac{9}{16}\right) = 24 - 9 = 15$

113. Simplifying: $-2(-14) + 3(-4) - 1(-10) = 28 - 12 + 10 = 26$

115. Simplifying: $1(0)(1) + 3(1)(4) + (-2)(2)(-1) = 0 + 12 + 4 = 16$

117. Simplifying: $-3(-1-1) + 4(-2+2) - 5\left[2 - (-2)\right] = -3(-2) + 4(0) - 5(4) = 6 + 0 - 20 = -14$

119. Simplifying: $4(-1)^2 + 3(-1) - 2 = 4(1) + 3(-1) - 2 = 4 - 3 - 2 = -1$

121. Simplifying: $5 \cdot 2^3 - 3 \cdot 2^2 + 4 \cdot 2 - 5 = 5 \cdot 8 - 3 \cdot 4 + 4 \cdot 2 - 5 = 40 - 12 + 8 - 5 = 31$

123. Simplifying: $\dfrac{0+6}{0-3} = \dfrac{6}{-3} = -2$ **125.** Simplifying: $\dfrac{6+6}{6-3} = \dfrac{12}{3} = 4$

127. Simplifying: $\dfrac{4-4}{4-2} = \dfrac{0}{2} = 0$ **129.** Simplifying: $\dfrac{3+6}{3-3} = \dfrac{9}{0}$, which is undefined

131. Simplifying: $\dfrac{-1-3}{3-(-3)} = \dfrac{-4}{6} = -\dfrac{2}{3}$

133. Completing the table:

a	b	SUM $a+b$	DIFFERENCE $a-b$	PRODUCT ab	QUOTIENT a/b
8	2	10	6	16	4
-8	2	-6	-10	-16	-4
8	-2	6	10	-16	-4
-8	-2	-10	-6	16	4

135. Computing the expression: $12 + \dfrac{1}{4} \cdot 12 = 15$

1.6 Properties of Real Numbers

1. Using the associative property: $4+(2+x)=(4+2)+x=6+x$

3. Using the associative property: $(a+3)+5=a+(3+5)=a+8$

5. Using the associative property: $5(3y)=(5\cdot 3)y=15y$

7. Using the associative property: $\frac{1}{3}(3x)=\left(\frac{1}{3}\cdot 3\right)x=x$

9. Using the associative property: $4\left(\frac{1}{4}a\right)=\left(4\cdot\frac{1}{4}\right)a=a$

11. Using the associative property: $\frac{2}{3}\left(\frac{3}{2}x\right)=\left(\frac{2}{3}\cdot\frac{3}{2}\right)x=x$

13. Applying the distributive property: $3(x+6)=3\cdot x+3\cdot 6=3x+18$

15. Applying the distributive property: $2(6x+4)=2\cdot 6x+2\cdot 4=12x+8$

17. Applying the distributive property: $5(3a+2b)=5\cdot 3a+5\cdot 2b=15a+10b$

19. Applying the distributive property: $\frac{1}{3}(4x+6)=\frac{1}{3}\cdot 4x+\frac{1}{3}\cdot 6=\frac{4}{3}x+2$

21. Applying the distributive property: $\frac{1}{5}(10+5y)=\frac{1}{5}\cdot 10+\frac{1}{5}\cdot 5y=2+y$

23. Applying the distributive property: $(5t+1)8=5t\cdot 8+1\cdot 8=40t+8$

25. Applying the distributive property: $3(3x+y-2z)=3\cdot 3x+3\cdot y-3\cdot 2z=9x+3y-6z$

27. Applying the distributive property: $10(0.3x+0.7y)=10\cdot 0.3x+10\cdot 0.7y=3x+7y$

29. Applying the distributive property: $100(0.06x+0.07y)=100\cdot 0.06x+100\cdot 0.07y=6x+7y$

31. Applying the distributive property: $3\left(x+\frac{1}{3}\right)=3\cdot x+3\cdot\frac{1}{3}=3x+1$

33. Applying the distributive property: $2\left(x-\frac{1}{2}\right)=2\cdot x-2\cdot\frac{1}{2}=2x-1$

35. Applying the distributive property: $x\left(1+\frac{2}{x}\right)=x\cdot 1+x\cdot\frac{2}{x}=x+2$

37. Applying the distributive property: $a\left(1-\frac{3}{a}\right)=a\cdot 1-a\cdot\frac{3}{a}=a-3$

39. Applying the distributive property: $8\left(\frac{1}{8}x+3\right)=8\cdot\frac{1}{8}x+8\cdot 3=x+24$

41. Applying the distributive property: $6\left(\frac{1}{2}x-\frac{1}{3}y\right)=6\cdot\frac{1}{2}x-6\cdot\frac{1}{3}y=3x-2y$

43. Applying the distributive property: $12\left(\frac{1}{4}x+\frac{2}{3}y\right)=12\cdot\frac{1}{4}x+12\cdot\frac{2}{3}y=3x+8y$

45. Applying the distributive property: $20\left(\frac{2}{5}x+\frac{1}{4}y\right)=20\cdot\frac{2}{5}x+20\cdot\frac{1}{4}y=8x+5y$

47. Applying the distributive property: $3(5x+2)+4=15x+6+4=15x+10$

49. Applying the distributive property: $4(2y+6)+8=8y+24+8=8y+32$

51. Applying the distributive property: $5(1+3t)+4=5+15t+4=15t+9$

53. Applying the distributive property: $3+(2+7x)4=3+8+28x=28x+11$

55. Simplifying the expression: $5a + 7 + 8a + a = (5a + 8a + a) + 7 = 14a + 7$

57. Simplifying the expression: $3y + y + 5 + 2y + 1 = (3y + y + 2y) + (5 + 1) = 6y + 6$

59. Simplifying the expression: $2(5x + 1) + 2x = 10x + 2 + 2x = 12x + 2$

61. Simplifying the expression: $7 + 2(4y + 2) = 7 + 8y + 4 = 8y + 11$

63. Simplifying the expression: $3 + 4(5a + 3) + 4a = 3 + 20a + 12 + 4a = 24a + 15$

65. Simplifying the expression: $5x + 2(3x + 8) + 4 = 5x + 6x + 16 + 4 = 11x + 20$

67. Commutative property of addition $\qquad$ **69.** Commutative property of multiplication

71. Additive inverse property $\qquad$ **73.** Commutative property of addition

75. Associative and commutative properties of multiplication

77. Commutative and associative properties of addition $\qquad$ **79.** Distributive property

81. Simplifying: $3(-4) - 2 = -12 - 2 = -14$

83. Simplifying: $4(-3) - 6(-5) = -12 + 30 = 18$

85. Simplifying: $2 - 5(-4) - 6 = 2 + 20 - 6 = 22 - 6 = 16$

87. Simplifying: $4 - 3(7 - 1) - 5 = 4 - 3(6) - 5 = 4 - 18 - 5 = -19$

89. Simplifying: $2(-3)^2 - 4(-2) = 2(9) - 4(-2) = 18 + 8 = 26$

91. Simplifying: $(2 - 8)^2 - (3 - 7)^2 = (-6)^2 - (-4)^2 = 36 - 16 = 20$

93. Simplifying: $7(3 - 5)^3 - 2(4 - 7)^3 = 7(-2)^3 - 2(-3)^3 = 7(-8) - 2(-27) = -56 + 54 = -2$

95. Simplifying: $-3(2 - 9) - 4(6 - 1) = -3(-7) - 4(5) = 21 - 20 = 1$

97. Simplifying: $2 - 4[3 - 5(-1)] = 2 - 4(3 + 5) = 2 - 4(8) = 2 - 32 = -30$

99. Simplifying: $(8 - 7)[4 - 7(-2)] = (1)(4 + 14) = (1)(18) = 18$

101. Simplifying: $-3 + 4[6 - 8(-3 - 5)] = -3 + 4[6 - 8(-8)] = -3 + 4(6 + 64) = -3 + 4(70) = -3 + 280 = 277$

103. Simplifying: $5 - 6[-3(2 - 9) - 4(8 - 6)] = 5 - 6[-3(-7) - 4(2)] = 5 - 6(21 - 8) = 5 - 6(13) = 5 - 78 = -73$

105. Simplifying: $1(-2) - 2(-16) + 1(9) = -2 + 32 + 9 = 39$

107. Simplifying: $1(1) - 3(-2) + (-2)(-2) = 1 + 6 + 4 = 11$

109. Applying the distributive property: $-1(a - b) = (-1) \cdot a - (-1) \cdot b = -a + b = b - a$

111. Applying the distributive property: $-1(6 - y) = (-1) \cdot 6 - (-1) \cdot y = -6 + y = y - 6$

113. Applying the distributive property: $-1(x - 2z) = (-1) \cdot x - (-1) \cdot 2z = -x + 2z$

115. Applying the distributive property: $6\left(\dfrac{x}{3} + 1\right) = 6 \cdot \dfrac{x}{3} + 6 \cdot 1 = 2x + 6$

117. Applying the distributive property: $15\left(\dfrac{a}{3} + 2\right) = 15 \cdot \dfrac{a}{3} + 15 \cdot 2 = 5a + 30$

119. Applying the distributive property: $10\left(\dfrac{x}{2} - 9\right) = 10 \cdot \dfrac{x}{2} - 10 \cdot 9 = 5x - 90$

121. Applying the distributive property: $12\left(\dfrac{y}{3} - \dfrac{y}{6} + \dfrac{y}{2}\right) = 12 \cdot \dfrac{y}{3} - 12 \cdot \dfrac{y}{6} + 12 \cdot \dfrac{y}{2} = 4y - 2y + 6y = 8y$

123. Simplifying: $4(7x + 3) - x = 28x + 12 - x = 27x + 12$

125. Simplifying: $3 - 5(5 - m) = 3 - 25 + 5m = 5m - 22$

127. Simplifying: $8 - 5(2x - 3) + 4x = 8 - 10x + 15 + 4x = -6x + 23$

129. Simplifying: $4(6y + 3) - (6y - 6) = 24y + 12 - 6y + 6 = 18y + 18$

131. Simplifying: $7(1-2x)-(4-10x) = 7-14x-4+10x = -4x+3$

133. Simplifying: $7-2(3x+5)-(2x-3) = 7-6x-10-2x+3 = -8x$

135. Simplifying: $\dfrac{6(-4)-5(-2)}{7-6} = \dfrac{-24+10}{1} = -14$

137. Simplifying: $-9-5\left[\dfrac{11(-1)-9}{4(-3)+2(5)}\right] = -9-5\left(\dfrac{-11-9}{-12+10}\right) = -9-5\left(\dfrac{-20}{-2}\right) = -9-5(10) = -9-50 = -59$

139. Simplifying: $8-(-7)\left[\dfrac{6-1(6-10)}{4-3(5-7)}\right] = 8-(-7)\left[\dfrac{6-1(-4)}{4-3(-2)}\right] = 8-(-7)\left(\dfrac{6+4}{4+6}\right) = 8-(-7)\left(\dfrac{10}{10}\right) = 8+7 = 15$

141. Completing the table:

x	$(x+1)^2$	x^2+1	x^2+2x+1
-2	1	5	1
-1	0	2	0
0	1	1	1
1	4	2	4
2	9	5	9

143. **a.** Evaluating when $a=3$, $b=-2$, and $c=4$: $b^2-4ac = (-2)^2-4(3)(4) = 4-48 = -44$

b. Evaluating when $a=1$, $b=-3$, and $c=-28$: $b^2-4ac = (-3)^2-4(1)(-28) = 9+112 = 121$

c. Evaluating when $a=1$, $b=-6$, and $c=9$: $b^2-4ac = (-6)^2-4(1)(9) = 36-36 = 0$

d. Evaluating when $a=0.1$, $b=-27$, and $c=1700$: $b^2-4ac = (-27)^2-4(0.1)(1700) = 729-680 = 49$

145. Using a calculator: $\dfrac{1.0792}{0.6990} \approx 1.5439$

147. Using a calculator: $\dfrac{1}{2}(-0.7948) = -0.3974$

149. Using a calculator: $\dfrac{1}{2}\left(\dfrac{1.3}{1.1}-1\right) \approx 0.0909$

151. Using a calculator: $\dfrac{(2.4)(1.8)}{1.2} = 3.6$

153. Using a calculator: $\dfrac{0.0006(400)}{(0.25)^2} = 3.84$

155. Using a calculator: $-500+27(170)-0.1(170)^2 = -500+4,590-2,890 = 1,200$

157. Using a calculator: $-0.04(130)^2+8.5(130)-210 = -676+1,105-210 = 219$

159. The outside rectangle area is $9(x+3) = 9x+27$, while the sum of the two areas is $9x+9(3) = 9x+27$. Note that the two areas are the same.

161. The outside rectangle area is $a(b+3) = ab+3a$, while the sum of the two areas is $ab+3a$. Note that the two areas are the same.

163. Subtracting: $29,028-(-35,840) = 64,868$ feet

1.7 Recognizing Patterns

1. The pattern is to add 1, so the next term is 5.

3. The pattern is to add 2, so the next term is 10.

5. These numbers are squares, so the next term is $5^2 = 25$.

7. The pattern is to add 7, so the next term is 29.

9. The pattern is to add 7, then add 6, then add 5, so the next term is $19 + 4 = 23$.

11. One possibility is: Δ

13. One possibility is: $\odot$

15. The pattern is to add 4, so the next two numbers are 17 and 21.

17. The pattern is to add -1, so the next two numbers are -2 and -3.

19. The pattern is to add -3, so the next two numbers are -4 and -7.

21. The pattern is to add $-\dfrac{1}{4}$, so the next two numbers are $-\dfrac{1}{2}$ and $-\dfrac{3}{4}$.

23. The pattern is to add $\dfrac{1}{2}$, so the next two numbers are $\dfrac{5}{2}$ and 3.

25. The pattern is to multiply by 3, so the next number is 27.

27. The pattern is to multiply by -3, so the next number is -270.

29. The pattern is to multiply by $\dfrac{1}{2}$, so the next number is $\dfrac{1}{8}$.

31. The pattern is to multiply by $\dfrac{1}{2}$, so the next number is $\dfrac{5}{2}$.

33. The pattern is to multiply by -5, so the next number is -625.

35. The pattern is to multiply by $-\dfrac{1}{5}$, so the next number is $-\dfrac{1}{125}$.

37. **a.** The pattern is to add 4, so the next number is 12.   **b.** The pattern is to multiply by 2, so the next number is 16.

39. The sequence is 1,1,2,3,5,8,13,21,34,55,89,144,233,..., so the 12th term is 144.

41. Three Fibonacci numbers that are prime numbers are 2, 3, and 5 (among others).

43. The even numbers are 2, 8, and 34.

45. Completing the table:

Two Numbers a and b	Their Product ab	Their Sum $a+b$
1, -24	-24	-23
$-1, 24$	-24	23
2, -12	-24	-10
$-2, 12$	-24	10
3, -8	-24	-5
$-3, 8$	-24	5
4, -6	-24	-2
$-4, 6$	-24	2

47. The sequence of air temperatures is: 41°, 37.5°, 34°, 30.5°, 27°, 23.5°. This is an arithmetic sequence.

49. The sequence of air temperatures is: 41°, 45.5°, 50°, 54.5°, 59°, 63.5°. This is an arithmetic sequence.

51. The patient on antidepressant 1 will have less medication in the body, approximately half as much. The patient on antidepressant 2 will still have more of the medication in the body, since a full half-life has not passed by the second day.

Hours Since Discontinuing	Concentration (ng/ml)
0	60
4	30
8	15
12	7.5
16	3.75

53. Completing the table:

Sketching the graph:

Elevation (ft)	Boiling Point (°F)
−2,000	215.6
−1,000	213.8
0	212
1,000	210.2
2,000	208.4
3,000	206.6

55. Completing the table:

57. **a.** The steep straight line segments show when the patient takes his medication.
b. The graph is falling off because the patient stops taking his medication.
c. The maximum medication is approximately 50 ng/ml.
d. Since the patient takes his medication every 4 hours, the values are $A = 4$ hours, $B = 8$ hours, and $C = 12$ hours.

Temperature (Fahrenheit)	Shelf Life (days)
32°	24
40°	10
50°	2
60°	1
70°	½

59. Completing the table:

Constructing a line graph:

Chapter 1 Review

1. Written in symbols, the expression is: $x + 2$

3. Written in symbols, the expression is: $\dfrac{x}{2}$

5. Written in symbols, the expression is: $2(x + y)$

7. Multiplying: $3^3 = 3 \cdot 3 \cdot 3 = 27$

9. Multiplying: $8^2 = 8 \cdot 8 = 64$

11. Multiplying: $2^5 = 2 \cdot 2 \cdot 2 \cdot 2 \cdot 2 = 32$

13. Using the rule for order of operations: $2 + 3 \cdot 5 = 2 + 15 = 17$

15. Using the rule for order of operations: $20 \div 2 + 3 = 10 + 3 = 13$

17. Using the rule for order of operations: $3 + 2(5 - 2) = 3 + 2(3) = 3 + 6 = 9$

19. Using the rule for order of operations: $3 \cdot 4^2 - 2 \cdot 3^2 = 3 \cdot 16 - 2 \cdot 9 = 48 - 18 = 30$

21. The set is $A \cup B = \{1, 2, 3, 4, 5, 6\}$.

23. The set is $\{5\}$.

25. Sketching a number line:

27. The integers are: $-7, 0, 5$

29. The irrational numbers are: $-\sqrt{3}, \pi$

31. Reducing the fraction: $\dfrac{4{,}356}{5{,}148} = \dfrac{396 \cdot 11}{396 \cdot 13} = \dfrac{11}{13}$

33. Multiplying the fractions: $\left(\dfrac{3}{4}\right)^3 = \dfrac{3}{4} \cdot \dfrac{3}{4} \cdot \dfrac{3}{4} = \dfrac{27}{64}$

35. Graphing the inequality:

37. Graphing the inequality:

39. The inequality is: $x \geq 4$

41. The inequality is: $0 < x < 8$

43. The opposite is -2 and the reciprocal is $\dfrac{1}{2}$.

45. Writing without absolute values: $|-3| = 3$

47. Writing without absolute values: $|-4| = 4$

49. Simplifying: $-2y + 4y = 2y$

51. Simplifying: $3x - 2 + 5x + 7 = (3x + 5x) + (-2 + 7) = 8x + 5$

53. Commutative property of addition (a)

55. Commutative property of addition (a)

57. Commutative and associative properties of addition (a,c)

59. Additive identity property (e)

61. Finding the difference: $5 - 3 = 2$

63. Computing: $7 + (-2) - 4 = 5 - 4 = 1$

65. Computing: $|-4| - |-3| + |-2| = 4 - 3 + 2 = 3$

67. Finding the differences: $6 - (-3) - 2 - 5 = 6 + 3 - 2 - 5 = 9 - 2 - 5 = 7 - 5 = 2$

69. Finding the differences: $-\dfrac{1}{12} - \dfrac{1}{6} - \dfrac{1}{4} - \dfrac{1}{3} = -\dfrac{1}{12} - \dfrac{2}{12} - \dfrac{3}{12} - \dfrac{4}{12} = -\dfrac{10}{12} = -\dfrac{5}{6}$

71. Finding the product: $6(-7) = -42$

73. Finding the product: $7(3x) = 21x$

75. Applying the distributive property: $-2(3x - 5) = -2(3x) - (-2)(5) = -6x + 10$

77. Applying the distributive property: $-\dfrac{1}{2}(2x - 6) = -\dfrac{1}{2}(2x) - \left(-\dfrac{1}{2}\right)(6) = -x + 3$

79. Dividing: $-\dfrac{5}{8} \div \dfrac{3}{4} = -\dfrac{5}{8} \cdot \dfrac{4}{3} = -\dfrac{5}{6}$

81. Dividing: $\dfrac{3}{5} \div 6 = \dfrac{3}{5} \cdot \dfrac{1}{6} = \dfrac{1}{10}$

83. Simplifying: $2(-5) - 3 = -10 - 3 = -13$

85. Simplifying: $6 + 3(-2) = 6 + (-6) = 0$

87. Simplifying: $-3(2) - 5(6) = -6 - 30 = -36$

89. Simplifying: $8 - 2(6 - 10) = 8 - 2(-4) = 8 + 8 = 16$

91. Simplifying: $\dfrac{3(-4) - 8}{-5 - 5} = \dfrac{-12 - 8}{-5 - 5} = \dfrac{-20}{-10} = 2$

93. Simplifying: $4 - (-2)\left[\dfrac{6 - 3(-4)}{1 + 5(-2)}\right] = 4 - (-2)\left(\dfrac{6 + 12}{1 - 10}\right) = 4 - (-2)\left(\dfrac{18}{-9}\right) = 4 - (-2)(-2) = 4 - 4 = 0$

95. Simplifying: $4(3x-1)-5(6x+2)=12x-4-30x-10=-18x-14$

97. The pattern is to add -3, so the next term is $2+(-3)=-1$. This is an arithmetic sequence.

99. The pattern is to add $-\dfrac{1}{2}$, so the next term is $-\dfrac{1}{2}-\dfrac{1}{2}=-1$. This is an arithmetic sequence.

Chapter 1 Test

1. Written in symbols, the expression is: $2(3x+4y)$

2. Written in symbols, the expression is: $2a-3b<2a+3b$

3. Using the rule for order of operations: $3\cdot2^2+5\cdot3^2=3\cdot4+5\cdot9=12+45=57$

4. Using the rule for order of operations: $6+2(4\cdot3-10)=6+2(12-10)=6+2(2)=6+4=10$

5. Using the rule for order of operations: $12-8\div4+2\cdot3=12-2+6=10+6=16$

6. Using the rule for order of operations:

$$20-4\left[3^2-2\left(2^3-6\right)\right]=20-4\left[9-2(8-6)\right]=20-4\left[9-2(2)\right]=20-4(9-4)=20-4(5)=20-20=0$$

7. The set is $A\cap B=\{2,4\}$.   **8.**   The set is $B\cap C=\varnothing$.

9. The integers are: $-5,0,1,4$

10. The rational numbers are: $-5,-4.1,-\dfrac{5}{6},0,1,1.8,4$

11. The irrational numbers are: $-\sqrt{2},\sqrt{3}$ **12.** The whole numbers are: $0,1,4$

13. Factoring: $585=9\cdot65=(3\cdot3)\cdot(5\cdot13)=3^2\cdot5\cdot13$ **14.** Factoring: $620=10\cdot62=(2\cdot5)\cdot(2\cdot31)=2^2\cdot5\cdot31$

15. Reducing the fraction: $\dfrac{585}{620}=\dfrac{5\cdot117}{5\cdot124}=\dfrac{117}{124}$

16. **a.** Simplifying: $-(-3)=3$ **b.** Simplifying: $-|-2|=-2$

17. The inequality is: $\{x\mid x>3\}$ **18.** The inequality is: $\{x\mid x\le-5\}$

19. The inequality is: $\{x\mid-5<x<3\}$ **20.** The inequality is: $\{x\mid x\le-5 \text{ or } x\ge3\}$

21. Graphing the inequality: **22.** Graphing the inequality:

23. This is the commutative property of addition.   **24.** This is the multiplicative identity property.

25. This is the associative property of multiplication.   **26.** This is the associative property of addition.

27. Simplifying: $\dfrac{-4(-1)-(-10)}{5-(-2)}=\dfrac{4+10}{5+2}=\dfrac{14}{7}=2$

28. Simplifying: $3-2\left[\dfrac{8(-1)-7}{-3(2)-4}\right]=3-2\left(\dfrac{-8-7}{-6-4}\right)=3-2\left(\dfrac{-15}{-10}\right)=3-2\left(\dfrac{3}{2}\right)=3-3=0$

29. Simplifying: $-\dfrac{3}{8}+\dfrac{5}{12}-\left(-\dfrac{7}{9}\right)=-\dfrac{3}{8}\cdot\dfrac{9}{9}+\dfrac{5}{12}\cdot\dfrac{6}{6}+\dfrac{7}{9}\cdot\dfrac{8}{8}=-\dfrac{27}{72}+\dfrac{30}{72}+\dfrac{56}{72}=\dfrac{59}{72}$

30. Simplifying: $-\dfrac{1}{2}(8x)=-4x$

31. Simplifying: $-4(3x+2)+7x=-12x-8+7x=-5x-8$

32. Simplifying: $5(2y-3)-(6y-5)=10y-15-6y+5=4y-10$

33. Simplifying: $3+4(2x-5)-5x=3+8x-20-5x=3x-17$

34. Simplifying: $2+5a+3(2a-4)=2+5a+6a-12=11a-10$

35. Computing: $-\dfrac{2}{3}+(-2)\left(\dfrac{5}{6}\right)=-\dfrac{2}{3}-\dfrac{5}{3}=-\dfrac{7}{3}$

36.   Computing: $-4\left(\dfrac{7}{16}\right) - \dfrac{3}{4} = -\dfrac{7}{4} - \dfrac{3}{4} = -\dfrac{10}{4} = -\dfrac{5}{2}$

37.   The pattern is to multiply by -4, so the next term is $80(-4) = -320$. This is a geometric sequence.

38.   The pattern is to add -1, so the next term is $-1 - 1 = -2$. This is an arithmetic sequence.

39.   The pattern is to add 1, then 2, then 3, so the next term is $13 + 4 = 17$.

40.   The pattern is to multiply by $\dfrac{1}{5}$, so the next term is $\dfrac{1}{25} \cdot \dfrac{1}{5} = \dfrac{1}{125}$. This is a geometric sequence.

41.   The number of users is: $31 + 16 + 27 = 74$ **42.**   The number of users is: $120 - 74 = 46$

Chapter 2
Equations and Inequalities in One Variable

Getting Ready for Chapter 2

1. Simplifying: $-6\left(\dfrac{2}{3}\right) = -4$

2. Simplifying: $-24\left(\dfrac{3}{8}\right) = -9$

3. Simplifying: $\dfrac{4}{3}(-9) = -12$

4. Simplifying: $\dfrac{2}{3}\left(-\dfrac{21}{16}\right) = -\dfrac{7}{8}$

5. Simplifying: $-\dfrac{3}{8} + \left(-\dfrac{1}{2}\right) = -\dfrac{3}{8} - \dfrac{1}{2} \cdot \dfrac{4}{4} = -\dfrac{3}{8} - \dfrac{4}{8} = -\dfrac{7}{8}$

6. Simplifying: $-\dfrac{7}{8} + \dfrac{1}{2} = -\dfrac{7}{8} + \dfrac{1}{2} \cdot \dfrac{4}{4} = -\dfrac{7}{8} + \dfrac{4}{8} = -\dfrac{3}{8}$

7. Simplifying: $\dfrac{1}{2}(2x) = x$

8. Simplifying: $\dfrac{4}{3}\left(\dfrac{3}{4}x\right) = x$

9. Simplifying: $24\left(\dfrac{2}{3}x\right) = 16x$

10. Simplifying: $24\left(\dfrac{2}{3}x + \dfrac{1}{2}\right) = 16x + 12$

11. Simplifying: $100(0.05x) = 5x$

12. Simplifying: $0.06x + 0.05(10{,}000 - x) = 0.06x + 500 - 0.05x = 0.01x + 500$

13. Simplifying: $3(2y - 1) + y = 6y - 3 + y = 7y - 3$

14. Simplifying: $8 - 3(4x - 2) + 5x = 8 - 12x + 6 + 5x = -7x + 14$

15.
   a. Evaluating when $x = 2$: $2x - 1 = 2(2) - 1 = 4 - 1 = 3$

   b. Evaluating when $x = 3$: $2x - 1 = 2(3) - 1 = 6 - 1 = 5$

   c. Evaluating when $x = 5$: $2x - 1 = 2(5) - 1 = 10 - 1 = 9$

16.
   a. Evaluating when $x = 1$: $\dfrac{1}{x+1} = \dfrac{1}{1+1} = \dfrac{1}{2}$

   b. Evaluating when $x = 2$: $\dfrac{1}{x+1} = \dfrac{1}{2+1} = \dfrac{1}{3}$

   c. Evaluating when $x = 3$: $\dfrac{1}{x+1} = \dfrac{1}{3+1} = \dfrac{1}{4}$

17. Translating into symbols: $2x - 3$

18. Translating into symbols: $4x - 10$

19. Translating into symbols: $x + y = 180$

20. Translating into symbols: $a + b = 90$

2.1 Linear Equations in One Variable

1. Solving the equation:
$$x - 5 = 3$$
$$x = 8$$

3. Solving the equation:
$$2x - 4 = 6$$
$$2x = 10$$
$$x = \frac{10}{2} = 5$$

5. Solving the equation:
$$7 = 4a - 1$$
$$8 = 4a$$
$$a = \frac{8}{4} = 2$$

7. Solving the equation:
$$3 - y = 10$$
$$-y = 7$$
$$y = -7$$

9. Solving the equation:
$$-3 - 4x = 15$$
$$-4x = 18$$
$$x = -\frac{9}{2}$$

11. Solving the equation:
$$-3 = 5 + 2x$$
$$-8 = 2x$$
$$x = -4$$

13. Solving the equation:
$$-300y + 100 = 500$$
$$-300y = 400$$
$$y = -\frac{4}{3}$$

15. Solving the equation:
$$160 = -50x - 40$$
$$200 = -50x$$
$$x = -4$$

17. Solving the equation:
$$-x = 2$$
$$x = -1 \cdot 2 = -2$$

19. Solving the equation:
$$-a = -\frac{3}{4}$$
$$a = -1 \cdot \left(-\frac{3}{4}\right) = \frac{3}{4}$$

21. Solving the equation:
$$\frac{2}{3}x = 8$$
$$x = \frac{3}{2} \cdot 8 = 12$$

23. Solving the equation:
$$-\frac{3}{5}a + 2 = 8$$
$$-\frac{3}{5}a = 6$$
$$a = -\frac{5}{3} \cdot 6 = -10$$

25. Solving the equation:
$$8 = 6 + \frac{2}{7}y$$
$$2 = \frac{2}{7}y$$
$$y = \frac{7}{2} \cdot 2 = 7$$

27. Solving the equation:
$$2x - 5 = 3x + 2$$
$$-x - 5 = 2$$
$$-x = 7$$
$$x = -7$$

29. Solving the equation:
$$-3a + 2 = -2a - 1$$
$$-a + 2 = -1$$
$$-a = -3$$
$$a = 3$$

31. Solving the equation:
$$5 - 2x = 3x + 1$$
$$5 - 5x = 1$$
$$-5x = -4$$
$$x = \frac{4}{5}$$

33. Solving the equation:
$$11x - 5 + 4x - 2 = 8x$$
$$15x - 7 = 8x$$
$$7x = 7$$
$$x = 1$$

35. Solving the equation:
$$6 - 7(m - 3) = -1$$
$$6 - 7m + 21 = -1$$
$$-7m + 27 = -1$$
$$-7m = -28$$
$$m = 4$$

37. Solving the equation:
$$7 + 3(x+2) = 4(x-1)$$
$$7 + 3x + 6 = 4x - 4$$
$$3x + 13 = 4x - 4$$
$$-x + 13 = -4$$
$$-x = -17$$
$$x = 17$$

39. Solving the equation:
$$5 = 7 - 2(3x-1) + 4x$$
$$5 = 7 - 6x + 2 + 4x$$
$$5 = -2x + 9$$
$$-2x = -4$$
$$x = 2$$

41. Solving the equation:
$$\frac{1}{2}x + \frac{1}{4} = \frac{1}{3}x + \frac{5}{4}$$
$$12\left(\frac{1}{2}x + \frac{1}{4}\right) = 12\left(\frac{1}{3}x + \frac{5}{4}\right)$$
$$6x + 3 = 4x + 15$$
$$2x + 3 = 15$$
$$2x = 12$$
$$x = 6$$

43. Solving the equation:
$$-\frac{2}{5}x + \frac{2}{15} = \frac{2}{3}$$
$$15\left(-\frac{2}{5}x + \frac{2}{15}\right) = 15\left(\frac{2}{3}\right)$$
$$-6x + 2 = 10$$
$$-6x = 8$$
$$x = -\frac{4}{3}$$

45. Solving the equation:
$$\frac{3}{4}(8x-4) = \frac{2}{3}(6x-9)$$
$$6x - 3 = 4x - 6$$
$$2x - 3 = -6$$
$$2x = -3$$
$$x = -\frac{3}{2}$$

47. Solving the equation:
$$\frac{1}{4}(12a+1) - \frac{1}{4} = 5$$
$$3a + \frac{1}{4} - \frac{1}{4} = 5$$
$$3a = 5$$
$$a = \frac{5}{3}$$

49. Solving the equation:
$$0.35x - 0.2 = 0.15x + 0.1$$
$$0.2x - 0.2 = 0.1$$
$$0.2x = 0.3$$
$$2x = 3$$
$$x = \frac{3}{2}$$

51. Solving the equation:
$$0.42 - 0.18x = 0.48x - 0.24$$
$$0.42 - 0.66x = -0.24$$
$$-0.66x = -0.66$$
$$x = 1$$

53. Solving the equation:
$$3x - 6 = 3(x+4)$$
$$3x - 6 = 3x + 12$$
$$-6 = 12$$
Since this statement is false, there is no solution.

55. Solving the equation:
$$4y + 2 - 3y + 5 = 3 + y + 4$$
$$y + 7 = y + 7$$
$$7 = 7$$
Since this statement is true, the solution is all real numbers.

57. Solving the equation:
$$2(4t-1) + 3 = 5t + 4 + 3t$$
$$8t - 2 + 3 = 8t + 4$$
$$8t + 1 = 8t + 4$$
$$1 = 4$$
Since this statement is false, there is no solution.

59. Solving the equation:
$$3x + 2 = 0$$
$$3x = -2$$
$$x = -\frac{2}{3}$$

61. Solving the equation:
$$0 = 6,400a + 70$$
$$-70 = 6,400a$$
$$a = -\frac{70}{6,400} = -\frac{7}{640}$$

63. Solving the equation:
$$x + 2 = 2x$$
$$2 = 2x - x$$
$$x = 2$$

65. Solving the equation:
$$0.07x = 1.4$$
$$x = \frac{1.4}{0.07} = 20$$

67. Solving the equation:
$$5(2x + 1) = 12$$
$$10x + 5 = 12$$
$$10x = 7$$
$$x = \frac{7}{10} = 0.7$$

69. Solving the equation:
$$50 = \frac{K}{48}$$
$$50 \cdot 48 = \frac{K}{48} \cdot 48$$
$$K = 2,400$$

71. Solving the equation:
$$100P = 2,400$$
$$P = \frac{2,400}{100} = 24$$

73. Solving the equation:
$$x + (3x + 2) = 26$$
$$4x + 2 = 26$$
$$4x = 24$$
$$x = 6$$

75. Solving the equation:
$$2x - 3(3x - 5) = -6$$
$$2x - 9x + 15 = -6$$
$$-7x + 15 = -6$$
$$-7x = -21$$
$$x = 3$$

77. Solving the equation:
$$2(2x - 3) + 2x = 45$$
$$4x - 6 + 2x = 45$$
$$6x - 6 = 45$$
$$6x = 51$$
$$x = \frac{51}{6} = \frac{17}{2} = 8.5$$

79. Solving the equation:
$$3x + (x - 2) \cdot 2 = 6$$
$$3x + 2x - 4 = 6$$
$$5x - 4 = 6$$
$$5x = 10$$
$$x = 2$$

81. Solving the equation:
$$15 - 3(x - 1) = x - 2$$
$$15 - 3x + 3 = x - 2$$
$$-3x + 18 = x - 2$$
$$-4x + 18 = -2$$
$$-4x = -20$$
$$x = 5$$

83. Solving the equation:
$$2(x + 3) + x = 4(x - 3)$$
$$2x + 6 + x = 4x - 12$$
$$3x + 6 = 4x - 12$$
$$-x + 6 = -12$$
$$-x = -18$$
$$x = 18$$

85. Solving the equation:
$$6(y - 3) - 5(y + 2) = 8$$
$$6y - 18 - 5y - 10 = 8$$
$$y - 28 = 8$$
$$y = 36$$

87. Solving the equation:
$$2(20 + x) = 3(20 - x)$$
$$40 + 2x = 60 - 3x$$
$$40 + 5x = 60$$
$$5x = 20$$
$$x = 4$$

89. Solving the equation:
$$2x + 1.5(75 - x) = 127.5$$
$$2x + 112.5 - 1.5x = 127.5$$
$$0.5x + 112.5 = 127.5$$
$$0.5x = 15$$
$$x = 30$$

91. Solving the equation:
$$0.08x + 0.09(9,000 - x) = 750$$
$$0.08x + 810 - 0.09x = 750$$
$$-0.01x + 810 = 750$$
$$-0.01x = -60$$
$$x = 6,000$$

93. Solving the equation:
$$0.12x + 0.10(15,000 - x) = 1,600$$
$$0.12x + 1,500 - 0.10x = 1,600$$
$$0.02x + 1,500 = 1,600$$
$$0.02x = 100$$
$$x = 5,000$$

95. Solving the equation:

$$5\left(\frac{19}{15}\right) + 5y = 9$$

$$\frac{19}{3} + 5y = 9$$

$$5y = \frac{8}{3}$$

$$y = \frac{8}{15}$$

97. Solving the equation:

$$2\left(\frac{29}{22}\right) - 3y = 4$$

$$\frac{29}{11} - 3y = 4$$

$$-3y = \frac{15}{11}$$

$$y = -\frac{5}{11}$$

99. **a.** Solving the equation:

$$8x - 5 = 0$$

$$8x = 5$$

$$x = \frac{5}{8}$$

b. Solving the equation:

$$8x - 5 = -5$$

$$8x = 0$$

$$x = 0$$

c. Adding: $(8x - 5) + (2x - 5) = 10x - 10$

d. Solving the equation:

$$8x - 5 = 2x - 5$$

$$6x - 5 = -5$$

$$6x = 0$$

$$x = 0$$

e. Multiplying: $8(x - 5) = 8x - 40$

f. Solving the equation:

$$8(x - 5) = 2(x - 5)$$

$$8x - 40 = 2x - 10$$

$$6x - 40 = -10$$

$$6x = 30$$

$$x = 5$$

101. **a.** The equation is $25.45 = 1.80n + 3.85$

b. Solving the equation:

$$25.45 = 1.80n + 3.85$$

$$21.6 = 1.8n$$

$$n = 12$$

The woman traveled 12 miles.

103. **a.** The equation is $1,125A = 3,937,000$.

b. Solving the equation:

$$1,125A = 3,937,000$$

$$A \approx 3,500$$

The area of Puerto Rico is approximately 3,500 square miles.

105. Let x represent the speed of the car. Solving the equation:

$$5x + 23 = 273$$

$$5x = 250$$

$$x = \frac{250}{5} = 50$$

The speed is 50 mph.

107. This is the commutative property of multiplication. **109.** This is the associative property of addition.

111. These are the commutative and associative properties of addition.

113. This is the multiplicative identity property. **115.** This is the commutative property of multiplication.

117. This is the additive identity property.

119. Simplifying: $2x - 3y + 5 + 5x - 7 = 2x + 5x - 3y + 5 - 7 = 7x - 3y - 2$

121. Simplifying: $3(2y - 4) + 7 - 5y = 6y - 12 + 7 - 5y = y - 5$

123. Simplifying: $7 - 4(x - 2) + 9 = 7 - 4x + 8 + 9 = -4x + 24$

125. Solving the equation:
$$x \cdot 42 = 21$$
$$x = \frac{21}{42} = 0.5$$

127. Solving the equation:
$$25 = 0.4x$$
$$x = \frac{25}{0.4} = 62.5$$

129. Solving the equation:
$$12 - 4y = 12$$
$$-4y = 0$$
$$y = 0$$

131. Solving the equation:
$$525 = 900 - 300p$$
$$-375 = -300p$$
$$p = \frac{-375}{-300} = 1.25$$

133. Solving the equation:
$$486.7 = 78.5 + 31.4h$$
$$408.2 = 31.4h$$
$$h = \frac{408.2}{31.4} = 13$$

135. **a.**   Evaluating when $x = 2$: $2x - 1 = 2(2) - 1 = 4 - 1 = 3$

   b.   Evaluating when $x = 3$: $2x - 1 = 2(3) - 1 = 6 - 1 = 5$

   c.   Evaluating when $x = 5$: $2x - 1 = 2(5) - 1 = 10 - 1 = 9$

137. Solving the equation:
$$\frac{x+4}{5} - \frac{x+3}{3} = -\frac{7}{15}$$
$$15\left(\frac{x+4}{5} - \frac{x+3}{3}\right) = 15\left(-\frac{7}{15}\right)$$
$$3(x+4) - 5(x+3) = -7$$
$$3x + 12 - 5x - 15 = -7$$
$$-2x - 3 = -7$$
$$-2x = -4$$
$$x = 2$$

139. Solving the equation:
$$\frac{1}{x} - \frac{2}{3} = \frac{2}{x}$$
$$3x\left(\frac{1}{x} - \frac{2}{3}\right) = 3x\left(\frac{2}{x}\right)$$
$$3 - 2x = 6$$
$$-2x = 3$$
$$x = -\frac{3}{2}$$

141. Solving the equation:
$$\frac{x+3}{2} - \frac{x-4}{4} = -\frac{1}{8}$$
$$8\left(\frac{x+3}{2} - \frac{x-4}{4}\right) = 8\left(-\frac{1}{8}\right)$$
$$4(x+3) - 2(x-4) = -1$$
$$4x + 12 - 2x + 8 = -1$$
$$2x + 20 = -1$$
$$2x = -21$$
$$x = -\frac{21}{2}$$

143. Solving the equation:
$$\frac{x-1}{2} - \frac{x+2}{3} = \frac{x+3}{6}$$
$$6\left(\frac{x-1}{2} - \frac{x+2}{3}\right) = 6\left(\frac{x+3}{6}\right)$$
$$3(x-1) - 2(x+2) = 1(x+3)$$
$$3x - 3 - 2x - 4 = x + 3$$
$$x - 7 = x + 3$$
$$-7 = 3$$

Since this statement is false, there is no solution.

2.2 Formulas

1. Substituting $x = 0$:

$$3(0) - 4y = 12$$
$$-4y = 12$$
$$y = -3$$

3. Substituting $x = 4$:

$$3(4) - 4y = 12$$
$$12 - 4y = 12$$
$$-4y = 0$$
$$y = 0$$

5. Substituting $y = 0$:
$$2x - 3 = 0$$
$$2x = 3$$
$$x = \frac{3}{2}$$

7. Substituting $y = 5$:
$$2x - 3 = 5$$
$$2x = 8$$
$$x = 4$$

9. Substituting $y = -\frac{6}{5}$:

$$x - 2\left(-\frac{6}{5}\right) = 4$$
$$x + \frac{12}{5} = 4$$
$$x = \frac{8}{5}$$

11. Substituting $x = 0$:

$$2(0) + 3y = 6$$
$$3y = 6$$
$$y = 2$$

13. Substituting $x = 160$ and $y = 0$:

$$0 = a(160 - 80)^2 + 70$$
$$-70 = a(80)^2$$
$$6,400\,a = -70$$
$$a = -\frac{70}{6,400} = -\frac{7}{640}$$

15. Substituting $p = 1.5$: $R = (900 - 300 \cdot 1.5)(1.5) = (450)(1.5) = 675$

17. **a.** Substituting $x = 100$: $P = -0.1(100)^2 + 27(100) - 1,700 = -1,000 + 2,700 - 1,700 = 0$

 b. Substituting $x = 170$: $P = -0.1(170)^2 + 27(170) - 1,700 = -2,890 + 4,590 - 1,700 = 0$

19. **a.** Substituting $t = \frac{1}{4}$: $h = 16 + 32\left(\frac{1}{4}\right) - 16\left(\frac{1}{4}\right)^2 = 16 + 8 - 1 = 23$

 b. Substituting $t = \frac{7}{4}$: $h = 16 + 32\left(\frac{7}{4}\right) - 16\left(\frac{7}{4}\right)^2 = 16 + 56 - 49 = 23$

21. Substituting $x = \frac{3}{2}$: $y = -2\left(\frac{3}{2}\right)^2 + 6\left(\frac{3}{2}\right) - 5 = -\frac{9}{2} + 9 - 5 = -\frac{1}{2}$

23. Substituting $x = 5$ and $y = 15$:

$$15 = K(5)$$
$$K = 3$$

25. Substituting $P = 48$ and $V = 50$:

$$50 = \frac{K}{48}$$
$$K = 50 \cdot 48 = 2,400$$

27. Solving for l:
$$A = lw$$
$$l = \frac{A}{w}$$

29. Solving for t:
$$I = prt$$
$$t = \frac{I}{pr}$$

31. Solving for T:
$$PV = nRT$$
$$T = \frac{PV}{nR}$$

33. Solving for x:
$$y = mx + b$$
$$y - b = mx$$
$$x = \frac{y - b}{m}$$

35. Solving for F:
$$C = \frac{5}{9}(F - 32)$$
$$\frac{9}{5}C = F - 32$$
$$F = \frac{9}{5}C + 32$$

37. Solving for v:
$$h = vt + 16t^2$$
$$h - 16t^2 = vt$$
$$v = \frac{h - 16t^2}{t}$$

39. Solving for d:
$$A = a + (n - 1)d$$
$$A - a = (n - 1)d$$
$$d = \frac{A - a}{n - 1}$$

41. Solving for y:
$$2x + 3y = 6$$
$$3y = -2x + 6$$
$$y = \frac{-2x + 6}{3}$$
$$y = -\frac{2}{3}x + 2$$

43. Solving for y:
$$-3x + 5y = 15$$
$$5y = 3x + 15$$
$$y = \frac{3x + 15}{5}$$
$$y = \frac{3}{5}x + 3$$

45. Solving for y:
$$2x - 6y + 12 = 0$$
$$-6y = -2x - 12$$
$$y = \frac{-2x - 12}{-6}$$
$$y = \frac{1}{3}x + 2$$

47. Solving for x:
$$ax + 4 = bx + 9$$
$$ax - bx + 4 = 9$$
$$ax - bx = 5$$
$$x(a - b) = 5$$
$$x = \frac{5}{a - b}$$

49. Solving for P:
$$A = P + Prt$$
$$A = P(1 + rt)$$
$$P = \frac{A}{1 + rt}$$

51. Solving for y:
$$\frac{x}{8} + \frac{y}{2} = 1$$
$$8\left(\frac{x}{8} + \frac{y}{2}\right) = 8(1)$$
$$x + 4y = 8$$
$$4y = -x + 8$$
$$y = -\frac{1}{4}x + 2$$

53. Solving for y:
$$\frac{x}{5} + \frac{y}{-3} = 1$$
$$15\left(\frac{x}{5} + \frac{y}{-3}\right) = 15(1)$$
$$3x - 5y = 15$$
$$-5y = -3x + 15$$
$$y = \frac{3}{5}x - 3$$

55. Solving for y:
$$x = 2y - 3$$
$$2y = x + 3$$
$$y = \frac{x + 3}{2} = \frac{1}{2}x + \frac{3}{2}$$

57. Solving for y:
$$y - 3 = -2(x + 4)$$
$$y - 3 = -2x - 8$$
$$y = -2x - 5$$

59. Solving for y:

$$y - 3 = -\frac{2}{3}(x + 3)$$
$$y - 3 = -\frac{2}{3}x - 2$$
$$y = -\frac{2}{3}x + 1$$

61. Solving for y:

$$y - 4 = -\frac{1}{2}(x + 1)$$
$$y - 4 = -\frac{1}{2}x - \frac{1}{2}$$
$$y = -\frac{1}{2}x + \frac{7}{2}$$

63. **a.** Solving for y:

$$\frac{y + 1}{x - 0} = 4$$
$$y + 1 = 4(x - 0)$$
$$y + 1 = 4x$$
$$y = 4x - 1$$

b. Solving for y:

$$\frac{y + 2}{x - 4} = -\frac{1}{2}$$
$$y + 2 = -\frac{1}{2}(x - 4)$$
$$y + 2 = -\frac{1}{2}x + 2$$
$$y = -\frac{1}{2}x$$

c. Solving for y:

$$\frac{y + 3}{x - 7} = 0$$
$$y + 3 = 0(x - 7)$$
$$y + 3 = 0$$
$$y = -3$$

65. Writing a linear equation and solving:

$$x = 0.54 \cdot 38$$
$$x = 20.52$$

67. Writing a linear equation and solving:

$$x \cdot 36 = 9$$
$$x = \frac{1}{4} = 25\%$$

69. Writing a linear equation and solving:

$$0.04 \cdot x = 37$$
$$x = \frac{37}{0.04} = 925$$

71. **a.** Solving the equation:

$$-4x + 5 = 20$$
$$-4x = 15$$
$$x = -\frac{15}{4} = -3.75$$

b. Substituting $x = 3$: $-4x + 5 = -4(3) + 5 = -7$

c. Solving for y:

$$-4x + 5y = 20$$
$$5y = 4x + 20$$
$$y = \frac{4}{5}x + 4$$

d. Solving for x:

$$-4x + 5y = 20$$
$$-4x = -5y + 20$$
$$x = \frac{5}{4}y - 5$$

73. Let p represent the percent. The equation is:

$$p \cdot 25 = 4$$
$$p = \frac{4}{25} = 0.16$$

The tip represented 16% of the cost.

75. Computing the sales tax: $0.065(50) = 3.25$. The sales tax was $3.25.

77. The discount was: $\$94.00 - \$56.40 = \$37.60$. Finding the percent discount:

$$p \cdot 94 = 37.60$$
$$p = \frac{37.6}{94} = 0.40$$

The percent discount was 40%.

79. Finding the area: $A = 0.072(286.3) \approx 20.6$ square miles

81. Substituting $x = 800$:
$$800 = 3,800 - 100p$$
$$-3,000 = -100p$$
$$p = 30$$
They should charge $30.00 per cartridge.

85. Let c represent the rate of the current. The equation is:
$$2(15 - c) = 18$$
$$30 - 2c = 18$$
$$-2c = -12$$
$$c = 6$$
The speed of the current is 6 mph.

89. Finding the rate: $\dfrac{220 \text{ miles}}{4 \text{ hours}} = 55$ miles per hour

91. Finding the rate: $\dfrac{252 \text{ kilometers}}{3 \text{ hours}} = 84$ kilometers per hour

93. The distance traveled by the rider is the circumference: $C = \pi(65) \approx (3.14)(65) \approx 204.1$ feet

Finding the rate: $\dfrac{204.1 \text{ feet}}{30 \text{ seconds}} \approx 6.8$ feet per second

95. Substituting $n = 1$, $y = 7$, and $z = 15$:
$$x^1 + 7^1 = 15^1$$
$$x + 7 = 15$$
$$x = 8$$

97. For Shar, $M = 220 - 46 = 174$ and $R = 60$: $T = R + 0.6(M - R) = 60 + 0.6(174 - 60) = 128.4$ beats per minute

For Sara, $M = 220 - 26 = 194$ and $R = 60$: $T = R + 0.6(M - R) = 60 + 0.6(194 - 60) = 140.4$ beats per minute

99. For Maria, $M = 220 - 35 = 185$ and $T = 137$:
$$137 = R + 0.6(185 - R)$$
$$137 = R + 111 - 0.6R$$
$$26 = 0.4R$$
$$R = 65$$
Her resting heart rate is 65 beats per minute.

101. Using the order of operations: $38 - 19 + 1 = 19 + 1 = 20$

103. Using the order of operations: $57 - 18 - 8 = 39 - 8 = 31$

105. Using the order of operations: $28 \div 7 \cdot 2 = 4 \cdot 2 = 8$

107. Using the order of operations: $125 \div 25 \div 5 = 5 \div 5 = 1$

109. Solving the equation:
$$3(2x - 9) + 15 - x = -27$$
$$6x - 27 + 15 - x = -27$$
$$5x - 12 = -27$$
$$5x = -15$$
$$x = -3$$

113. Solving the equation:
$$\frac{x - 4}{3} - 5 + 2x = 24$$
$$x - 4 - 15 + 6x = 72$$
$$7x - 19 = 72$$
$$7x = 91$$
$$x = 13$$

115. Translating into symbols: $2x - 3$

83. Substituting $x = 400$:
$$400 = 3,800 - 100p$$
$$-3,400 = -100p$$
$$p = 34$$
They should charge $34.00 per cartridge.

87. Let w represent the rate of the wind. The equation is:
$$4(258 - w) = 864$$
$$1032 - 4w = 864$$
$$-4w = -168$$
$$w = 42$$
The speed of the wind is 42 mph.

111. Solving the equation:
$$8x - 7 + 5x - 12 = 7$$
$$13x - 19 = 7$$
$$13x = 26$$
$$x = 2$$

117. Translating into symbols: $x + y = 180$

119. Solving the equation:

$$x + 2x = 90$$
$$3x = 90$$
$$x = 30$$

121. Solving the equation:

$$2(2x - 3) + 2x = 45$$
$$4x - 6 + 2x = 45$$
$$6x - 6 = 45$$
$$6x = 51$$
$$x = \frac{51}{6} = 8.5$$

123. Solving the equation:

$$6x + 5(10,000 - x) = 56,000$$
$$6x + 50,000 - 5x = 56,000$$
$$x + 50,000 = 56,000$$
$$x = 6,000$$

125. Solving the equation:

$$\frac{x}{a} + \frac{y}{b} = 1$$
$$ab\left(\frac{x}{a} + \frac{y}{b}\right) = ab(1)$$
$$bx + ay = ab$$
$$bx = -ay + ab$$
$$x = -\frac{a}{b}y + a$$

127. Solving the equation for a:

$$\frac{1}{a} + \frac{1}{b} = \frac{1}{c}$$
$$abc\left(\frac{1}{a} + \frac{1}{b}\right) = abc\left(\frac{1}{c}\right)$$
$$bc + ac = ab$$
$$bc = ab - ac$$
$$bc = a(b - c)$$
$$a = \frac{bc}{b - c}$$

129. Solving the equation for b:

$$\frac{2}{a} + \frac{2}{b} = \frac{2}{c}$$
$$abc\left(\frac{2}{a} + \frac{2}{b}\right) = abc\left(\frac{2}{c}\right)$$
$$2bc + 2ac = 2ab$$
$$bc + ac = ab$$
$$bc = ab - ac$$
$$bc = a(b - c)$$
$$a = \frac{bc}{b - c}$$

131. Solving the equation for a:

$$\frac{2}{a} + \frac{4}{b} = \frac{2}{c}$$
$$abc\left(\frac{2}{a} + \frac{4}{b}\right) = abc\left(\frac{2}{c}\right)$$
$$2bc + 4ac = 2ab$$
$$bc + 2ac = ab$$
$$bc = ab - 2ac$$
$$bc = a(b - 2c)$$
$$a = \frac{bc}{b - 2c}$$

2.3 Applications

1. Let w represent the width and $2w$ represent the length. Using the perimeter formula:
$$2w + 2(2w) = 60$$
$$2w + 4w = 60$$
$$6w = 60$$
$$w = 10$$
The dimensions are 10 feet by 20 feet.

3. Let s represent the side of the square. Using the perimeter formula:
$$4s = 28$$
$$s = 7$$
The length of each side is 7 feet.

5. Let x represent the shortest side, $x + 3$ represent the medium side, and $2x$ represent the longest side. Using the perimeter formula:
$$x + x + 3 + 2x = 23$$
$$4x + 3 = 23$$
$$4x = 20$$
$$x = 5$$
The shortest side is 5 inches.

7. Let w represent the width and $2w - 3$ represent the length. Using the perimeter formula:
$$2w + 2(2w - 3) = 18$$
$$2w + 4w - 6 = 18$$
$$6w - 6 = 18$$
$$6w = 24$$
$$w = 4$$
The width is 4 meters.

9. Let w represent the width and $2w$ represent the length. Using the perimeter formula:
$$2w + 2(2w) = 48$$
$$2w + 4w = 48$$
$$6w = 48$$
$$w = 8$$
The width is 8 feet and the length is 16 feet. Finding the cost: $C = 1.75(32) + 2.25(16) = 56 + 36 = 92$
The cost to build the pen is $92.00.

11. Let b represent the amount of money Shane had at the beginning of the trip. Using the percent increase:
$$b + 0.50b = 300$$
$$1.5b = 300$$
$$b = 200$$
Shane had $200.00 at the beginning of the trip.

13. Let I represent the accountant's monthly income before the raise. The equation is:
$$I + 0.055I = 3440$$
$$1.055I = 3440$$
$$I \approx 3260.66$$
The accountant's monthly income before the raise was approximately $3,260.66.

15. Let R represent the total box office receipts. The equation is:
$$0.53R = 52.8$$
$$R \approx 99.6$$
The total receipts were approximately $99.6 million.

17. Let x represent one angle and $8x$ represent the other angle. Since the angles are supplementary:
$$x + 8x = 180$$
$$9x = 180$$
$$x = 20$$
The two angles are 20° and 160°.

19. **a.** Let x represent one angle and $4x - 12$ represent the other angle. Since the angles are complementary:

$$x + 4x - 12 = 90$$
$$5x - 12 = 90$$
$$5x = 102$$
$$x = 20.4$$
$$4x - 12 = 4(20.4) - 12 = 69.6$$

The two angles are 20.4° and 69.6°.

b. Let x represent one angle and $4x - 12$ represent the other angle. Since the angles are supplementary:

$$x + 4x - 12 = 180$$
$$5x - 12 = 180$$
$$5x = 192$$
$$x = 38.4$$
$$4x - 12 = 4(38.4) - 12 = 141.6$$

The two angles are 38.4° and 141.6°.

21. Let x represent the smallest angle, $3x$ represent the largest angle, and $3x - 9$ represent the third angle. The equation is:

$$x + 3x + 3x - 9 = 180$$
$$7x - 9 = 180$$
$$7x = 189$$
$$x = 27$$

The three angles are 27°, 72° and 81°.

23. Let x represent the largest angle, $\frac{1}{3}x$ represent the smallest angle, and $\frac{1}{3}x + 10$ represent the third angle.

The equation is:

$$\frac{1}{3}x + x + \frac{1}{3}x + 10 = 180$$
$$\frac{5}{3}x + 10 = 180$$
$$\frac{5}{3}x = 170$$
$$x = 102$$

The three angles are 34°, 44° and 102°.

25. Let x represent the measure of the two base angles, and $2x + 8$ represent the third angle. The equation is:

$$x + x + 2x + 8 = 180$$
$$4x + 8 = 180$$
$$4x = 172$$
$$x = 43$$

The three angles are 43°, 43° and 94°.

27. Let x represent the amount invested at 8% and $9,000 - x$ represent the amount invested at 9%. The equation is:

$$0.08x + 0.09(9,000 - x) = 750$$
$$0.08x + 810 - 0.09x = 750$$
$$-0.01x + 810 = 750$$
$$-0.01x = -60$$
$$x = 6,000$$

She invested $6,000 at 8% and $3,000 at 9%.

29. Let x represent the amount invested at 12% and $15,000 - x$ represent the amount invested at 10%. The equation is:

$$0.12x + 0.10(15,000 - x) = 1,600$$
$$0.12x + 1,500 - 0.10x = 1,600$$
$$0.02x + 1,500 = 1,600$$
$$0.02x = 100$$
$$x = 5,000$$

The investment was $5,000 at 12% and $10,000 at 10%.

31.   Let x represent the amount invested at 8% and $6{,}000 - x$ represent the amount invested at 9%. The equation is:

$$0.08x + 0.09(6{,}000 - x) = 500$$
$$0.08x + 540 - 0.09x = 500$$
$$-0.01x + 540 = 500$$
$$-0.01x = -40$$
$$x = 4{,}000$$

Stacy invested $4,000 at 8% and $2,000 at 9%.

33.   Let x represent the number of adult tickets sold and $x + 6$ the number of children's tickets sold. The equation is:

$$6(x) + 4(x + 6) = 184$$
$$6x + 4x + 24 = 184$$
$$10x + 24 = 184$$
$$10x = 160$$
$$x = 16$$

Miguel sold 16 adult tickets and 22 children's tickets.

35.   The total money collected is: $\$1204 - \$250 = \$954$

Let x represent the amount of her sales (not including tax). Since this amount includes the tax collected, the equation is:

$$x + 0.06x = 954$$
$$1.06x = 954$$
$$x = 900$$

Her sales were $900, so the sales tax is: $0.06(900) = \$54$

37.   Completing the table:

t	0	$\frac{1}{4}$	1	$\frac{7}{4}$	2
h	0	7	16	7	0

39.   Completing the table:

Year	Bets (millions of dollars)
1985	20.2
1990	34.4
1995	44.8
2000	65.4
2005	104
2010	163

41.   Completing the table:

Year	Sales (billions of dollars)
2005	7
2006	7.5
2007	8
2008	8.6
2009	9.2

43.   Completing the table:

Speed (miles per hour)	Distance (miles)
20	10
30	15
40	20
50	25
60	30
70	35

45.   Completing the table:

Time (hours)	Distance Upstream (miles)	Distance Downstream (miles)
1	6	14
2	12	28
3	18	42
4	24	56
5	30	70
6	36	84

47. Completing the table:

Age (years)	Maximum Heart Rate (beats per minute)
18	202
19	201
20	200
21	199
22	198
23	197

49. Completing the table:

w (ft)	l (ft)	A (ft^2)
2	22	44
4	20	80
6	18	108
8	16	128
10	14	140
12	12	144

51. Graphing the inequality:

53. Graphing the inequality:

55. Graphing the inequality:

57. Graphing the inequality:

59. Graphing the inequality:

61. Graphing the inequality:

63. Solving the equation:

$$-2x - 3 = 7$$
$$-2x = 10$$
$$x = -5$$

65. Solving the equation:

$$3(2x - 4) - 7x = -3x$$
$$6x - 12 - 7x = -3x$$
$$-x - 12 = -3x$$
$$-12 = -2x$$
$$x = 6$$

2.4 Linear Inequalities in One Variable

1. Solving the inequality:

$$2x \le 3$$
$$x \le \frac{3}{2}$$

Graphing the solution set:

3. Solving the inequality:

$$\frac{1}{2}x > 2$$
$$x > 4$$

Graphing the solution set:

5. Solving the inequality:

$$-5x \le 25$$
$$x \ge -5$$

Graphing the solution set:

7. Solving the inequality:

$$-\frac{3}{2}x > -6$$
$$-3x > -12$$
$$x < 4$$

Graphing the solution set:

9. Solving the inequality:

$$-12 \le 2x$$
$$x \ge -6$$

Graphing the solution set:

11. Solving the inequality:

$$-1 \ge -\frac{1}{4}x$$
$$x \ge 4$$

Graphing the solution set:

13. Solving the inequality:

$$-3x+1>10$$
$$-3x>9$$
$$x<-3$$

Graphing the solution set:

15. Solving the inequality:

$$\frac{1}{2}-\frac{m}{12}\le\frac{7}{12}$$
$$12\left(\frac{1}{2}-\frac{m}{12}\right)\le 12\left(\frac{7}{12}\right)$$
$$6-m\le 7$$
$$-m\le 1$$
$$m\ge -1$$

Graphing the solution set:

17. Solving the inequality:

$$\frac{1}{2}\ge -\frac{1}{6}-\frac{2}{9}x$$
$$18\left(\frac{1}{2}\right)\ge 18\left(-\frac{1}{6}-\frac{2}{9}x\right)$$
$$9\ge -3-4x$$
$$12\ge -4x$$
$$x\ge -3$$

Graphing the solution set:

19. Solving the inequality:

$$-40\le 30-20y$$
$$-70\le -20y$$
$$y\le\frac{7}{2}$$

Graphing the solution set:

21. Solving the inequality:

$$\frac{2}{3}x-3<1$$
$$\frac{2}{3}x<4$$
$$2x<12$$
$$x<6$$

Graphing the solution set:

23. Solving the inequality:

$$10-\frac{1}{2}y\le 36$$
$$-\frac{1}{2}y\le 26$$
$$y\ge -52$$

Graphing the solution set:

25. Solving the inequality:

$$2(3y+1)\le -10$$
$$6y+2\le -10$$
$$6y\le -12$$
$$y\le -2$$

The solution set is $(-\infty,-2]$.

27. Solving the inequality:

$$-(a+1)-4a\le 2a-8$$
$$-a-1-4a\le 2a-8$$
$$-5a-1\le 2a-8$$
$$-7a\le -7$$
$$a\ge 1$$

The solution set is $[1,\infty)$.

29. Solving the inequality:

$$\frac{1}{3}t-\frac{1}{2}(5-t)<0$$
$$6\left(\frac{1}{3}t-\frac{1}{2}(5-t)\right)<6(0)$$
$$2t-3(5-t)<0$$
$$2t-15+3t<0$$
$$5t-15<0$$
$$5t<15$$
$$t<3$$

The solution set is $(-\infty,3)$.

31. Solving the inequality:

$$-2\le 5-7(2a+3)$$
$$-2\le 5-14a-21$$
$$-2\le -16-14a$$
$$14\le -14a$$
$$a\le -1$$

The solution set is $(-\infty,-1]$.

33. Solving the inequality:

$$-\frac{1}{3}(x+5) \le -\frac{2}{9}(x-1)$$
$$9\left[-\frac{1}{3}(x+5)\right] \le 9\left[-\frac{2}{9}(x-1)\right]$$
$$-3(x+5) \le -2(x-1)$$
$$-3x-15 \le -2x+2$$
$$-x-15 \le 2$$
$$-x \le 17$$
$$x \ge -17$$

The solution set is $[-17, \infty)$.

35. Solving the inequality:

$$20x + 9,300 > 18,000$$
$$20x > 8,700$$
$$x > 435$$

37. Solving the inequality:

$$-2 \le m - 5 \le 2$$
$$3 \le m \le 7$$

The solution set is $[3, 7]$. Graphing the solution set:

39. Solving the inequality:

$$-60 < 20a + 20 < 60$$
$$-80 < 20a < 40$$
$$-4 < a < 2$$

The solution set is $(-4, 2)$. Graphing the solution set:

41. Solving the inequality:

$$0.5 \le 0.3a - 0.7 \le 1.1$$
$$1.2 \le 0.3a \le 1.8$$
$$4 \le a \le 6$$

The solution set is $[4, 6]$. Graphing the solution set:

43. Solving the inequality:

$$3 < \frac{1}{2}x + 5 < 6$$
$$-2 < \frac{1}{2}x < 1$$
$$-4 < x < 2$$

The solution set is $(-4, 2)$. Graphing the solution set:

45. Solving the inequality:

$$4 < 6 + \frac{2}{3}x < 8$$
$$-2 < \frac{2}{3}x < 2$$
$$-6 < 2x < 6$$
$$-3 < x < 3$$

The solution set is $(-3, 3)$. Graphing the solution set:

47. Solving the inequality:

$$x + 5 \le -2 \quad \text{or} \quad x + 5 \ge 2$$
$$x \le -7 \quad \text{or} \quad x \ge -3$$

The solution set is $(-\infty, -7] \cup [-3, \infty)$. Graphing the solution set:

49. Solving the inequality:

$$5y+1 \le -4 \qquad \text{or} \qquad 5y+1 \ge 4$$
$$5y \le -5 \qquad \text{or} \qquad 5y \ge 3$$
$$y \le -1 \qquad \text{or} \qquad y \ge \frac{3}{5}$$

The solution set is $\left(-\infty,-1\right] \cup \left[\frac{3}{5},\infty\right)$. Graphing the solution set:

51. Solving the inequality:

$$2x+5 < 3x-1 \qquad \text{or} \qquad x-4 > 2x+6$$
$$-x+5 < -1 \qquad \text{or} \qquad -x-4 > 6$$
$$-x < -6 \qquad \text{or} \qquad -x > 10$$
$$x > 6 \qquad \text{or} \qquad x < -10$$

The solution set is $\left(-\infty,-10\right) \cup \left(6,\infty\right)$. Graphing the solution set:

53. Writing as an inequality: $-2 < x \le 4$ **55.** Writing as an inequality: $x < -4$ or $x \ge 1$

57. **a.** Writing as an inequality: $x > 0$ **b.** Writing as an inequality: $x \ge 0$

       c. Writing as an inequality: $x \ge 0$

59. There is no solution, since $x^2 \ge 0$ for all values of x.

61. This inequality is true for all values of x, since $x^2 \ge 0$ is always true.

63. There is no solution, since $\dfrac{1}{x^2} > 0$ for all values of x.

65. **a.** Evaluating when $x = 0$: $-\dfrac{1}{2}x+1 = -\dfrac{1}{2}(0)+1 = 1$

       b. Solving the equation:

$$-\frac{1}{2}x+1 = -7$$
$$-\frac{1}{2}x = -8$$
$$x = 16$$

       c. Substituting $x = 0$: $-\dfrac{1}{2}x+1 = -\dfrac{1}{2}(0)+1 = 1$

          No, 0 is not a solution to the inequality.

       d. Solving the inequality:

$$-\frac{1}{2}x+1 < -7$$
$$-\frac{1}{2}x < -8$$
$$x > 16$$

67. Written as an inequality: $31 < d < 49$

69. Solving the inequality:

$$2000 - 200p \ge 300$$
$$-200p \ge -1700$$
$$p \le 8.5$$

They should charge \$8.50 per pad or less.

71. Solving the inequality:

$$2000 - 200p < 550$$
$$-200p < -1450$$
$$p > 7.25$$

They should charge more than \$7.25 per pad.

73. Solving the inequality:
$$22149 - 399x > 20500$$
$$-399x > -1649$$
$$x < 4.13$$
In the years 1990 through 1994 Amtrak had more than 20,500 million passengers.

75. For adults, the inequality is $0.72 - 0.11 \le r \le 0.72 + 0.11$, or $0.61 \le r \le 0.83$. The survival rate for adults is between 61% and 83%. For juveniles, the inequality is $0.13 - 0.07 \le r \le 0.13 + 0.07$, or $0.06 \le r \le 0.20$. The survival rate for juveniles is between 6% and 20%.

77. **a.** Solving the inequality:
$$95 \le \frac{9}{5}C + 32 \le 113$$
$$63 \le \frac{9}{5}C \le 81$$
$$315 \le 9C \le 405$$
$$35° \le C \le 45°$$

   b. Solving the inequality:
$$68 \le \frac{9}{5}C + 32 \le 86$$
$$36 \le \frac{9}{5}C \le 54$$
$$180 \le 9C \le 270$$
$$20° \le C \le 30°$$

   c. Solving the inequality:
$$-13 \le \frac{9}{5}C + 32 \le 14$$
$$-45 \le \frac{9}{5}C \le -18$$
$$-225 \le 9C \le -90$$
$$-25° \le C \le -10°$$

   d. Solving the inequality:
$$-4 \le \frac{9}{5}C + 32 \le 23$$
$$-36 \le \frac{9}{5}C \le -9$$
$$-180 \le 9C \le -45$$
$$20° \le C \le -5°$$

79. Simplifying: $|-3| = 3$

81. Simplifying: $-|-3| = -3$

83. $|x|$ represents the distance from x to 0 on the number line.

85. Solving the equation:
$$2a - 1 = -7$$
$$2a = -6$$
$$a = -3$$

87. Solving the equation:
$$\frac{2}{3}x - 3 = 7$$
$$\frac{2}{3}x = 10$$
$$x = 15$$

89. Solving the equation:
$$x - 5 = x - 7$$
$$-5 = -7$$

   The equation has no solution $(\varnothing)$.

91. Solving the equation:
$$x - 5 = -x - 7$$
$$2x - 5 = -7$$
$$2x = -2$$
$$x = -1$$

93. Solving the inequality:
$$ax + b < c$$
$$ax < c - b$$
$$x < \frac{c - b}{a}$$

95. Solving the inequality:
$$-c < ax + b < c$$
$$-c - b < ax < c - b$$
$$\frac{-c - b}{a} < x < \frac{c - b}{a}$$

2.5 Equations with Absolute Value

1. Solving the equation:
$$|x| = 4$$
$$x = -4, 4$$

3. Solving the equation:
$$2 = |a|$$
$$a = -2, 2$$

5. The equation $|x| = -3$ has no solution, or $\varnothing$.

7. Solving the equation:
$$|a| + 2 = 3$$
$$|a| = 1$$
$$a = -1, 1$$

9. Solving the equation:
$$|y| + 4 = 3$$
$$|y| = -1$$

The equation $|y| = -1$ has no solution, or $\varnothing$.

11. Solving the equation:
$$4 = |x| - 2$$
$$|x| = 6$$
$$x = -6, 6$$

13. Solving the equation:
$$|x - 2| = 5$$
$$x - 2 = -5, 5$$
$$x = -3, 7$$

15. Solving the equation:
$$|a - 4| = \frac{5}{3}$$
$$a - 4 = -\frac{5}{3}, \frac{5}{3}$$
$$a = \frac{7}{3}, \frac{17}{3}$$

17. Solving the equation:
$$1 = |3 - x|$$
$$3 - x = -1, 1$$
$$-x = -4, -2$$
$$x = 2, 4$$

19. Solving the equation:
$$\left|\frac{3}{5}a + \frac{1}{2}\right| = 1$$
$$\frac{3}{5}a + \frac{1}{2} = -1, 1$$
$$\frac{3}{5}a = -\frac{3}{2}, \frac{1}{2}$$
$$a = -\frac{5}{2}, \frac{5}{6}$$

21. Solving the equation:
$$60 = |20x - 40|$$
$$20x - 40 = -60, 60$$
$$20x = -20, 100$$
$$x = -1, 5$$

23. Since $|2x + 1| = -3$ is impossible, there is no solution, or $\varnothing$.

25. Solving the equation:
$$\left|\frac{3}{4}x - 6\right| = 9$$
$$\frac{3}{4}x - 6 = -9, 9$$
$$\frac{3}{4}x = -3, 15$$
$$3x = -12, 60$$
$$x = -4, 20$$

27. Solving the equation:
$$\left|1 - \frac{1}{2}a\right| = 3$$
$$1 - \frac{1}{2}a = -3, 3$$
$$-\frac{1}{2}a = -4, 2$$
$$a = -4, 8$$

29. Solving the equation:
$$|3x + 4| + 1 = 7$$
$$|3x + 4| = 6$$
$$3x + 4 = -6, 6$$
$$3x = -10, 2$$
$$x = -\frac{10}{3}, \frac{2}{3}$$

31. Solving the equation:

$$|3 - 2y| + 4 = 3$$
$$|3 - 2y| = -1$$

Since this equation is impossible, there is no solution, or $\varnothing$.

33. Solving the equation:

$$3 + |4t - 1| = 8$$
$$|4t - 1| = 5$$
$$4t - 1 = -5, 5$$
$$4t = -4, 6$$
$$t = -1, \frac{3}{2}$$

35. Solving the equation:

$$\left|9 - \frac{3}{5}x\right| + 6 = 12$$
$$\left|9 - \frac{3}{5}x\right| = 6$$
$$9 - \frac{3}{5}x = -6, 6$$
$$-\frac{3}{5}x = -15, -3$$
$$-3x = -75, -15$$
$$x = 5, 25$$

37. Solving the equation:

$$5 = \left|\frac{2x}{7} + \frac{4}{7}\right| - 3$$
$$\left|\frac{2x}{7} + \frac{4}{7}\right| = 8$$
$$\frac{2x}{7} + \frac{4}{7} = -8, 8$$
$$2x + 4 = -56, 56$$
$$2x = -60, 52$$
$$x = -30, 26$$

39. Solving the equation:

$$2 = -8 + \left|4 - \frac{1}{2}y\right|$$
$$\left|4 - \frac{1}{2}y\right| = 10$$
$$4 - \frac{1}{2}y = -10, 10$$
$$-\frac{1}{2}y = -14, 6$$
$$y = -12, 28$$

41. Solving the equation:

$$|3a + 1| = |2a - 4|$$

$3a + 1 = 2a - 4$	or	$3a + 1 = -2a + 4$
$a + 1 = -4$		$5a = 3$
$a = -5$		$a = \frac{3}{5}$

43. Solving the equation:

$$\left|x - \frac{1}{3}\right| = \left|\frac{1}{2}x + \frac{1}{6}\right|$$

$x - \frac{1}{3} = \frac{1}{2}x + \frac{1}{6}$	or	$x - \frac{1}{3} = -\frac{1}{2}x - \frac{1}{6}$
$6x - 2 = 3x + 1$		$6x - 2 = -3x - 1$
$3x - 2 = 1$		$9x - 2 = -1$
$3x = 3$		$9x = 1$
$x = 1$		$x = \frac{1}{9}$

45. Solving the equation:

$$|y - 2| = |y + 3|$$

$y - 2 = y + 3$	or	$y - 2 = -y - 3$
$-2 = -3$		$2y = -1$
$y = $ impossible		$y = -\frac{1}{2}$

47.   Solving the equation:

$$|3x-1|=|3x+1|$$

$$3x-1=3x+1 \qquad \text{or} \qquad 3x-1=-3x-1$$
$$-1=1 \qquad\qquad\qquad\qquad 6x=0$$
$$x=\text{impossible} \qquad\qquad\qquad x=0$$

49.   Solving the equation:

$$|3-m|=|m+4|$$

$$3-m=m+4 \qquad \text{or} \qquad 3-m=-m-4$$
$$-2m=1 \qquad\qquad\qquad\qquad 3=-4$$
$$m=-\frac{1}{2} \qquad\qquad\qquad m=\text{impossible}$$

51.   Solving the equation:

$$|0.03-0.01x|=|0.04+0.05x|$$

$$0.03-0.01x=0.04+0.05x \qquad \text{or} \qquad 0.03-0.01x=-0.04-0.05x$$
$$-0.06x=0.01 \qquad\qquad\qquad\qquad 0.04x=-0.07$$
$$x=-\frac{1}{6} \qquad\qquad\qquad\qquad x=-\frac{7}{4}$$

53.   Since $|x-2|=|2-x|$ is always true, the solution set is all real numbers.

55.   Since $\left|\frac{x}{5}-1\right|=\left|1-\frac{x}{5}\right|$ is always true, the solution set is all real numbers.

57.   **a.**   Solving the equation:

$$4x-5=0$$
$$4x=5$$
$$x=\frac{5}{4}=1.25$$

b.   Solving the equation:

$$|4x-5|=0$$
$$4x-5=0$$
$$4x=5$$
$$x=\frac{5}{4}=1.25$$

c.   Solving the equation:

$$4x-5=3$$
$$4x=8$$
$$x=2$$

d.   Solving the equation:

$$|4x-5|=3$$
$$4x-5=-3,3$$
$$4x=2,8$$
$$x=\frac{1}{2},2$$

e.   Solving the equation:

$$|4x-5|=|2x+3|$$

$$4x-5=2x+3 \qquad \text{or} \qquad 4x-5=-2x-3$$
$$2x-5=3 \qquad\qquad\qquad\qquad 6x-5=-3$$
$$2x=8 \qquad\qquad\qquad\qquad\qquad 6x=2$$
$$x=4 \qquad\qquad\qquad\qquad\qquad x=\frac{1}{3}$$

59.   The distance is: $|1,671-1,614|=|57|=57$ feet

61.   Graphing the inequality:

$$-2 \qquad\qquad\qquad\qquad 8$$

63.   Graphing the inequality:

$$-2 \qquad\qquad\qquad\qquad 1$$

65.   Simplifying: $\dfrac{38}{30}=\dfrac{2\cdot 19}{2\cdot 15}=\dfrac{19}{15}$

67.   Simplifying: $\dfrac{240}{6}=\dfrac{6\cdot 40}{6\cdot 1}=40$

69.   Simplifying: $\dfrac{0+6}{0-3}=\dfrac{6}{-3}=-2$

71.   Simplifying: $\dfrac{4-4}{4-2}=\dfrac{0}{2}=0$

73. Solving the inequality:
$$2x - 5 < 3$$
$$2x < 8$$
$$x < 4$$

75. Solving the inequality:
$$-4 \le 3a + 7$$
$$-11 \le 3a$$
$$a \ge -\frac{11}{3}$$

77. Solving the inequality:
$$4t - 3 \le -9$$
$$4t \le -6$$
$$t \le -\frac{3}{2}$$

79. Solving the inequality:
$$|x - a| = b$$
$$x - a = -b, b$$
$$x = a - b \ \text{ or } \ x = a + b$$

81. Solving the equation:
$$|ax + b| = c$$
$$ax + b = -c, c$$
$$ax = -b - c, -b + c$$
$$x = \frac{-b - c}{a} \ \text{ or } \ x = \frac{-b + c}{a}$$

83. Solving the equation:
$$\left|\frac{x}{a} + \frac{y}{b}\right| = 1$$
$$\frac{x}{a} + \frac{y}{b} = -1, 1$$
$$\frac{x}{a} = -\frac{y}{b} - 1, -\frac{y}{b} + 1$$
$$x = -\frac{a}{b}y - a \ \text{ or } \ x = -\frac{a}{b}y + a$$

2.6 Inequalities Involving Absolute Value

1. Solving the inequality:
$$|x| < 3$$
$$-3 < x < 3$$
Graphing the solution set:

3. Solving the inequality:
$$|x| \ge 2$$
$$x \le -2 \text{ or } x \ge 2$$
Graphing the solution set:

5. Solving the inequality:
$$|x| + 2 < 5$$
$$|x| < 3$$
$$-3 < x < 3$$
Graphing the solution set:

7. Solving the inequality:
$$|t| - 3 > 4$$
$$|t| > 7$$
$$t < -7 \text{ or } t > 7$$
Graphing the solution set:

9. Since the inequality $|y| < -5$ is never true, there is no solution, or $\varnothing$. Graphing the solution set:

11. Since the inequality $|x| \ge -2$ is always true, the solution set is all real numbers.

Graphing the solution set:

13. Solving the inequality:
$$|x - 3| < 7$$
$$-7 < x - 3 < 7$$
$$-4 < x < 10$$
Graphing the solution set:

15. Solving the inequality:
$$|a + 5| \ge 4$$
$$a + 5 \le -4 \text{ or } a + 5 \ge 4$$
$$a \le -9 \text{ or } \quad a \ge -1$$
Graphing the solution set:

17. Since the inequality $|a - 1| < -3$ is never true, there is no solution, or $\varnothing$. Graphing the solution set:

19. Solving the inequality:
$$|2x - 4| < 6$$
$$-6 < 2x - 4 < 6$$
$$-2 < 2x < 10$$
$$-1 < x < 5$$
Graphing the solution set:

21. Solving the inequality:
$$|3y + 9| \geq 6$$

$$3y + 9 \leq -6 \qquad \text{or} \qquad 3y + 9 \geq 6$$
$$3y \leq -15 \qquad\qquad\qquad 3y \geq -3$$
$$y \leq -5 \qquad\qquad\qquad y \geq -1$$
Graphing the solution set:

23. Solving the inequality:
$$|2k + 3| \geq 7$$
$$2k + 3 \leq -7 \qquad \text{or} \qquad 2k + 3 \geq 7$$
$$2k \leq -10 \qquad\qquad\qquad 2k \geq 4$$
$$k \leq -5 \qquad\qquad\qquad k \geq 2$$
Graphing the solution set:

25. Solving the inequality:
$$|x - 3| + 2 < 6$$
$$|x - 3| < 4$$
$$-4 < x - 3 < 4$$
$$-1 < x < 7$$
Graphing the solution set:

27. Solving the inequality:
$$|2a + 1| + 4 \geq 7$$
$$|2a + 1| \geq 3$$
$$2a + 1 \leq -3 \qquad \text{or} \qquad 2a + 1 \geq 3$$
$$2a \leq -4 \qquad\qquad\qquad 2a \geq 2$$
$$a \leq -2 \qquad\qquad\qquad a \geq 1$$

Graphing the solution set:

29. Solving the inequality:
$$|3x + 5| - 8 < 5$$
$$|3x + 5| < 13$$
$$-13 < 3x + 5 < 13$$
$$-18 < 3x < 8$$
$$-6 < x < \frac{8}{3}$$
Graphing the solution set:

31. Solving the inequality:

$$|5 - x| > 3$$
$$5 - x < -3 \qquad \text{or} \qquad 5 - x > 3$$
$$-x < -8 \qquad\qquad\qquad -x > -2$$
$$x > 8 \qquad\qquad\qquad x < 2$$

Graphing the solution set:

33. Solving the inequality:

$$\left|3 - \frac{2}{3}x\right| \geq 5$$
$$3 - \frac{2}{3}x \leq -5 \qquad \text{or} \qquad 3 - \frac{2}{3}x \geq 5$$
$$-\frac{2}{3}x \leq -8 \qquad\qquad\qquad -\frac{2}{3}x \geq 2$$
$$-2x \leq -24 \qquad\qquad\qquad -2x \geq 6$$
$$x \geq 12 \qquad\qquad\qquad x \leq -3$$

Graphing the solution set:

35. Solving the inequality:

$$\left|2 - \frac{1}{2}x\right| > 1$$

$$2 - \frac{1}{2}x < -1 \quad \text{or} \quad 2 - \frac{1}{2}x > 1$$

$$-\frac{1}{2}x < -3 \qquad\qquad -\frac{1}{2}x > -1$$

$$x > 6 \qquad\qquad\qquad x < 2$$

Graphing the solution set:

37. Solving the inequality:

$$|x - 1| < 0.01$$

$$-0.01 < x - 1 < 0.01$$

$$0.99 < x < 1.01$$

39. Solving the inequality:

$$|2x + 1| \geq \frac{1}{5}$$

$$2x + 1 \leq -\frac{1}{5} \quad \text{or} \quad 2x + 1 \geq \frac{1}{5}$$

$$2x \leq -\frac{6}{5} \qquad\qquad 2x \geq -\frac{4}{5}$$

$$x \leq -\frac{3}{5} \qquad\qquad x \geq -\frac{2}{5}$$

41. Solving the inequality:

$$\left|\frac{3x - 2}{5}\right| \leq \frac{1}{2}$$

$$-\frac{1}{2} \leq \frac{3x - 2}{5} \leq \frac{1}{2}$$

$$-\frac{5}{2} \leq 3x - 2 \leq \frac{5}{2}$$

$$-\frac{1}{2} \leq 3x \leq \frac{9}{2}$$

$$-\frac{1}{6} \leq x \leq \frac{3}{2}$$

43. Solving the inequality:

$$\left|2x - \frac{1}{5}\right| < 0.3$$

$$-0.3 < 2x - 0.2 < 0.3$$

$$-0.1 < 2x < 0.5$$

$$-0.05 < x < 0.25$$

45. Writing as an absolute value inequality: $|x - 5| \leq 1$

47. Evaluating when $x = 0$: $|5x + 3| = |5(0) + 3| = |3| = 3$

No, 0 is not a solution.

49. Solving the inequality:

$$|5x + 3| > 7$$

$$5x + 3 < -7 \quad \text{or} \quad 5x + 3 > 7$$

$$5x < -10 \qquad\qquad 5x > 4$$

$$x < -2 \qquad\qquad x > \frac{4}{5}$$

51. Simplifying: $-9 \div \frac{3}{2} = -9 \cdot \frac{2}{3} = -6$

53. Simplifying: $3 - 7(-6 - 3) = 3 - 7(-9) = 3 + 63 = 66$

55. Simplifying: $-4(-2)^3 - 5(-3)^2 = -4(-8) - 5(9) = 32 - 45 = -13$

57. Simplifying: $\dfrac{2(-3) - 5(-6)}{-1 - 2 - 3} = \dfrac{-6 + 30}{-1 - 2 - 3} = \dfrac{24}{-6} = -4$

59. Simplifying: $6(1) - 1(-5) + 1(2) = 6 + 5 + 2 = 13$

61. Simplifying: $1(0)(1) + 3(1)(4) + (-2)(2)(-1) = 0 + 12 + 4 = 16$

63. Simplifying: $4(-1)^2 + 3(-1) - 2 = 4(1) + 3(-1) - 2 = 4 - 3 - 2 = -1$

65. Solving the inequality:

$$|x - a| < b$$
$$-b < x - a < b$$
$$a - b < x < a + b$$

67. Solving the inequality:

$$|ax - b| > c$$

$$ax - b < -c \qquad \text{or} \qquad ax - b > c$$
$$ax < b - c \qquad\qquad\qquad ax > b + c$$
$$x < \frac{b - c}{a} \qquad\qquad\qquad x > \frac{b + c}{a}$$

69. Solving the inequality:

$$|ax + b| \leq c$$
$$-c \leq ax + b \leq c$$
$$-c - b \leq ax \leq c - b$$
$$\frac{-c - b}{a} \leq x \leq \frac{c - b}{a}$$

Chapter 2 Review

1. Solving the equation:

$$x - 3 = 7$$
$$x = 10$$

3. Solving the equation:

$$400 - 100a = 200$$
$$-100a = -200$$
$$a = 2$$

5. Solving the equation:

$$4x - 2 = 7x + 7$$
$$-3x - 2 = 7$$
$$-3x = 9$$
$$x = -3$$

7. Solving the equation:

$$7y - 5 - 2y = 2y - 3$$
$$5y - 5 = 2y - 3$$
$$3y - 5 = -3$$
$$3y = 2$$
$$y = \frac{2}{3}$$

9. Solving the equation:

$$3(2x + 1) = 18$$
$$6x + 3 = 18$$
$$6x = 15$$
$$x = \frac{5}{2}$$

11. Solving the equation:

$$8 - 3(2t + 1) = 5(t + 2)$$
$$8 - 6t - 3 = 5t + 10$$
$$-6t + 5 = 5t + 10$$
$$-11t + 5 = 10$$
$$-11t = 5$$
$$t = -\frac{5}{11}$$

13. Solving for h:

$$40 = 2(3) + 2h$$
$$40 = 6 + 2h$$
$$2h = 34$$
$$h = 17$$

15. Solving for p:

$$I = prt$$
$$p = \frac{I}{rt}$$

17. Solving for y:

$$4x - 3y = 12$$
$$-3y = -4x + 12$$
$$y = \frac{4}{3}x - 4$$

19. Solving for y:

$$5x + 3y - 6 = 0$$
$$3y = -5x + 6$$
$$y = -\frac{5}{3}x + 2$$

21. Solving for y:

$$y + 3 = -2(x - 1)$$
$$y + 3 = -2x + 2$$
$$y = -2x - 1$$

23. Let w represent the width and $3w$ represent the length. Using the perimeter formula:
$$2w + 2(3w) = 32$$
$$2w + 6w = 32$$
$$8w = 32$$
$$w = 4$$
The dimensions are 4 feet by 12 feet.

25. **a.** Substituting $L = 45$ and $H = 12$: $N = 7(45)(12) = 3,780$ bricks

   b. Substituting $N = 35,000$ and $H = 8$:
$$35000 = 7L(8)$$
$$35000 = 56L$$
$$L = 625$$
The wall could be 625 feet long.

27. Solving the inequality:

$$-8a > -4$$
$$a < \frac{1}{2}$$

The solution set is $\left(-\infty, \frac{1}{2}\right)$.

29. Solving the inequality:
$$\frac{3}{4}x + 1 \le 10$$
$$\frac{3}{4}x \le 9$$
$$3x \le 36$$
$$x \le 12$$
The solution set is $(-\infty, 12]$.

31. Solving the inequality:
$$\frac{1}{3} \le \frac{1}{6}x \le 1$$
$$2 \le x \le 6$$

The solution set is $[2, 6]$.

33. Solving the inequality:
$$5t + 1 \le 3t - 2 \quad \text{or} \quad -7t \le -21$$
$$2t \le -3 \qquad\qquad t \ge 3$$
$$t \le -\frac{3}{2} \qquad\qquad t \ge 3$$
The solution set is $\left(-\infty, -\frac{3}{2}\right] \cup [3, \infty)$.

35. Solving the equation:
$$|x| = 2$$
$$x = -2, 2$$

37. Solving the equation:
$$|x - 3| = 1$$
$$x - 3 = -1, 1$$
$$x = 2, 4$$

39. Solving the equation:
$$|4x - 3| + 2 = 11$$
$$|4x - 3| = 9$$
$$4x - 3 = -9, 9$$
$$4x = -6, 12$$
$$x = -\frac{3}{2}, 3$$

41. Solving the equation:

$$|5t - 3| = |3t - 5|$$
$$5t - 3 = 3t - 5 \quad \text{or} \quad 5t - 3 = -3t + 5$$
$$2t = -2 \qquad\qquad 8t = 8$$
$$t = -1 \qquad\qquad t = 1$$

43. Solving the inequality:
$$|x| < 5$$
$$-5 < x < 5$$
Graphing the solution set:

45. Since the inequality $|x| < 0$ is never true, there is no solution, or $\varnothing$. Graphing the solution set:

47.   Solving the equation:
$$2x - 3 = 2(x - 3)$$
$$2x - 3 = 2x - 6$$
$$-3 = -6$$
Since this statement is false, there is no solution, or $\varnothing$.

49.   Since $|4y + 8| = -1$ is never true, there is no solution, or $\varnothing$.

51.   Solving the inequality:
$$|5 - 8t| + 4 \le 1$$
$$|5 - 8t| \le -3$$

Since this statement is never true, there is no solution, or $\varnothing$.

Chapter 2 Cumulative Review

1.   Simplifying: $7^2 = 7 \cdot 7 = 49$

3.   Simplifying: $-2 \cdot 5 + 4 = -10 + 4 = -6$

5.   Simplifying: $6 + 2 \cdot 5 - 3 = 6 + 10 - 3 = 13$

7.   Simplifying: $3 + 2(7 + 4) = 3 + 2(11) = 3 + 22 = 25$

9.   Simplifying: $-(6 - 4) - (3 - 8) = -2 - (-5) = -2 + 5 = 3$

11.   Simplifying: $\dfrac{-7 - 7}{-3 - 4} = \dfrac{-14}{-7} = 2$

13.   Reducing the fraction: $\dfrac{630}{735} = \dfrac{2 \cdot 3 \cdot 3 \cdot 5 \cdot 7}{3 \cdot 5 \cdot 7 \cdot 7} = \dfrac{2 \cdot 3}{7} = \dfrac{6}{7}$

15.   Adding the fractions: $\dfrac{11}{30} + \dfrac{13}{42} = \dfrac{11}{30} \cdot \dfrac{7}{7} + \dfrac{13}{42} \cdot \dfrac{5}{5} = \dfrac{77}{210} + \dfrac{65}{210} = \dfrac{142}{210} = \dfrac{2 \cdot 71}{2 \cdot 105} = \dfrac{71}{105}$

17.   Finding the difference: $-3 - 6 = -9$

19.   The set is $A \cup C = \{2, 3, 4, 6, 8, 9\}$.

21.   The set is $\{4, 5\}$.

23.   The rational numbers are: $-13, -6.7, 0, \dfrac{1}{2}, 2, \dfrac{5}{2}$

25.   The irrational numbers are: $-\sqrt{5}, \pi, \sqrt{13}$

27.   The pattern is to multiply by –3, so the next number is: $18(-3) = -54$

29.   Simplifying: $6\left(\dfrac{x}{3} - 2\right) = 6 \cdot \dfrac{x}{3} - 6 \cdot 2 = 2x - 12$

31.   Simplifying: $x\left(1 - \dfrac{4}{x}\right) = x \cdot 1 - x \cdot \dfrac{4}{x} = x - 4$

33.   Solving the equation:
$$7 - 3y = 14$$
$$-3y = 7$$
$$y = -\dfrac{7}{3}$$

35.   Solving the equation:
$$3 - \dfrac{4}{5}a = -5$$
$$5\left(3 - \dfrac{4}{5}a\right) = 5(-5)$$
$$15 - 4a = -25$$
$$-4a = -40$$
$$a = 10$$

37.   Solving the equation:
$$|y| - 2 = 3$$
$$|y| = 5$$
$$y = -5, 5$$

39.   Solving for t:
$$1000 = 500 + 500(0.1)t$$
$$1000 = 500 + 50t$$
$$500 = 50t$$
$$t = 10$$

41.   Solving for x:
$$4x - 3y = 12$$
$$4x = 3y + 12$$
$$x = \dfrac{3}{4}y + 3$$

43.   Solving for h:
$$A = \dfrac{1}{2}bh$$
$$bh = 2A$$
$$h = \dfrac{2A}{b}$$

45.   Let x and $2x + 15$ represent the two angles. Since they are complementary, the equation is:

$$x + 2x + 15 = 90$$
$$3x + 15 = 90$$
$$3x = 75$$
$$x = 25$$

The angles are 25° and 65°.

47.   Solving the inequality:

$$600 - 300x < 900$$
$$-300x < 300$$
$$x > -1$$

The solution set is $(-1, \infty)$.

49.   Solving the inequality:

$$|5x - 1| > 3$$

$$\begin{array}{ccc} 5x - 1 < -3 & \text{or} & 5x - 1 > 3 \\ 5x < -2 & & 5x > 4 \\ x < -\dfrac{2}{5} & & x > \dfrac{4}{5} \end{array}$$

Graphing the solution set:

$-2/5$ $\qquad\qquad$ $4/5$

51.   Finding the difference: $\$3.03 - \$2.69 = \$0.34$

Chapter 2 Test

1.   Solving the equation:

$$x - 5 = 7$$
$$x = 12$$

2.   Solving the equation:

$$3y = -4$$
$$y = -\frac{4}{3}$$

3.   Solving the equation:

$$5(x - 1) - 2(2x + 3) = 5x - 4$$
$$5x - 5 - 4x - 6 = 5x - 4$$
$$x - 11 = 5x - 4$$
$$-4x = 7$$
$$x = -\frac{7}{4}$$

4.   Solving the equation:

$$0.07 - 0.02(3x + 1) = -0.04x + 0.01$$
$$0.07 - 0.06x - 0.02 = -0.04x + 0.01$$
$$-0.06x + 0.05 = -0.04x + 0.01$$
$$-0.02x = -0.04$$
$$x = 2$$

5.   Substituting $x = 0$:

$$4(0) - 6y = 12$$
$$0 - 6y = 12$$
$$-6y = 12$$
$$y = -2$$

6.   Substituting $x = 3$:

$$4(3) - 6y = 12$$
$$12 - 6y = 12$$
$$-6y = 0$$
$$y = 0$$

7.   Substituting $y = 0$:

$$-2x + 4(0) = -8$$
$$-2x + 0 = -8$$
$$-2x = -8$$
$$x = 4$$

8.   Substituting $y = 2$:

$$-2x + 4(2) = -8$$
$$-2x + 8 = -8$$
$$-2x = -16$$
$$x = 8$$

9.   Solving for w:

$$P = 2l + 2w$$
$$P - 2l = 2w$$
$$w = \frac{P - 2l}{2}$$

10.   Solving for B:

$$A = \frac{1}{2}h(b + B)$$
$$2A = h(b + B)$$
$$2A = hb + hB$$
$$2A - hb = hB$$
$$B = \frac{2A - hb}{h}$$

11. Solving for y:
$$3x - 2y = -4$$
$$-2y = -3x - 4$$
$$y = \frac{3}{2}x + 2$$

12. Solving for y:
$$5x + 3y = 6$$
$$3y = -5x + 6$$
$$y = -\frac{5}{3}x + 2$$

13. Solving for y:
$$y - 3 = \frac{2}{3}(x + 6)$$
$$y - 3 = \frac{2}{3}x + 4$$
$$y = \frac{2}{3}x + 7$$

14. Solving for y:
$$y + 4 = \frac{1}{2}(6x - 4)$$
$$y + 4 = 3x - 2$$
$$y = 3x - 6$$

15. Finding the percent:
$$p \bullet 25 = 15$$
$$p = \frac{15}{25} = 60\%$$

16. Finding the percent:
$$p \bullet 45 = 20$$
$$p = \frac{20}{45} = 44\frac{4}{9}\%$$

17. Finding the amount: $0.25(220) = 55$

18. Finding the amount: $1.30(20) = 26$

19. Let w represent the width and $4w$ represent the length. Using the perimeter formula:
$$2(w) + 2(4w) = 60$$
$$2w + 8w = 60$$
$$10w = 60$$
$$w = 6$$
$$4w = 24$$
The width is 6 cm and the length of 24 cm.

20. Adding up the sides of the triangle:
$$x + 3x + (x + 5) = 35$$
$$5x + 5 = 35$$
$$5x = 30$$
$$x = 6 \text{ meters}$$

21. Let x and $2x + 15$ represent the two angles. Since they are supplementary:
$$x + 2x + 15 = 180$$
$$3x + 15 = 180$$
$$3x = 165$$
$$x = 55$$
$$2x + 15 = 125$$
The angles are 55° and 125°.

22. Since the angles are complementary:
$$3x + 6x = 90$$
$$9x = 90$$
$$x = 10$$
$$3x = 30$$
$$6x = 60$$
The angles are 30° and 60°.

23. The amount collected for the day is $881.25 – $75 = $806.25. Since this amount includes the sales s plus the tax, the equation is:
$$s + 0.075s = 806.25$$
$$1.075s = 806.25$$
$$s = 750$$
The portion that is sales tax is: $0.075(750) = 56.25$. The sales tax collected is $56.25.

24. Completing the table:

	Dollars Invested at 11%	Dollars Invested at 12%
Dollars	x	$x + 4000$
Interest	$0.11(x)$	$0.12(x + 4000)$

The equation is:
$$0.11(x) + 0.12(x + 4000) = 940$$
$$0.11x + 0.12x + 480 = 940$$
$$0.23x + 480 = 940$$
$$0.23x = 460$$
$$x = 2000$$
$$x + 4000 = 6000$$

I have \$2,000 invested at 11% and \$6,000 invested at 12%.

25. Solving the inequality:

$$-5t \le 30$$
$$t \ge -6$$

The solution set is $[-6, \infty)$. Graphing the solution set:

26. Solving the inequality:

$$5 - \frac{3}{2}x > -1$$
$$-\frac{3}{2}x > -6$$
$$-3x > -12$$
$$x < 4$$

The solution set is $(-\infty, 4)$. Graphing the solution set:

27. Solving the inequality:
$$1.6x - 2 < 0.8x + 2.8$$
$$0.8x - 2 < 2.8$$
$$0.8x < 4.8$$
$$x < 6$$

The solution set is $(-\infty, 6)$. Graphing the solution set:

28. Solving the inequality:
$$3(2y + 4) \ge 5(y - 8)$$
$$6y + 12 \ge 5y - 40$$
$$y + 12 \ge -40$$
$$y \ge -52$$

The solution set is $[-52, \infty)$. Graphing the solution set:

29. Solving the equation:

$$\left|\frac{1}{4}x - 1\right| = \frac{1}{2}$$
$$\frac{1}{4}x - 1 = -\frac{1}{2}, \frac{1}{2}$$
$$\frac{1}{4}x = \frac{1}{2}, \frac{3}{2}$$
$$x = 2, 6$$

30. Solving the equation:
$$\left|\frac{2}{3}a + 4\right| = 6$$
$$\frac{2}{3}a + 4 = -6, 6$$
$$\frac{2}{3}a = -10, 2$$
$$2a = -30, 6$$
$$a = -15, 3$$

31. Solving the equation:
$$|3 - 2x| + 5 = 2$$
$$|3 - 2x| = -3$$

Since this statement is false, there is no solution, or $\varnothing$.

32. Solving the equation:
$$5 = |3y + 6| - 4$$
$$9 = |3y + 6|$$
$$3y + 6 = -9, 9$$
$$3y = -15, 3$$
$$y = -5, 1$$

33.   Solving the inequality:

$$|6x-1|>7$$
$$6x-1<-7 \qquad \text{or} \qquad 6x-1>7$$
$$6x<-6 \qquad\qquad\qquad\qquad 6x>8$$
$$x<-1 \qquad\qquad\qquad\qquad x>\frac{4}{3}$$

Graphing the solution set:

$$\xleftarrow{\hspace{2cm}}\,)\,\xrightarrow{\hspace{1.5cm}}(\,\xrightarrow{\hspace{1.5cm}}$$
$$\qquad\qquad -1 \qquad\qquad\qquad 4/3$$

34.   Solving the inequality:

$$|3x-5|-4\le 3$$
$$|3x-5|\le 7$$
$$-7\le 3x-5\le 7$$
$$-2\le 3x\le 12$$
$$-\frac{2}{3}\le x\le 4$$

Graphing the solution set:

$$\xleftarrow{\hspace{2cm}}[\,\rule{3cm}{0.4pt}\,]\xrightarrow{\hspace{1.5cm}}$$
$$\qquad\qquad -2/3 \qquad\qquad\qquad 4$$

35.   Since the inequality is always true, the solution set is all real numbers. Graphing the solution set:

$$\xleftarrow{\hspace{3cm}}\rule{0pt}{0pt}\big|\rule{0pt}{0pt}\xrightarrow{\hspace{3cm}}$$
$$0$$

36.   Since $|4t-1|<-3$ is never true, there is no solution, or $\varnothing$. Graphing the solution set:

$$\xleftarrow{\hspace{3cm}}\rule{0pt}{0pt}\big|\rule{0pt}{0pt}\xrightarrow{\hspace{3cm}}$$
$$0$$

Chapter 3
Equations and Inequalities in Two Variables

Getting Ready for Chapter 3

1. Simplifying: $16(4.5)^2 = 16(20.25) = 324$

2. Simplifying: $\dfrac{0.2(0.5)^2}{100} = \dfrac{0.2(0.25)}{100} = \dfrac{0.05}{100} = 0.0005$

3. Simplifying: $-\dfrac{1}{3}(-3) + 2 = 1 + 2 = 3$

4. Simplifying: $4(-2) - 1 = -8 - 1 = -9$

5. Simplifying: $\dfrac{4-(-2)}{3-1} = \dfrac{6}{2} = 3$

6. Simplifying: $\dfrac{-3-(-3)}{2-(-1)} = \dfrac{-3+3}{2+1} = 0$

7. Simplifying: $\dfrac{4}{3}(3.14) \cdot 3^3 = \dfrac{4}{3}(3.14)(27) \approx 113$

8. Simplifying: $3(-2)^2 + 2(-2) - 1 = 12 - 4 - 1 = 7$

9. Substituting $y = 0$:
$$2x + 3(0) = 6$$
$$2x = 6$$
$$x = 3$$

10. Substituting $t = 15$:
$$s = \dfrac{60}{15}$$
$$s = 4$$

11. Writing as a fraction: $-0.06 = -\dfrac{6}{100} = -\dfrac{3}{50}$

12. Writing as a fraction: $-0.6 = -\dfrac{6}{10} = -\dfrac{3}{5}$

13. Solving for y:

$$2x - 3y = 5$$
$$-3y = -2x + 5$$
$$y = \dfrac{2}{3}x - \dfrac{5}{3}$$

14. Solving for y:
$$y - 4 = -\dfrac{1}{2}(x+1)$$
$$y - 4 = -\dfrac{1}{2}x - \dfrac{1}{2}$$
$$y = -\dfrac{1}{2}x + \dfrac{7}{2}$$

15. Substituting each point:

(0,0): $0 + 0 \geq 5$ (false)

(5,0): $5 + 0 \geq 5$ (true)

(4,3): $4 + 3 \geq 5$ (true)

So (5,0) and (4,3) are solutions.

16. Substituting each point:

(0,0): $0 < \dfrac{1}{2}(0) + 3$ (true)

(2,0): $0 < \dfrac{1}{2}(2) + 3$ (true)

(−2,4): $4 < \dfrac{1}{2}(-2) + 3$ (false)

So (0,0) and (2,0) are solutions.

17.   Completing the table:

t	0	1	2	3	4
h	0	48	64	48	0

18.   Completing the table:

t	4	6	8	10
s	15	10	7.5	6

19.   Substituting $x = 3$ and $y = 45$:

$$45 = \frac{K}{3}$$
$$K = 135$$

20.   Substituting $x = 2$, $y = 81$, and $z = 9$:

$$81 = K(2)(9)^2$$
$$81 = 162K$$
$$K = \frac{1}{2}$$

3.1   Paired Data and the Rectangular Coordinate System

1.   Plotting the points:

3.
A.   (4,1)   B.   (–4,3)
C.   (–2,–5)   D.   (2,–2)
E.   (0,5)   F.   (–4,0)
G.   (1,0)

5.   The x-intercept is 3 and the y-intercept is –2.

7.   The x-intercept is 5 and the y-intercept is –4.

9.   The *x*-intercept is $-\dfrac{3}{2}$ and the *y*-intercept is 3.

11.   The *x*-intercept is –4 and the *y*-intercept is 6.

13.   The *x*-intercept is $\dfrac{10}{3}$ and the *y*-intercept is –4.

15.   The *x*-intercept is $\dfrac{5}{3}$ and the *y*-intercept is –5.

17.   The *x*-intercept is 2 and the *y*-intercept is 3.

19.   Table b, since its values match the equation.

21.   Graphing the line:

23.   Graphing the line:

25. Graphing the line:

27. Graphing the line:

29. Since the x-intercept is 3 and the y-intercept is -2, this is the graph of b.

31. **a.** Solving the equation:

$$4x + 12 = -16$$
$$4x = -28$$
$$x = -7$$

b. Substituting $y = 0$:

$$4x + 12(0) = -16$$
$$4x = -16$$
$$x = -4$$

c. Substituting $x = 0$:

$$4(0) + 12y = -16$$
$$12y = -16$$
$$y = -\frac{4}{3}$$

d. Graphing the line:

e. Solving for y:

$$4x + 12y = -16$$
$$12y = -4x - 16$$
$$y = -\frac{1}{3}x - \frac{4}{3}$$

33. **a.** Graphing the line:

b. Graphing the line:

c.   Graphing the line:

35. **a.**   Graphing the line:

b.   Graphing the line:

c.   Graphing the line:

37.   Graphing the line:

39.   Graphing the line:

41. **a.** Three ordered pairs are (5,40), (10,80), and (20,160).
  b. She will earn (8)(40) = $320 for working 40 hours.
  c. Solving the equation:

$$8x = 240$$
$$x = 30$$

   She worked 30 hours that week.
  d. She should earn (8)(35) = $280 for working 35 hours, not $260. No, the paycheck was not the correct amount.

43. Constructing a line graph:

45. Six ordered pairs are (1985, 20.2), (1990, 34.4), (1995, 44.8), (2000, 65.4), (2005, 104), and (2010, 163).

47. Completing the table:

x (year)	1970	1975	1980	1985	1990	1995	2000	2005	2006
y (age)	21.5	21.75	22.75	23.75	24.25	24.5	24.75	25	24.75

49. The coordinates of A are (6 − 5,2) = (1,2) and of B are (6,2+5) = (6,7).

51. The coordinates of A are (8 − 5,3) = (3,3), of B are (3,3 + 3) = (3,6), and of C are (8,3 + 3) = (8,6).

53. **a.** Yes, the point (2000, 7500) is on the graph.
  b. No, the point (2004, 15000) is not on the graph.
  c. Yes, the point (2005, 15000) is on the graph.

55. Solving the equation:

$$5x - 4 = -3x + 12$$
$$8x - 4 = 12$$
$$8x = 16$$
$$x = 2$$

57. Solving the equation:

$$\frac{1}{2} - \frac{1}{8}(3t - 4) = -\frac{7}{8}t$$
$$8\left(\frac{1}{2} - \frac{1}{8}(3t - 4)\right) = 8\left(-\frac{7}{8}t\right)$$
$$4 - (3t - 4) = -7t$$
$$4 - 3t + 4 = -7t$$
$$4t + 8 = 0$$
$$4t = -8$$
$$t = -2$$

59. Solving the equation:

$$50 = \frac{K}{24}$$
$$K = 24 \cdot 50 = 1,200$$

61. Solving the equation:

$$2(1) + y = 4$$
$$2 + y = 4$$
$$y = 2$$

63. Solving the equation:

$$4\left(\frac{19}{15}\right) - 2y = 4$$
$$\frac{76}{15} - 2y = 4$$
$$-2y = -\frac{16}{15}$$
$$y = \frac{8}{15}$$

65. Writing as a fraction: $-0.06 = -\frac{6}{100}$

67. Substituting $x = 2$:
$$y = 2(2) - 3$$
$$y = 4 - 3$$
$$y = 1$$

69. Simplifying: $\dfrac{1 - (-3)}{-5 - (-2)} = \dfrac{4}{-3} = -\dfrac{4}{3}$

71. Simplifying: $\dfrac{-1 - 4}{3 - 3} = \dfrac{-5}{0}$, which is undefined

73. **a.** The number is $\dfrac{3}{2}$, since $\dfrac{2}{3} \cdot \dfrac{3}{2} = 1$.

b. The number is $-\dfrac{3}{2}$, since $\dfrac{2}{3} \cdot \left(-\dfrac{3}{2}\right) = -1$.

75. Setting $y = 0$, the x-intercept is $\dfrac{c}{a}$. Setting $x = 0$, the y-intercept is $\dfrac{c}{b}$.

77. Setting $y = 0$, the x-intercept is a. Setting $x = 0$, the y-intercept is b.

3.2 The Slope of a Line

1. The slope is $\dfrac{3}{2}$.

3. There is no slope (undefined).

5. The slope is $\dfrac{2}{3}$.

7. Finding the slope: $m = \dfrac{4 - 1}{4 - 2} = \dfrac{3}{2}$

Sketching the graph:

9. Finding the slope: $m = \dfrac{2 - 4}{5 - 1} = \dfrac{-2}{4} = -\dfrac{1}{2}$

Sketching the graph:

11. Finding the slope: $m = \dfrac{2 - (-3)}{4 - 1} = \dfrac{2 + 3}{3} = \dfrac{5}{3}$

Sketching the graph:

13. Finding the slope: $m = \dfrac{3 - (-2)}{1 - (-3)} = \dfrac{3 + 2}{1 + 3} = \dfrac{5}{4}$

Sketching the graph:

15. Finding the slope: $m = \dfrac{-2-2}{3-(-3)} = \dfrac{-4}{3+3} = -\dfrac{2}{3}$

Sketching the graph:

17. Finding the slope: $m = \dfrac{-2-(-5)}{3-2} = \dfrac{-2+5}{1} = 3$

Sketching the graph:

19. Using the slope formula:

$$\dfrac{3-6}{a-2} = -1$$
$$-3 = -1(a-2)$$
$$-3 = -a+2$$
$$-5 = -a$$
$$a = 5$$

21. Using the slope formula:

$$\dfrac{4b-b}{-1-2} = -2$$
$$3b = -2(-3)$$
$$3b = 6$$
$$b = 2$$

23. Completing the table:

x	y
0	2
3	0

Finding the slope: $m = \dfrac{2-0}{0-3} = -\dfrac{2}{3}$

25. Completing the table:

x	y
0	-5
3	-3

Finding the slope: $m = \dfrac{-5-(-3)}{0-3} = \dfrac{-5+3}{-3} = \dfrac{2}{3}$

27. Graphing the line:

The slope is $m = -\dfrac{1}{2}$.

29. Finding the slope of this line: $m = \dfrac{-3-5}{5-2} = -\dfrac{8}{3}$. Since the parallel slope is the same, its slope is $-\dfrac{8}{3}$.

31. Finding the slope of this line: $m = \dfrac{1-4}{-3-3} = \dfrac{-3}{-6} = \dfrac{1}{2}$

Since the perpendicular slope is the negative reciprocal, its slope is –2.

33. Finding the slope of this line: $m = \dfrac{-6-(-4)}{-2-3} = \dfrac{-2}{-5} = \dfrac{2}{5}$. Since the parallel slope is the same, its slope is $\dfrac{2}{5}$.

35. Finding the slope of this line: $m = \dfrac{7-(-3)}{-2-6} = \dfrac{10}{-8} = -\dfrac{5}{4}$

Since the perpendicular slope is the negative reciprocal, its slope is $\dfrac{4}{5}$.

37. **a.** Since the slopes between each successive pairs of points is 2, this could represent ordered pairs from a line.
b. Since the slopes between each successive pairs of points is not the same, this could not represent ordered pairs from a line.

39. Using the slope formula:

$$\frac{\Delta y}{12} = \frac{2}{3}$$
$$3\Delta y = 24$$
$$\Delta y = 8$$

41. **a.** Finding the slope: $m = \dfrac{105-0}{6-0} = 17.5$ miles/hour

b. Finding the slope: $m = \dfrac{80-0}{2-0} = 40$ kilometers/hour

c. Finding the slope: $m = \dfrac{3600-0}{30-0} = 120$ feet/second

d. Finding the slope: $m = \dfrac{420-0}{15-0} = 28$ meters/minute

43. For line segment A, the slope is: $m = \dfrac{250-300}{2007-2006} = -50$ million cameras/year

For line segment B, the slope is: $m = \dfrac{175-250}{2008-2007} = -75$ million cameras/year

For line segment C, the slope is: $m = \dfrac{125-150}{2010-2009} = -25$ million cameras/year

45. **a.** Let d represent the distance. Using the slope formula:

$$\frac{0-1106}{d-0} = -\frac{7}{100}$$
$$-7d = -110600$$
$$d = 15800$$

The distance to point A is 15,800 feet.

b. The slope is $-\dfrac{7}{100} = -0.07$.

47. Computing the slope: $m \approx \dfrac{17{,}000-8{,}000}{2008-1997} = \dfrac{9{,}000}{11} = 818.2$

Between 1997 and 2008, the number of solar thermal collector shipments increased an average of 818.2 shipments per year.

49. **a.** Computing the slope: $m \approx \dfrac{34.4-20.2}{1990-1985} = \dfrac{14.2}{5} = 2.84$

Between 1985 and 1990, the amount of money bet on horse racing increased at an average of $2.84 million per year.

b. Computing the slope: $m \approx \dfrac{104-65.4}{2005-2000} = \dfrac{38.6}{5} = 7.72$

Between 2000 and 2005, the amount of money bet on horse racing increased at an average of $7.72 million per year.

51. Substituting $x = 4$:

$$3(4)+2y = 12$$
$$12+2y = 12$$
$$2y = 0$$
$$y = 0$$

53. Solving for y:

$$3x+2y = 12$$
$$2y = -3x+12$$
$$y = -\frac{3}{2}x+6$$

55. Solving for t:

$$A = P + Prt$$
$$A - P = Prt$$
$$t = \frac{A - P}{Pr}$$

57. Simplifying: $2\left(-\dfrac{1}{2}\right) = -1$

59. Simplifying: $\dfrac{5 - (-3)}{2 - 6} = \dfrac{8}{-4} = -2$

61. Solving for y:

$$\frac{y - b}{x - 0} = m$$
$$y - b = mx$$
$$y = mx + b$$

63. Solving for y:

$$y - 3 = -2(x + 4)$$
$$y - 3 = -2x - 8$$
$$y = -2x - 5$$

65. Solving for y: $y = -\dfrac{4}{3}(0) + 5 = 0 + 5 = 5$

3.3 The Equation of a Line

1. Using the slope-intercept formula: $y = 2x + 3$

3. Using the slope-intercept formula: $y = x - 5$

5. Using the slope-intercept formula: $y = \dfrac{1}{2}x + \dfrac{3}{2}$

7. Using the slope-intercept formula: $y = 4$

9. The slope is 3, the y-intercept is -2, and the perpendicular slope is $-\dfrac{1}{3}$.

11. The slope is $\dfrac{2}{3}$, the y-intercept is -4, and the perpendicular slope is $-\dfrac{3}{2}$.

13. The slope is $-\dfrac{4}{5}$, the y-intercept is 4, and the perpendicular slope is $\dfrac{5}{4}$.

$4x + 5y = 20$

15. The slope is $\dfrac{1}{2}$ and the y-intercept is -4. Using the slope-intercept form, the equation is $y = \dfrac{1}{2}x - 4$.

17. The slope is $-\dfrac{2}{3}$ and the y-intercept is 3. Using the slope-intercept form, the equation is $y = -\dfrac{2}{3}x + 3$.

19. Using the point-slope formula:

$$y - (-5) = 2(x - (-2))$$
$$y + 5 = 2(x + 2)$$
$$y + 5 = 2x + 4$$
$$y = 2x - 1$$

21. Using the point-slope formula:

$$y - 1 = -\dfrac{1}{2}(x - (-4))$$
$$y - 1 = -\dfrac{1}{2}(x + 4)$$
$$y - 1 = -\dfrac{1}{2}x - 2$$
$$y = -\dfrac{1}{2}x - 1$$

23. Using the point-slope formula:

$$y - 2 = -3\left(x - \left(-\dfrac{1}{3}\right)\right)$$
$$y - 2 = -3\left(x + \dfrac{1}{3}\right)$$
$$y - 2 = -3x - 1$$
$$y = -3x + 1$$

25. Using the point-slope formula:

$$y - (-2) = \dfrac{2}{3}(x - (-4))$$
$$y + 2 = \dfrac{2}{3}(x + 4)$$
$$y + 2 = \dfrac{2}{3}x + \dfrac{8}{3}$$
$$y = \dfrac{2}{3}x + \dfrac{2}{3}$$

27. Using the point-slope formula:

$$y - 2 = -\dfrac{1}{4}(x - (-5))$$
$$y - 2 = -\dfrac{1}{4}(x + 5)$$
$$y - 2 = -\dfrac{1}{4}x - \dfrac{5}{4}$$
$$y = -\dfrac{1}{4}x + \dfrac{3}{4}$$

29. First find the slope: $m = \dfrac{-1 - (-4)}{1 - (-2)} = \dfrac{-1 + 4}{1 + 2} = \dfrac{3}{3} = 1$. Using the point-slope formula:

$$y - (-1) = 1(x - 1)$$
$$y + 1 = x - 1$$
$$-x + y = -2$$
$$x - y = 2$$

31.   First find the slope: $m = \dfrac{1-(-5)}{2-(-1)} = \dfrac{1+5}{2+1} = \dfrac{6}{3} = 2$. Using the point-slope formula:

$$y - 1 = 2(x - 2)$$
$$y - 1 = 2x - 4$$
$$-2x + y = -3$$
$$2x - y = 3$$

33.   First find the slope: $m = \dfrac{-1-\left(-\dfrac{1}{5}\right)}{-\dfrac{1}{3}-\dfrac{1}{3}} = \dfrac{-1+\dfrac{1}{5}}{-\dfrac{2}{3}} = \dfrac{-\dfrac{4}{5}}{-\dfrac{2}{3}} = \dfrac{4}{5} \cdot \dfrac{3}{2} = \dfrac{6}{5}$. Using the point-slope formula:

$$y - (-1) = \frac{6}{5}\left(x - \left(-\frac{1}{3}\right)\right)$$
$$y + 1 = \frac{6}{5}\left(x + \frac{1}{3}\right)$$
$$5(y + 1) = 6\left(x + \frac{1}{3}\right)$$
$$5y + 5 = 6x + 2$$
$$6x - 5y = 3$$

35.   **a.**   The parallel slope is 3.   **b.**   The perpendicular slope is $-\dfrac{1}{3}$.

37.   Solving for y: **39.**   Solving for y:
$$3x + y = -2$$ $$2x + 5y = -11$$
$$y = -3x - 2$$ $$5y = -2x - 11$$
$$y = -\frac{2}{5}x - \frac{11}{5}$$

a.   The parallel slope is –3.   **a.**   The parallel slope is $-\dfrac{2}{5}$.

b.   The perpendicular slope is $\dfrac{1}{3}$.   **b.**   The perpendicular slope is $\dfrac{5}{2}$.

41.   **a.**   For the x-intercept, substitute $y = 0$:   For the y-intercept, substitute $x = 0$:
$$3x - 2(0) = 10$$ $$3(0) - 2y = 10$$
$$3x = 10$$ $$-2y = 10$$
$$x = \frac{10}{3}$$ $$y = -5$$

b.   Substituting $y = 1$:
$$3x - 2(1) = 10$$
$$3x - 2 = 10$$
$$3x = 12$$
$$x = 4$$
Another solution is (4,1). Other answers are possible.

c.   Solving for y:
$$3x - 2y = 10$$
$$-2y = -3x + 10$$
$$y = \frac{3}{2}x - 5$$

d.   Substituting $x = 2$: $y = \dfrac{3}{2}(2) - 5 = 3 - 5 = -2$. No, the point (2,2) is not a solution to the equation.

43. **a.** For the x-intercept, substitute $y = 0$:

$$\frac{3x}{4} - \frac{0}{2} = 1$$
$$\frac{3x}{4} = 1$$
$$3x = 4$$
$$x = \frac{4}{3}$$

For the y-intercept, substitute $x = 0$:

$$\frac{3(0)}{4} - \frac{y}{2} = 1$$
$$-\frac{y}{2} = 1$$
$$-y = 2$$
$$y = -2$$

b. Substituting $x = 2$:

$$\frac{3(2)}{4} - \frac{y}{2} = 1$$
$$\frac{3}{2} - \frac{y}{2} = 1$$
$$-\frac{y}{2} = -\frac{1}{2}$$
$$y = 1$$

Another solution is (2,1). Other answers are possible.

c. Solving for y:

$$\frac{3x}{4} - \frac{y}{2} = 1$$
$$4\left(\frac{3x}{4} - \frac{y}{2}\right) = 4(1)$$
$$3x - 2y = 4$$
$$-2y = -3x + 4$$
$$y = \frac{3}{2}x - 2$$

d. Substituting $x = 1$: $y = \frac{3}{2}(1) - 2 = \frac{3}{2} - 2 = -\frac{1}{2}$. No, the point (1,2) is not a solution to the equation.

45. **a.** Solving for x:
$$-2x + 1 = -3$$
$$-2x = -4$$
$$x = 2$$

b. Substituting $y = 0$:
$$-2x + 0 = -3$$
$$-2x = -3$$
$$x = \frac{3}{2}$$

c. Substituting $x = 0$:
$$-2(0) + y = -3$$
$$y = -3$$

d. Sketching the graph:

e. Solving for y:
$$-2x + y = -3$$
$$y = 2x - 3$$

47.   Two points on the line are $(0,-4)$ and $(2,0)$. Finding the slope: $m = \dfrac{0-(-4)}{2-0} = \dfrac{4}{2} = 2$

Using the slope-intercept form, the equation is $y = 2x - 4$.

49.   Two points on the line are $(0,4)$ and $(-2,0)$. Finding the slope: $m = \dfrac{0-4}{-2-0} = \dfrac{-4}{-2} = 2$

Using the slope-intercept form, the equation is $y = 2x + 4$.

51.   The slope is 0 and the y-intercept is -2.

53.   First find the slope:
$$3x - y = 5$$
$$-y = -3x + 5$$
$$y = 3x - 5$$

So the slope is 3. Using $(-1,4)$ in the point-slope formula:
$$y - 4 = 3(x - (-1))$$
$$y - 4 = 3(x + 1)$$
$$y - 4 = 3x + 3$$
$$y = 3x + 7$$

55.   First find the slope:
$$2x - 5y = 10$$
$$-5y = -2x + 10$$
$$y = \frac{2}{5}x - 2$$

So the perpendicular slope is $-\dfrac{5}{2}$. Using $(-4,-3)$ in the point-slope formula:

$$y - (-3) = -\frac{5}{2}(x - (-4))$$
$$y + 3 = -\frac{5}{2}(x + 4)$$
$$y + 3 = -\frac{5}{2}x - 10$$
$$y = -\frac{5}{2}x - 13$$

57.   The perpendicular slope is $\dfrac{1}{4}$. Using $(-1,0)$ in the point-slope formula:

$$y - 0 = \frac{1}{4}(x - (-1))$$
$$y = \frac{1}{4}(x + 1)$$
$$y = \frac{1}{4}x + \frac{1}{4}$$

59.   Using the points $(3,0)$ and $(0,2)$, first find the slope: $m = \dfrac{2-0}{0-3} = -\dfrac{2}{3}$

Using the slope-intercept formula, the equation is: $y = -\dfrac{2}{3}x + 2$

61.   Using the points $(2000, 65.4)$ and $(2005, 104)$, first find the slope: $m = \dfrac{104 - 65.4}{2005 - 2000} = \dfrac{38.6}{5} = 7.72$

Using the point-slope formula:
$$y - 104 = 7.72(x - 2005)$$
$$y - 104 = 7.72x - 15,478.6$$
$$y = 7.72x - 15,374.6$$

63.   a.   Using the points $(0,32)$ and $(25,77)$, first find the slope: $m = \dfrac{77 - 32}{25 - 0} = \dfrac{45}{25} = \dfrac{9}{5}$

Using the slope-intercept formula, the equation is: $F = \dfrac{9}{5}C + 32$

b.   Substituting $C = 30$: $F = \dfrac{9}{5}(30) + 32 = 54 + 32 = 86°$

65.   a.   Substituting $n = 10,000$: $C = 215,000 + 9.5(10,000) = \$310,000$

b.   Finding the average cost: $\dfrac{\$310,000}{10,000} = \$31$ per textbook

c.   Since each textbook costs \$9.50 in materials, this is the cost to produce the next textbook.

67.   a.   If he sells 20 TV's, he will earn \$1,500.   **b.**   If he sells 40 TV's, he will earn \$2,000.

c.   Using the slope formula: $m = \dfrac{2000 - 1500}{40 - 20} = \dfrac{500}{20} = 25$

d.   He makes \$25 for each TV he sells.   **e.**   The equation of the line is $y = 25x + 1,000$.

69.   a.   After 5 years the car is worth \$5,000.

b.   Using the slope formula: $m = \dfrac{5000 - 20000}{5 - 0} = \dfrac{-15000}{5} = -3000$

Since $(0, 20000)$ is the y-intercept, the equation is $y = -3,000x + 20,000$

c.   Substituting $x = 10$: $y = -3,000(10) + 20,000 = -30,000 + 20,000 = -10,000$

This can't be possible since the value is negative.

71.   Let w represent the width and $4w + 3$ represent the length. Using the perimeter formula:
$$2w + 2(4w + 3) = 56$$
$$2w + 8w + 6 = 56$$
$$10w = 50$$
$$w = 5$$
The width is 5 inches and the length is 23 inches.

73.   The total amount collected is: \$732.50 − \$66 = \$666.50

Let x represent the sales. Since this amount includes the sales tax:
$$x + 0.075x = 666.50$$
$$1.075x = 666.50$$
$$x = 620$$
The amount which is sales tax is therefore: \$666.50 − \$620 = \$46.50

75.   Since $0 + 0 \le 4$ and $4 + 0 \le 4$, but $2 + 3 > 4$, the points $(0,0)$ and $(4,0)$ are solutions.

77.   Since $0 \le \dfrac{1}{2}(0)$ and $0 \le \dfrac{1}{2}(2)$, but $0 > \dfrac{1}{2}(-2)$, the points $(0,0)$ and $(2,0)$ are solutions.

79. First find the midpoint: $\left(\dfrac{1+7}{2}, \dfrac{4+8}{2}\right) = \left(\dfrac{8}{2}, \dfrac{12}{2}\right) = (4, 6)$. Now find the slope: $m = \dfrac{8-4}{7-1} = \dfrac{4}{6} = \dfrac{2}{3}$

So the perpendicular slope is $-\dfrac{3}{2}$. Using $(4,6)$ in the point-slope formula:

$$y - 6 = -\dfrac{3}{2}(x - 4)$$

$$y - 6 = -\dfrac{3}{2}x + 6$$

$$y = -\dfrac{3}{2}x + 12$$

81. First find the midpoint: $\left(\dfrac{-5-1}{2}, \dfrac{1+4}{2}\right) = \left(\dfrac{-6}{2}, \dfrac{5}{2}\right) = \left(-3, \dfrac{5}{2}\right)$. Now find the slope: $m = \dfrac{4-1}{-1+5} = \dfrac{3}{4}$

So the perpendicular slope is $-\dfrac{4}{3}$. Using $\left(-3, \dfrac{5}{2}\right)$ in the point-slope formula:

$$y - \dfrac{5}{2} = -\dfrac{4}{3}(x + 3)$$

$$y - \dfrac{5}{2} = -\dfrac{4}{3}x - 4$$

$$y = -\dfrac{4}{3}x - \dfrac{3}{2}$$

3.4 Linear Inequalities in Two Variables

1. Graphing the solution set:

3. Graphing the solution set:

5. Graphing the solution set:

7. Graphing the solution set:

9.   Graphing the solution set:

11.   Graphing the solution set:

13.   Graphing the solution set:

15.   Graphing the solution set:

17.   Graphing the solution set:

19.   Graphing the solution set:

21.   Graphing the solution set:

23.   Graphing the solution set:

25. Graphing the solution set:

27. Graphing the solution set:

29. Graphing the solution set:

31. Graphing the solution set:

33. Graphing the solution set:

35. Graphing the solution set:

37. The inequality is $x + y > 4$.

39. The inequality is $-x + 2y \le 4$ or $y \le \dfrac{1}{2}x + 2$.

41. The inequality is $y \ge \dfrac{2}{3}x + 2$.

43. **a.** Solving the inequality:

$$\frac{1}{3} + \frac{y}{2} < 1$$
$$\frac{y}{2} < \frac{2}{3}$$
$$y < \frac{4}{3}$$

b. Solving the inequality:

$$\frac{1}{3} - \frac{y}{2} < 1$$
$$-\frac{y}{2} < \frac{2}{3}$$
$$y > -\frac{4}{3}$$

c. Solving for y:

$$\frac{x}{3} + \frac{y}{2} = 1$$
$$6\left(\frac{x}{3} + \frac{y}{2}\right) = 6(1)$$
$$2x + 3y = 6$$
$$3y = -2x + 6$$
$$y = -\frac{2}{3}x + 2$$

d. Graphing the inequality:

$y = -\frac{4}{3}x + 4$

45. Graphing the region:

$x + y = 200$

47. Graphing the region:

$22x + 12y = 7920$

49.   The inequality is $440x + 715y < 11,440$, which simplifies to $8x + 13y < 208$. Sketching the graph:

51.   Solving the inequality:

$$\frac{1}{3} + \frac{y}{5} \le \frac{26}{15}$$

$$15\left(\frac{1}{3} + \frac{y}{5}\right) \le 15\left(\frac{26}{15}\right)$$

$$5 + 3y \le 26$$

$$3y \le 21$$

$$y \le 7$$

53.   Solving the inequality:

$$5t - 4 > 3t - 8$$

$$2t - 4 > -8$$

$$2t > -4$$

$$t > -2$$

55.   Solving the inequality:

$$-9 < -4 + 5t < 6$$

$$-5 < 5t < 10$$

$$-1 < t < 2$$

57.   Completing the table:

x	y
0	0
10	75
20	150

59.   Completing the table:

x	y
0	0
1	1
1	−1

61.   Graphing the inequality:

63.   Graphing the inequality:

65.   Graphing the inequality:

67.   Graphing the inequality:

69.   Shading the region:

3.5 Introduction to Functions

1.   The domain is {1,2,4} and the range is {1,3,5}. This is a function.

3.   The domain is {−1,1,2} and the range is {−5,3}. This is a function.

5.   The domain is {3,7} and the range is {−1,4}. This is not a function.

7.   The domain is {a,b,c,d} and the range is {3,4,5}. This is a function.

9.   The domain is {a} and the range is {1,2,3,4}. This is not a function.

11.   Yes, since it passes the vertical line test.   **13.**   No, since it fails the vertical line test.

15.   No, since it fails the vertical line test.   **17.**   Yes, since it passes the vertical line test.

19.   Yes, since it passes the vertical line test.

21.   The domain is $\{x \mid -5 \leq x \leq 5\}$ and the range is $\{y \mid 0 \leq y \leq 5\}$.

23.   The domain is $\{x \mid -5 \leq x \leq 3\}$ and the range is $\{y \mid y = 4\}$.

25.   The domain is all real numbers and the range is $\{y \mid y \geq -1\}$. This is a function.

27.   The domain is all real numbers and the range is $\{y \mid y \geq 4\}$. This is a function.

29.   The domain is $\{x \mid x \geq -1\}$ and the range is all real numbers. This is not a function.

31.   The domain is $\{x \mid x \geq 4\}$ and the range is all real numbers. This is not a function.

33.   The domain is all real numbers and the range is $\{y \mid y \geq 0\}$. This is a function.

35.   The domain is all real numbers and the range is $\{y \mid y \geq -2\}$. This is a function.

37.   The domain is {2004, 2005, 2006, 2007, 2008, 2009, 2010} and the range is {680, 730, 800, 900, 920, 990, 1030}.

39.   **a.**   The equation is $y = 8.5x$ for $10 \leq x \leq 40$.

Hours Worked	Function Rule	Gross Pay ($)
x	$y = 8.5x$	y
10	$y = 8.5(10) = 85$	85
20	$y = 8.5(20) = 170$	170
30	$y = 8.5(30) = 255$	255
40	$y = 8.5(40) = 340$	340

   b.   Completing the table:

   c.   Constructing a line graph:

   d.   The domain is $\{x \mid 10 \leq x \leq 40\}$ and the range is $\{y \mid 85 \leq y \leq 340\}$.

   e.   The minimum is $85 and the maximum is $340.

Time (sec)	Function Rule	Height (ft)
t	$h = 16t - 16t^2$	h
0	$h = 16(0) - 16(0)^2$	0
0.1	$h = 16(0.1) - 16(0.1)^2$	1.44
0.2	$h = 16(0.2) - 16(0.2)^2$	2.56
0.3	$h = 16(0.3) - 16(0.3)^2$	3.36
0.4	$h = 16(0.4) - 16(0.4)^2$	3.84
0.5	$h = 16(0.5) - 16(0.5)^2$	4
0.6	$h = 16(0.6) - 16(0.6)^2$	3.84
0.7	$h = 16(0.7) - 16(0.7)^2$	3.36
0.8	$h = 16(0.8) - 16(0.8)^2$	2.56
0.9	$h = 16(0.9) - 16(0.9)^2$	1.44
1	$h = 16(1) - 16(1)^2$	0

41.   a.   Completing the table:

b.   The domain is $\{t \mid 0 \le t \le 1\}$ and the range is $\{h \mid 0 \le h \le 4\}$.

c.   Graphing the function:

43.   a.   Graphing the function:

b.   The domain is $\{r \mid 0 \le r \le 3\}$ and the range is $\{A \mid 0 \le A \le 9\pi\}$.

45.   a.   Yes, since it passes the vertical line test.

b.   The domain is $\{t \mid 0 \le t \le 6\}$ and the range is $\{h \mid 0 \le h \le 60\}$.

c.   At time $t = 3$ the ball reaches its maximum height.

d.   The maximum height is $h = 60$.   **e.**   At time $t = 6$ the ball hits the ground.

47.   a.   Figure III   **b.**   Figure I

c.   Figure II   **d.**   Figure IV

49. Substituting $x = 4$: $y = 3(4) - 2 = 12 - 2 = 10$

51. Substituting $x = -4$: $y = 3(-4) - 2 = -12 - 2 = -14$

53. Substituting $x = 2$: $y = (2)^2 - 3 = 4 - 3 = 1$

55. Substituting $x = 0$: $y = (0)^2 - 3 = 0 - 3 = -3$

57. Solving for y:

$$\frac{8}{5} - 2y = 4$$

$$-2y = \frac{12}{5}$$

$$y = -\frac{6}{5}$$

59. Substituting $x = 0$ and $y = 0$:

$$0 = a(0 - 80)^2 + 70$$
$$0 = 6400a + 70$$
$$6400a = -70$$
$$a = -\frac{7}{640}$$

61. Simplifying: $7.5(20) = 150$

63. Simplifying: $4(3.14)(9) \approx 113$

65. Simplifying: $4(-2) - 1 = -8 - 1 = -9$

67. **a.** Substituting $t = 10$: $s = \frac{60}{10} = 6$ **b.** Substituting $t = 8$: $s = \frac{60}{8} = 7.5$

69. **a.** Substituting $x = 5$: $(5)^2 + 2 = 25 + 2 = 27$ **b.** Substituting $x = -2$: $(-2)^2 + 2 = 4 + 2 = 6$

71. The domain is all real numbers and the range is $\{y \mid y \leq 5\}$. This is a function.

73. The domain is $\{x \mid x \geq 3\}$ and the range is all real numbers. This is not a function.

75. The domain is $\{x \mid -4 \leq x \leq 4\}$ and the range is $\{y \mid -4 \leq y \leq 4\}$. This is not a function.

3.6 Function Notation

1. Evaluating the function: $f(2) = 2(2) - 5 = 4 - 5 = -1$

3. Evaluating the function: $f(-3) = 2(-3) - 5 = -6 - 5 = -11$

5. Evaluating the function: $g(-1) = (-1)^2 + 3(-1) + 4 = 1 - 3 + 4 = 2$

7. Evaluating the function: $g(-3) = (-3)^2 + 3(-3) + 4 = 9 - 9 + 4 = 4$

9. First evaluate each function:

$$g(4) = (4)^2 + 3(4) + 4 = 16 + 12 + 4 = 32 \qquad f(4) = 2(4) - 5 = 8 - 5 = 3$$

Now evaluating: $g(4) + f(4) = 32 + 3 = 35$

11. First evaluate each function:

$$f(3) = 2(3) - 5 = 6 - 5 = 1 \qquad\qquad g(2) = (2)^2 + 3(2) + 4 = 4 + 6 + 4 = 14$$

Now evaluating: $f(3) - g(2) = 1 - 14 = -13$

13. Evaluating the function: $f(0) = 3(0)^2 - 4(0) + 1 = 0 - 0 + 1 = 1$

15. Evaluating the function: $g(-4) = 2(-4) - 1 = -8 - 1 = -9$

17. Evaluating the function: $f(-1) = 3(-1)^2 - 4(-1) + 1 = 3 + 4 + 1 = 8$

19. Evaluating the function: $g(10) = 2(10) - 1 = 20 - 1 = 19$

21. Evaluating the function: $f(3) = 3(3)^2 - 4(3) + 1 = 27 - 12 + 1 = 16$

23. Evaluating the function: $g\left(\dfrac{1}{2}\right) = 2\left(\dfrac{1}{2}\right) - 1 = 1 - 1 = 0$ **25.** Evaluating the function: $f(a) = 3a^2 - 4a + 1$

27. $f(1) = 4$ **29.** $g\left(\dfrac{1}{2}\right) = 0$

31. $g(-2) = 2$ **33.** Evaluating the function: $f(0) = 2(0)^2 - 8 = 0 - 8 = -8$

35. Evaluating the function: $g(-4) = \dfrac{1}{2}(-4) + 1 = -2 + 1 = -1$

37. Evaluating the function: $f(a) = 2a^2 - 8$ **39.** Evaluating the function: $f(b) = 2b^2 - 8$

41. Graphing the function:

43.   Finding where $f(x) = x$:

$$\frac{1}{2}x + 2 = x$$

$$2 = \frac{1}{2}x$$

$$x = 4$$

45.   Graphing the function:

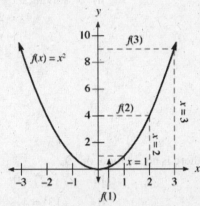

47.   Evaluating: $V(3) = 150 \cdot 2^{3/3} = 150 \cdot 2 = 300$; The painting is worth $300 in 3 years.

Evaluating: $V(6) = 150 \cdot 2^{6/3} = 150 \cdot 4 = 600$; The painting is worth $600 in 6 years.

49.   **a.**   True   **b.**   False

          c.   True

51.   **a.**   True   **b.**   True

          c.   False   **d.**   True

53.   **a.**   Evaluating: $V(3.75) = -3300(3.75) + 18000 = \$5,625$

          b.   Evaluating: $V(5) = -3300(5) + 18000 = \$1,500$

          c.   The domain of this function is $\{t \mid 0 \le t \le 5\}$.

          d.   Sketching the graph:

          e.   The range of this function is $\{V(t) \mid 1,500 \le V(t) \le 18,000\}$.

          f.   Solving $V(t) = 10000$:

$$-3300t + 18000 = 10000$$

$$-3300t = -8000$$

$$t \approx 2.42$$

          The copier will be worth $10,000 after approximately 2.42 years.

55.   Simplifying: $(35x - 0.1x^2) - (8x + 500) = 35x - 0.1x^2 - 8x - 500 = -0.1x^2 + 27x - 500$

57. Simplifying: $\left(4x^2+3x+2\right)+\left(2x^2-5x-6\right)=4x^2+3x+2+2x^2-5x-6=6x^2-2x-4$

59. Simplifying: $\left(4x^2+3x+2\right)-\left(2x^2-5x-6\right)=4x^2+3x+2-2x^2+5x+6=2x^2+8x+8$

61. Simplifying: $0.6\left(M-70\right)=0.6M-42$

63. Simplifying: $\left(4x-3\right)\left(x-1\right)=4x^2-4x-3x+3=4x^2-7x+3$

65. Solving the equation:

$$|3x-5|=7$$
$$3x-5=-7,7$$
$$3x=-2,12$$
$$x=-\frac{2}{3},4$$

67. Solving the equation:

$$|4y+2|-8=-2$$
$$|4y+2|=6$$
$$4y+2=-6,6$$
$$4y=-8,4$$
$$y=-2,1$$

69. Solving the equation:

$$5+|6t+2|=3$$
$$|6t+2|=-2$$

Since this last equation is impossible, there is no solution, or $\varnothing$.

71. **a.** $f(2)=2$ **b.** $f(-4)=0$

     c. $g(0)=1$ **d.** $g(3)=4$

73. **a.** Completing the table:

Weight (ounces)	0.6	1.0	1.1	2.5	3.0	4.8	5.0	5.3
Cost (cents)	39	39	63	87	87	135	135	159

     b. The letter weighs over 2 ounces, but not over 3 ounces. As an inequality, this can be written as $2<x\le3$.

     c. The domain is $\{x\,|\,0<x\le6\}$.   **d.** The range is $\{39,63,87,111,135,159\}$.

3.7 Algebra and Composition with Functions

1. Writing the formula: $f+g=f(x)+g(x)=\left(4x-3\right)+\left(2x+5\right)=6x+2$

3. Writing the formula: $g-f=g(x)-f(x)=\left(2x+5\right)-\left(4x-3\right)=-2x+8$

5. Writing the formula: $g+f=g(x)+f(x)=\left(x-2\right)+\left(3x-5\right)=4x-7$

7. Writing the formula: $g+h=g(x)+h(x)=\left(x-2\right)+\left(3x^2\right)=3x^2+x-2$

9. Writing the formula: $g-f=g(x)-f(x)=\left(x-2\right)-\left(3x-5\right)=-2x+3$

11. Writing the formula: $fh=f(x)\bullet h(x)=\left(3x-5\right)\left(3x^2\right)=9x^3-15x^2$

13. Writing the formula: $h/f=\dfrac{h(x)}{f(x)}=\dfrac{3x^2}{3x-5}$ **15.** Writing the formula: $f/h=\dfrac{f(x)}{h(x)}=\dfrac{3x-5}{3x^2}$

17. Writing the formula: $f+g+h=f(x)+g(x)+h(x)=\left(3x-5\right)+\left(x-2\right)+\left(3x^2\right)=3x^2+4x-7$

19. Evaluating: $(f+g)(2)=f(2)+g(2)=\left(2\bullet2+1\right)+\left(4\bullet2+2\right)=5+10=15$

21. Evaluating: $(fg)(3)=f(3)\bullet g(3)=\left(2\bullet3+1\right)\left(4\bullet3+2\right)=7\bullet14=98$

23. Evaluating: $(h/g)(1)=\dfrac{h(1)}{g(1)}=\dfrac{4(1)^2+4(1)+1}{4(1)+2}=\dfrac{9}{6}=\dfrac{3}{2}$

25. Evaluating: $(fh)(0)=f(0)\bullet h(0)=\left(2(0)+1\right)\left(4(0)^2+4(0)+1\right)=(1)(1)=1$

27. Evaluating: $(f+g+h)(2)=f(2)+g(2)+h(2)=\left(2(2)+1\right)+\left(4(2)+2\right)+\left(4(2)^2+4(2)+1\right)=5+10+25=40$

29. Evaluating:

$$(h+fg)(3)=h(3)+f(3)\bullet g(3)=\left(4(3)^2+4(3)+1\right)+(2(3)+1)\bullet(4(3)+2)=49+7\bullet14=49+98=147$$

31. **a.** Evaluating: $(f\circ g)(5)=f(g(5))=f(5+4)=f(9)=9^2=81$

b. Evaluating: $(g\circ f)(5)=g(f(5))=g\left(5^2\right)=g(25)=25+4=29$

c. Evaluating: $(f\circ g)(x)=f(g(x))=f(x+4)=(x+4)^2$

d. Evaluating: $(g\circ f)(x)=g(f(x))=g\left(x^2\right)=x^2+4$

33. **a.** Evaluating: $(f\circ g)(0)=f(g(0))=f(4\bullet0-1)=f(-1)=(-1)^2+3(-1)=1-3=-2$

b. Evaluating: $(g\circ f)(0)=g(f(0))=g\left(0^2+3\bullet0\right)=g(0)=4(0)-1=-1$

c. Evaluating: $(g\circ f)(x)=g(f(x))=g\left(x^2+3x\right)=4\left(x^2+3x\right)-1=4x^2+12x-1$

35. Evaluating each composition:

$$(f\circ g)(x)=f(g(x))=f\left(\frac{x+4}{5}\right)=5\left(\frac{x+4}{5}\right)-4=x+4-4=x$$

$$(g\circ f)(x)=g(f(x))=g(5x-4)=\frac{5x-4+4}{5}=\frac{5x}{5}=x$$

Thus $(f\circ g)(x)=(g\circ f)(x)=x$.

37. Using the graph: $f(2)+5=1+5=6$

39. Using the graph: $f(-3)+g(-3)=3+(-1)=2$

41. Using the graph: $(f\circ g)(0)=f(g(0))=f(-3)=3$

43. Using the graph: $f(x)=-3$ when $x=-8$

45. Using the graph: $f(-3)+2=4+2=6$

47. Using the graph: $f(2)+g(2)=3+2=5$

49. Using the graph: $(f\circ g)(0)=f(g(0))=f(2)=3$

51. Using the graph: $f(x)=1$ when $x=-6$

53. **a.** Finding the revenue: $R(x)=x(11.5-0.05x)=11.5x-0.05x^2$

b. Finding the cost: $C(x)=2x+200$

c. Finding the profit: $P(x)=R(x)-C(x)=\left(11.5x-0.05x^2\right)-(2x+200)=-0.05x^2+9.5x-200$

d. Finding the average cost: $\bar{C}(x)=\dfrac{C(x)}{x}=\dfrac{2x+200}{x}=2+\dfrac{200}{x}$

55. **a.** The function is $M(x)=220-x$.   **b.** Evaluating: $M(24)=220-24=196$ beats per minute

c. The training heart rate function is: $T(M)=62+0.6(M-62)=0.6M+24.8$

Finding the composition: $T(M(x))=T(220-x)=0.6(220-x)+24.8=156.8-0.6x$

Evaluating: $T(M(24))=156.8-0.6(24)\approx142$ beats per minute

d. Evaluating: $T(M(36))=156.8-0.6(36)\approx135$ beats per minute

e. Evaluating: $T(M(48))=156.8-0.6(48)\approx128$ beats per minute

57. Solving the inequality:

$$|x-3|<1$$
$$-1<x-3<1$$
$$2<x<4$$

59. Solving the inequality:

$$|6-x|>2$$

$$6-x<-2 \quad\text{or}\quad 6-x>2$$
$$-x<-8 \qquad\qquad -x>-4$$
$$x>8 \qquad\qquad\quad x<4$$

61. Solving the inequality:

$$|7x-1| \le 6$$
$$-6 \le 7x-1 \le 6$$
$$-5 \le 7x \le 7$$
$$-\frac{5}{7} \le x \le 1$$

63. Simplifying: $16(3.5)^2 = 16(12.25) = 196$

65. Simplifying: $\frac{180}{45} = 4$

67. Simplifying: $\frac{0.0005(200)}{(0.25)^2} = \frac{0.1}{0.0625} = 1.6$

69. Solving for K:

$$15 = K(5)$$
$$K = 3$$

71. Solving for K:

$$50 = \frac{K}{48}$$
$$K = 50 \cdot 48 = 2,400$$

3.8 Variation

1. Direct variation
3. Direct variation
5. Direct variation
7. Direct variation
9. Direct variation
11. Inverse variation
13. The variation equation is $y = Kx$. Substituting $x = 2$ and $y = 10$:

$$10 = K \cdot 2$$
$$K = 5$$

So $y = 5x$. Substituting $x = 6$: $y = 5 \cdot 6 = 30$

15. The variation equation is $y = Kx$. Substituting $x = 4$ and $y = -32$:

$$-32 = K \cdot 4$$
$$K = -8$$

So $y = -8x$. Substituting $y = -40$:

$$-40 = -8x$$
$$x = 5$$

17. The variation equation is $r = \frac{K}{s}$. Substituting $s = 4$ and $r = -3$:

$$-3 = \frac{K}{4}$$
$$K = -12$$

So $r = \frac{-12}{s}$. Substituting $s = 2$: $r = \frac{-12}{2} = -6$

19. The variation equation is $r = \frac{K}{s}$. Substituting $s = 3$ and $r = 8$:

$$8 = \frac{K}{3}$$
$$K = 24$$

So $r = \frac{24}{s}$. Substituting $r = 48$:

$$48 = \frac{24}{s}$$
$$48s = 24$$
$$s = \frac{1}{2}$$

21. The variation equation is $d = Kr^2$. Substituting $r = 5$ and $d = 10$:

$$10 = K \cdot 5^2$$
$$10 = 25K$$
$$K = \frac{2}{5}$$

So $d = \frac{2}{5}r^2$. Substituting $r = 10$: $d = \frac{2}{5}(10)^2 = \frac{2}{5} \cdot 100 = 40$

23. The variation equation is $d = Kr^2$. Substituting $r = 2$ and $d = 100$:

$$100 = K \cdot 2^2$$
$$100 = 4K$$
$$K = 25$$

So $d = 25r^2$. Substituting $r = 3$: $d = 25(3)^2 = 25 \cdot 9 = 225$

25. The variation equation is $y = K\sqrt{x}$. Substituting $x = 9$ and $y = 6$:

$$6 = K\sqrt{9}$$
$$6 = 3K$$
$$K = 2$$

So $y = 2\sqrt{x}$. Substituting $x = 25$: $y = 2\sqrt{25} = 2 \cdot 5 = 10$

27. The variation equation is $y = K\sqrt{x}$. Substituting $x = 25$ and $y = 20$:

$$20 = K\sqrt{25}$$
$$20 = 5K$$
$$K = 4$$

So $y = 4\sqrt{x}$. Substituting $x = 16$: $y = 4\sqrt{16} = 4 \cdot 4 = 16$

29. The variation equation is $y = \dfrac{K}{x^2}$. Substituting $x = 3$ and $y = 45$:

$$45 = \frac{K}{3^2}$$
$$45 = \frac{K}{9}$$
$$K = 405$$

So $y = \dfrac{405}{x^2}$. Substituting $x = 5$: $y = \dfrac{405}{5^2} = \dfrac{405}{25} = \dfrac{81}{5}$

31. The variation equation is $y = \dfrac{K}{x^2}$. Substituting $x = 3$ and $y = 18$:

$$18 = \frac{K}{3^2}$$
$$18 = \frac{K}{9}$$
$$K = 162$$

So $y = \dfrac{162}{x^2}$. Substituting $x = 2$: $y = \dfrac{162}{2^2} = \dfrac{162}{4} = \dfrac{81}{2}$

33. The variation equation is $z = Kxy^2$. Substituting $x = 3$, $y = 3$, and $z = 54$:

$$54 = K(3)(3)^2$$
$$54 = 27K$$
$$K = 2$$

So $z = 2xy^2$. Substituting $x = 2$ and $y = 4$: $z = 2(2)(4)^2 = 64$

35. The variation equation is $z = Kxy^2$. Substituting $x = 1$, $y = 4$, and $z = 64$:

$$64 = K(1)(4)^2$$
$$64 = 16K$$
$$K = 4$$

So $z = 4xy^2$. Substituting $z = 32$ and $y = 1$:

$$32 = 4x(1)^2$$
$$32 = 4x$$
$$x = 8$$

37. Let l represent the length and f represent the force. The variation equation is $l = Kf$. Substituting $f = 5$ and $l = 3$:

$$3 = K \cdot 5$$
$$K = \frac{3}{5}$$

So $l = \frac{3}{5}f$. Substituting $l = 10$:

$$10 = \frac{3}{5}f$$
$$50 = 3f$$
$$f = \frac{50}{3}$$

The force required is $\frac{50}{3}$ pounds.

39. Let s represent the speed in miles per hour and v represent the speed in feet per second. The variation equation is $s = Kv$. Substituting $s = 273$ and $v = 400\frac{2}{5}$:

$$273 = K \cdot 400\frac{2}{5}$$
$$273 = \frac{2002}{5}K$$
$$1365 = 2002K$$
$$K = \frac{1365}{2002}$$

So $s = \frac{1365}{2002}v$. Substituting $s = 253$:

$$253 = \frac{1365}{2002}v$$
$$506,506 = 1365v$$
$$v = \frac{506,506}{1365} = 371\frac{1}{15} \approx 371.1$$

The speed is $371\frac{1}{15} \approx 371.1$ feet per second.

41. **a.** The variation equation is $T = 4P$.

  b. Graphing the equation:

  c. Substituting $T = 280$:
  $$280 = 4P$$
  $$P = 70$$
  The pressure is 70 pounds per square inch.

43. Let v represent the volume and p represent the pressure. The variation equation is $v = \dfrac{K}{p}$.

  Substituting $p = 36$ and $v = 25$:
  $$25 = \frac{K}{36}$$
  $$K = 900$$

  The equation is $v = \dfrac{900}{p}$. Substituting $v = 75$:

  $$75 = \frac{900}{p}$$
  $$75p = 900$$
  $$p = 12$$

  The pressure is 12 pounds per square inch.

45. **a.** The variation equation is $f = \dfrac{80}{d}$.

  b. Graphing the equation:

  c. Substituting $d = 10$:
  $$f = \frac{80}{10}$$
  $$f = 8$$
  The f-stop is 8.

47.   Let A represent the surface area, h represent the height, and r represent the radius. The variation equation is $A = Khr$.
Substituting $A = 94$, $r = 3$, and $h = 5$:

$$94 = K(3)(5)$$
$$94 = 15K$$
$$K = \frac{94}{15}$$

The equation is $A = \frac{94}{15}hr$. Substituting $r = 2$ and $h = 8$: $A = \frac{94}{15}(8)(2) = \frac{1504}{15}$.

The surface area is $\frac{1504}{15}$ square inches

49.   Let R represent the resistance, l represent the length, and d represent the diameter. The variation equation is $R = \frac{Kl}{d^2}$.

Substituting $R = 10$, $l = 100$, and $d = 0.01$:

$$10 = \frac{K(100)}{(0.01)^2}$$
$$0.001 = 100K$$
$$K = 0.00001$$

The equation is $R = \frac{0.00001l}{d^2}$. Substituting $l = 60$ and $d = 0.02$: $R = \frac{0.00001(60)}{(0.02)^2} = 1.5$. The resistance is 1.5 ohms.

51.   Let p represent the pitch and w represent the wavelength. The variation equation is $p = \frac{K}{w}$.

Substituting $p = 420$ and $w = 2.2$:

$$420 = \frac{K}{2.2}$$
$$K = 924$$

The equation is $p = \frac{924}{w}$. Substituting $p = 720$:

$$720 = \frac{924}{w}$$
$$720w = 924$$
$$w \approx 1.28$$

The wavelength is approximately 1.28 meters.

53.   The variation equation is $F = \frac{Gm_1m_2}{d^2}$.

55.   Solving the equation:

$$x - 5 = 7$$
$$x = 12$$

57.   Solving the equation:

$$5 - \frac{4}{7}a = -11$$
$$7\left(5 - \frac{4}{7}a\right) = 7(-11)$$
$$35 - 4a = -77$$
$$-4a = -112$$
$$a = 28$$

59.   Solving the equation:

$$5(x - 1) - 2(2x + 3) = 5x - 4$$
$$5x - 5 - 4x - 6 = 5x - 4$$
$$x - 11 = 5x - 4$$
$$-4x = 7$$
$$x = -\frac{7}{4}$$

61.   Solving for w:

$$P = 2l + 2w$$
$$P - 2l = 2w$$
$$w = \frac{P - 2l}{2}$$

63. Solving the inequality:

$$-5t \le 30$$
$$t \ge -6$$

The solution set is $[-6, \infty)$. Graphing the solution set:

65. Solving the inequality:
$$1.6x - 2 < 0.8x + 2.8$$
$$0.8x - 2 < 2.8$$
$$0.8x < 4.8$$
$$x < 6$$

The solution set is $(-\infty, 6)$. Graphing the solution set:

67. Solving the equation:

$$\left| \frac{1}{4}x - 1 \right| = \frac{1}{2}$$
$$\frac{1}{4}x - 1 = -\frac{1}{2}, \frac{1}{2}$$
$$\frac{1}{4}x = \frac{1}{2}, \frac{3}{2}$$
$$x = 2, 6$$

69. Solving the equation:
$$|3 - 2x| + 5 = 2$$
$$|3 - 2x| = -3$$

Since this statement is false, there is no solution, or $\varnothing$.

71. Solving the inequality:

$$\left| \frac{x}{5} + 1 \right| \ge \frac{4}{5}$$

$$\frac{x}{5} + 1 \le -\frac{4}{5} \qquad \text{or} \qquad \frac{x}{5} + 1 \ge \frac{4}{5}$$
$$x + 5 \le -4 \qquad\qquad\qquad x + 5 \ge 4$$
$$x \le -9 \qquad\qquad\qquad\quad x \ge -1$$

Graphing the solution set:

73. Since $|3 - 4t| > -5$ is always true, the solution set is all real numbers. Graphing the solution set:

75. **a.**   The distance appears to vary directly with the square of the speed.

b.   Let $d = Ks^2$. Substituting $d = 108$ and $s = 40$:

$$108 = K(40)^2$$
$$1600K = 108$$
$$K = 0.0675$$

So the equation is $d = 0.0675s^2$.

c.   Substituting $s = 55$: $d = 0.0675(55)^2 \approx 204.2$ feet away

d.   Substituting $s = 56$: $d = 0.0675(56)^2 \approx 211.7$ feet. This is approximately 7.5 feet further away.

Chapter 3 Review

1. Graphing the line:

3. Graphing the line:

5. Finding the slope: $m = \dfrac{2-2}{3-(-4)} = \dfrac{0}{7} = 0$

7. Solving for x:

$$\frac{x-7}{2+4} = -\frac{1}{3}$$
$$\frac{x-7}{6} = -\frac{1}{3}$$
$$x-7 = -2$$
$$x = 5$$

9. Solving for y:

$$\frac{y-3y}{2-5} = 4$$
$$\frac{-2y}{-3} = 4$$
$$-2y = -12$$
$$y = 6$$

11. Using the slope-intercept formula, the equation is $y = -2x$.

13. Solving for y:

$$2x - 3y = 9$$
$$-3y = -2x + 9$$
$$y = \frac{2}{3}x - 3$$

The slope is $m = \dfrac{2}{3}$ and the y-intercept is $b = -3$.

15. Using the point-slope formula:

$$y - 1 = -\frac{1}{3}(x+3)$$
$$y - 1 = -\frac{1}{3}x - 1$$
$$y = -\frac{1}{3}x$$

17. First find the slope: $m = \dfrac{7-7}{4-(-3)} = \dfrac{0}{7} = 0$. Since the line is horizontal, its equation is $y = 7$.

19. First find the slope by solving for y:

$$2x - y = 4$$
$$-y = -2x + 4$$
$$y = 2x - 4$$

The parallel slope is also $m = 2$. Now using the point-slope formula:

$$y + 3 = 2(x-2)$$
$$y + 3 = 2x - 4$$
$$y = 2x - 7$$

21.   Graphing the inequality:

23.   The domain is $\{2, 3, 4\}$ and the range is $\{2, 3, 4\}$. This is a function.

25. $f(-3) = 0$

27.   Evaluating the function: $f(0) = 2(0)^2 - 4(0) + 1 = 0 - 0 + 1 = 1$

29.   Evaluating the function: $f(g(0)) = f(3(0) + 2) = f(2) = 2(2)^2 - 4(2) + 1 = 8 - 8 + 1 = 1$

31.   The variation equation is $y = Kx$. Substituting $x = 2$ and $y = 6$:

$$6 = K \bullet 2$$
$$K = 3$$

The equation is $y = 3x$. Substituting $x = 8$: $y = 3 \bullet 8 = 24$

33.   The variation equation is $y = \dfrac{K}{x^2}$. Substituting $x = 2$ and $y = 9$:

$$9 = \frac{K}{2^2}$$
$$9 = \frac{K}{4}$$
$$K = 36$$

The equation is $y = \dfrac{36}{x^2}$. Substituting $x = 3$: $y = \dfrac{36}{3^2} = \dfrac{36}{9} = 4$

35.   The variation equation is $t = Kd$. Substituting $t = 42$ and $d = 2$:
$$42 = K \bullet 2$$
$$K = 21$$

The equation is $t = 21d$. Substituting $d = 4$: $t = 21 \bullet 4 = 84$. The tension is 84 pounds.

Chapter 3 Cumulative Review

1.   Simplifying: $-5^2 = -5 \bullet 5 = -25$ **3.**   Simplifying: $5^2 + 7^2 = 25 + 49 = 74$

5.   Simplifying: $48 \div 8 \bullet 6 = 6 \bullet 6 = 36$ **7.**   Simplifying: $30 - 15 \div 3 + 2 = 30 - 5 + 2 = 27$

9.   Evaluating when $x = 4$: $x^2 - 10x + 25 = (4)^2 - 10(4) + 25 = 16 - 40 + 25 = 1$

11.   Simplifying: $\dfrac{336}{432} = \dfrac{2 \bullet 2 \bullet 2 \bullet 2 \bullet 3 \bullet 7}{2 \bullet 2 \bullet 2 \bullet 2 \bullet 3 \bullet 3 \bullet 3} = \dfrac{7}{3 \bullet 3} = \dfrac{7}{9}$

13.   Subtracting: $\dfrac{3}{15} - \dfrac{2}{20} = \dfrac{3 \bullet 4}{15 \bullet 4} - \dfrac{2 \bullet 3}{20 \bullet 3} = \dfrac{12}{60} - \dfrac{6}{60} = \dfrac{6}{60} = \dfrac{1}{10}$

15.   Evaluating: $-\dfrac{3}{4} + (-3)\left(\dfrac{5}{12}\right) = -\dfrac{3}{4} + \left(-\dfrac{5}{4}\right) = -\dfrac{8}{4} = -2$

17.   Simplifying: $8\left(\dfrac{3}{4}x - \dfrac{1}{2}y\right) = 8 \bullet \dfrac{3}{4}x - 8 \bullet \dfrac{1}{2}y = 6x - 4y$

19. Solving the equation:

$$\frac{2}{3}a - 4 = 6$$

$$\frac{2}{3}a = 10$$

$$2a = 30$$

$$a = 15$$

21. Solving the equation:

$$-4 + 3(3x + 2) = 7$$

$$-4 + 9x + 6 = 7$$

$$9x + 2 = 7$$

$$9x = 5$$

$$x = \frac{5}{9}$$

23. Solving the equation:

$$|2x - 3| - 7 = 1$$

$$|2x - 3| = 8$$

$$2x - 3 = -8, 8$$

$$2x = -5, 11$$

$$x = -\frac{5}{2}, \frac{11}{2}$$

25. Solving for y:

$$y - 5 = -3(x + 2)$$

$$y - 5 = -3x - 6$$

$$y = -3x - 1$$

27. Solving the inequality:

$$|3x - 2| \le 7$$

$$-7 \le 3x - 2 \le 7$$

$$-5 \le 3x \le 9$$

$$-\frac{5}{3} \le x \le 3$$

Graphing the solution set:

29. Graphing the equation:

$6x - 5y = 30$

31. Graphing the equation:

$y = \frac{1}{2}x - 1$

33. Finding the slope: $m = \dfrac{3 - (-2)}{-5 - (-3)} = \dfrac{3 + 2}{-5 + 3} = -\dfrac{5}{2}$

35. Since this line is horizontal, its slope is 0.

37. Solving for y:

$$4x - 5y = 20$$

$$-5y = -4x + 20$$

$$y = \frac{4}{5}x - 4$$

The slope is $m = \dfrac{4}{5}$ and the y-intercept is $b = -4$.

39.   Solving for y:
$$5x + 2y = -1$$
$$2y = -5x - 1$$
$$y = -\frac{5}{2}x - \frac{1}{2}$$

Using the slope of $-\frac{5}{2}$ in the point-slope formula:

$$y + 2 = -\frac{5}{2}(x + 4)$$
$$y + 2 = -\frac{5}{2}x - 10$$
$$y = -\frac{5}{2}x - 12$$

41.   Evaluating the function: $h(0) = 3(0) - 1 = -1$

43.   Evaluating the function: $g\big[f(-2)\big] = g\big((-2)^2 - 3(-2)\big) = g(4 + 6) = g(10) = 10 - 1 = 9$

45.   The domain is $\{-1, 2, 3\}$ and the range is $\{-1, 3\}$. This relation is a function.

47.   Finding the percent:
$$p \cdot 30 = 6$$
$$p = \frac{6}{30} = 20\%$$

49.   The difference is: $143,700 - 115,400 = 28,300$

Expressed as a percent: $\dfrac{28,300}{115,400} \approx 25\%$

It costs approximately 25% more to live in Tokyo.

Chapter 3 Test

1.   The x-intercept is 3, the y-intercept is 6, and the slope is -2.

2.   The x-intercept is $-\dfrac{3}{2}$, the y-intercept is -3, and the slope is -2.

3.   The x-intercept is $-\dfrac{8}{3}$, the y-intercept is 4, and the slope is $\dfrac{3}{2}$.

4.   The x-intercept is -2, there is no y-intercept, and there is no slope.

5.   The slope is 3 and the y-intercept is 3, so the slope-intercept form is $y = 3x + 3$.

6.   The slope is $-\dfrac{5}{3}$ and the y-intercept is 5, so the slope-intercept form is $y = -\dfrac{5}{3}x + 5$.

7.   Using the point-slope formula:

$$y - 3 = 2(x + 1)$$
$$y - 3 = 2x + 2$$
$$y = 2x + 5$$

8. First find the slope: $m = \dfrac{-1-2}{4-(-3)} = \dfrac{-3}{4+3} = -\dfrac{3}{7}$. Using the point-slope formula:

$$y - 2 = -\frac{3}{7}(x+3)$$
$$y - 2 = -\frac{3}{7}x - \frac{9}{7}$$
$$y = -\frac{3}{7}x + \frac{5}{7}$$

9. First solve for y to find the slope:
$$2x - 5y = 10$$
$$-5y = -2x + 10$$
$$y = \frac{2}{5}x - 2$$

The parallel line will also have a slope of $\dfrac{2}{5}$. Now using the point-slope formula:

$$y - (-3) = \frac{2}{5}(x - 5)$$
$$y + 3 = \frac{2}{5}x - 2$$
$$y = \frac{2}{5}x - 5$$

10. The perpendicular slope is $-\dfrac{1}{3}$. Using the point-slope formula:

$$y - (-2) = -\frac{1}{3}(x - (-1))$$
$$y + 2 = -\frac{1}{3}x - \frac{1}{3}$$
$$y = -\frac{1}{3}x - \frac{7}{3}$$

11. The inequality is $y \geq -x + 3$.

12. The inequality is $y < -x + 3$.

13. Graphing the inequality:

14. Graphing the inequality:

15. The domain is $\{-3, -2\}$ and the range is $\{0, 1\}$. This is not a function.

16. The domain is all real numbers and the range is $\{y \mid y \geq -9\}$. This is a function.

17. The domain is $\{0, 1, 2\}$ and the range is $\{0, 3, 5\}$. This is a function.

18. The domain is all real numbers and the range is all real numbers. This is a function.

19. The domain is $\{x \mid -4 \leq x \leq 4\}$ and the range is $\{y \mid -1 \leq y \leq 3\}$.

20. The domain is $\{x \mid -3 \leq x \leq 3\}$ and the range is $\{y \mid -4 \leq y \leq 4\}$.

21. Evaluating the function: $f(3)+g(2)=[3-2]+[3\bullet 2+4]=1+10=11$

22. Evaluating the function: $h(0)+g(0)=\left[3\bullet 0^2-2\bullet 0-8\right]+[3\bullet 0+4]=-8+4=-4$

23. Evaluating the function: $f(g(2))=f(3\bullet 2+4)=f(10)=10-2=8$

24. Evaluating the function: $g(f(2))=g(2-2)=g(0)=3\bullet 0+4=4$

25. Evaluating the function: $f(-6)+g(-6)=1+4=5$

26. Evaluating the function: $(f-g)(1)=f(1)-g(1)=2-2=0$

27. Evaluating the function: $(f\circ g)(5)=f[g(5)]=f(0)=3$

28. Evaluating the function: $(g\circ f)(5)=g[f(5)]=g(5)=0$

29. The variation equation is $y=Kx^2$. Substituting $x=5$ and $y=50$:
$$50=K(5)^2$$
$$50=25K$$
$$K=2$$
The equation is $y=2x^2$. Substituting $x=3$: $y=2(3)^2=2\bullet 9=18$

30. The variation equation is $z=Kxy^3$. Substituting $x=5$, $y=2$, and $z=15$:
$$15=K(5)(2)^3$$
$$15=40K$$
$$K=\frac{3}{8}$$
The equation is $z=\frac{3}{8}xy^3$. Substituting $x=2$ and $y=3$: $z=\frac{3}{8}(2)(3)^3=\frac{3}{8}\bullet 54=\frac{81}{4}$

Chapter 4
Systems of Linear Equations and Inequalities

Getting ready for Chapter 4

1. Simplifying: $0.09(6,000) = 540$

2. Simplifying: $1.5(500) = 750$

3. Simplifying: $3(11) - 5(7) = 33 - 35 = -2$

4. Simplifying: $6 - 12 = -6$

5. Simplifying: $3(3) - 5 = 9 - 5 = 4$

6. Simplifying: $4(5) - 3(-3) = 20 + 9 = 29$

7. Simplifying: $1 + 2(2) - 3(3) = 1 + 4 - 9 = -4$

8. Simplifying: $-4(-1) + 1(2) + 2(6) = 4 + 2 + 12 = 18$

9. Applying the distributive property: $-3(2x - 3y) = -6x + 9y$

10. Applying the distributive property: $3(4x + 5y) = 12x + 15y$

11. Applying the distributive property: $6\left(\dfrac{1}{2}x - \dfrac{1}{3}y\right) = 3x - 2y$

12. Applying the distributive property: $12\left(\dfrac{1}{4}x + \dfrac{2}{3}y\right) = 3x + 8y$

13. Applying the distributive property: $10(0.3x + 0.7y) = 3x + 7y$

14. Applying the distributive property: $100(0.06x + 0.07y) = 6x + 7y$

15. Solving the equation:

    $$2(1) + y = 4$$
    $$2 + y = 4$$
    $$y = 2$$

16. Solving the equation:

    $$2x - 3(3x - 5) = -6$$
    $$2x - 9x + 15 = -6$$
    $$-7x + 15 = -6$$
    $$-7x = -21$$
    $$x = 3$$

17. Solving the equation:

    $$2(2y + 6) + 3y = 5$$
    $$4y + 12 + 3y = 5$$
    $$7y + 12 = 5$$
    $$7y = -7$$
    $$y = -1$$

18. Solving the equation:

    $$5\left(\dfrac{19}{15}\right) + 5y = 9$$
    $$\dfrac{19}{3} + 5y = 9$$
    $$19 + 15y = 27$$
    $$15y = 8$$
    $$y = \dfrac{8}{15}$$

19. Solving the equation:
$$4x - 2x = 8$$
$$2x = 8$$
$$x = 4$$

20. Solving the inequality:
$$20x + 9,300 > 18,000$$
$$20x > 8,700$$
$$x > 435$$

4.1 Systems of Linear Equations in Two Variables

1. The intersection point is (4,3).

3. The intersection point is (−5,−6).

5. The intersection point is (4,2).

7. The lines are parallel. There is no solution to the system.

9. The lines coincide. Any solution to one of the equations is a solution to the other.

11. Solving the two equations:
$$x + y = 5$$
$$3x - y = 3$$
Adding yields:
$$4x = 8$$
$$x = 2$$
The solution is (2,3).

13. Multiply the first equation by -1:
$$-3x - y = -4$$
$$4x + y = 5$$
Adding yields: $x = 1$. The solution is (1,1).

15. Multiply the first equation by -2:
$$-6x + 4y = -12$$
$$6x - 4y = 12$$
Adding yields $0 = 0$, so the lines coincide. The solution is $\{(x,y) \mid 3x - 2y = 6\}$.

17. Multiply the first equation by 3:
$$3x + 6y = 0$$
$$2x - 6y = 5$$
Adding yields:
$$5x = 5$$
$$x = 1$$
The solution is $\left(1, -\dfrac{1}{2}\right)$.

19. Multiply the first equation by -2:
$$-4x + 10y = -32$$
$$4x - 3y = 11$$
Adding yields:
$$7y = -21$$
$$y = -3$$
The solution is $\left(\dfrac{1}{2}, -3\right)$.

21. Multiply the first equation by 3 and the second equation by -2:
$$18x + 9y = -3$$
$$-18x - 10y = -2$$
Adding yields:
$$-y = -5$$
$$y = 5$$
The solution is $\left(-\dfrac{8}{3}, 5\right)$.

23. Multiply the first equation by 2 and the second equation by 3:
$$8x + 6y = 28$$
$$27x - 6y = 42$$
Adding yields:
$$35x = 70$$
$$x = 2$$
The solution is (2,2).

25. Multiply the first equation by 2:
$$4x - 10y = 6$$
$$-4x + 10y = 3$$
Adding yields $0 = 9$, which is false (parallel lines). There is no solution ($\varnothing$).

27. To clear each equation of fractions, multiply the first equation by 12 and the second equation by 30:
$$3x - 2y = -24$$
$$-5x + 6y = 120$$
Multiply the first equation by 3:
$$9x - 6y = -72$$
$$-5x + 6y = 120$$
Adding yields $4x = 48$, so $x = 12$. Substituting into the first equation:
$$36 - 2y = -24$$
$$-2y = -60$$
$$y = 30$$
The solution is $(12, 30)$.

29. To clear each equation of fractions, multiply the first equation by 6 and the second equation by 20:
$$3x + 2y = 78$$
$$8x + 5y = 200$$
Multiply the first equation by 5 and the second equation by -2:
$$15x + 10y = 390$$
$$-16x - 10y = -400$$
Adding yields:
$$-x = -10$$
$$x = 10$$
The solution is $(10, 24)$.

31. Substituting into the first equation:
$$7(2y + 9) - y = 24$$
$$14y + 63 - y = 24$$
$$13y = -39$$
$$y = -3$$

The solution is $(3, -3)$.

33. Substituting into the first equation:
$$6x - \left(-\frac{3}{4}x - 1\right) = 10$$
$$6x + \frac{3}{4}x + 1 = 10$$
$$\frac{27}{4}x = 9$$
$$27x = 36$$
$$x = \frac{4}{3}$$
The solution is $\left(\frac{4}{3}, -2\right)$.

35. Multiply the second equation by -4:
$$3y + 4z = 23$$
$$-24y - 4z = -128$$
Adding yields:
$$-21y = -105$$
$$y = 5$$
Substituting into the second equation:
$$6(5) + z = 32$$
$$z = 2$$
The solution is $y = 5$, $z = 2$.

37. Substituting into the first equation:
$$4x - 4 = 3x - 2$$
$$x - 4 = -2$$
$$x = 2$$
The solution is $(2, 4)$.

39.   Solving the first equation for y yields $y = 2x - 5$. Substituting into the second equation:

$$4x - 2(2x - 5) = 10$$
$$4x - 4x + 10 = 10$$
$$10 = 10$$

Since this statement is true, the two lines coincide. The solution is $\{(x, y) \mid 2x - y = 5\}$.

41.   Substituting into the first equation:

$$\frac{1}{3}\left(\frac{3}{2}y\right) - \frac{1}{2}y = 0$$
$$\frac{1}{2}y - \frac{1}{2}y = 0$$
$$0 = 0$$

Since this statement is true, the two lines coincide. The solution is $\left\{(x, y) \mid x = \frac{3}{2}y\right\}$.

43.   Multiply the first equation by 2 and the second equation by 7:

$$8x - 14y = 6$$
$$35x + 14y = -21$$

Adding yields:

$$43x = -15$$
$$x = -\frac{15}{43}$$

Substituting into the original second equation:

$$5\left(-\frac{15}{43}\right) + 2y = -3$$
$$-\frac{75}{43} + 2y = -3$$
$$2y = -\frac{54}{43}$$
$$y = -\frac{27}{43}$$

The solution is $\left(-\frac{15}{43}, -\frac{27}{43}\right)$.

45.   Multiply the first equation by 3 and the second equation by 8:

$$27x - 24y = 12$$
$$16x + 24y = 48$$

Adding yields:

$$43x = 60$$
$$x = \frac{60}{43}$$

Substituting into the original second equation:

$$2\left(\frac{60}{43}\right) + 3y = 6$$
$$\frac{120}{43} + 3y = 6$$
$$3y = \frac{138}{43}$$
$$y = \frac{46}{43}$$

The solution is $\left(\frac{60}{43}, \frac{46}{43}\right)$.

47. Multiply the first equation by 2 and the second equation by 5:

$$6x - 10y = 4$$
$$35x + 10y = 5$$

Adding yields:

$$41x = 9$$
$$x = \frac{9}{41}$$

Substituting into the original second equation:

$$7\left(\frac{9}{41}\right) + 2y = 1$$
$$\frac{63}{41} + 2y = 1$$
$$2y = -\frac{22}{41}$$
$$y = -\frac{11}{41}$$

The solution is $\left(\dfrac{9}{41}, -\dfrac{11}{41}\right)$.

49. Multiply the second equation by 3:

$$x - 3y = 7$$
$$6x + 3y = -18$$

Adding yields:

$$7x = -11$$
$$x = -\frac{11}{7}$$

Substituting into the original second equation:

$$2\left(-\frac{11}{7}\right) + y = -6$$
$$-\frac{22}{7} + y = -6$$
$$y = -\frac{20}{7}$$

The solution is $\left(-\dfrac{11}{7}, -\dfrac{20}{7}\right)$.

51. Substituting into the first equation:

$$-\frac{1}{3}x + 2 = \frac{1}{2}x + \frac{1}{3}$$
$$6\left(-\frac{1}{3}x + 2\right) = 6\left(\frac{1}{2}x + \frac{1}{3}\right)$$
$$-2x + 12 = 3x + 2$$
$$-5x = -10$$
$$x = 2$$

Substituting into the first equation: $y = \dfrac{1}{2}(2) + \dfrac{1}{3} = 1 + \dfrac{1}{3} = \dfrac{4}{3}$. The solution is $\left(2, \dfrac{4}{3}\right)$.

53. Substituting into the first equation:

$$3\left(\tfrac{2}{3}y - 4\right) - 4y = 12$$
$$2y - 12 - 4y = 12$$
$$-2y - 12 = 12$$
$$-2y = 24$$
$$y = -12$$

Substituting into the second equation: $x = \dfrac{2}{3}(-12) - 4 = -8 - 4 = -12$. The solution is $(-12, -12)$.

55. Multiply the first equation by 2:

$$8x - 6y = -14$$
$$-8x + 6y = -11$$
$$0 = -25$$

Since this statement is false, there is no solution ($\varnothing$).

57. Multiply the first equation by -20:

$$-60y - 20z = -340$$
$$5y + 20z = 65$$

Adding yields:

$$-55y = -275$$
$$y = 5$$

Substituting into the first equation:

$$3(5) + z = 17$$
$$15 + z = 17$$
$$z = 2$$

The solution is $y = 5$, $z = 2$.

59. Substitute into the first equation:

$$\frac{3}{4}x - \frac{1}{3}\left(\frac{1}{4}x\right) = 1$$
$$\frac{3}{4}x - \frac{1}{12}x = 1$$
$$\frac{2}{3}x = 1$$
$$x = \frac{3}{2}$$

Substituting into the second equation: $y = \dfrac{1}{4}\left(\dfrac{3}{2}\right) = \dfrac{3}{8}$. The solution is $\left(\dfrac{3}{2}, \dfrac{3}{8}\right)$.

61. To clear each equation of fractions, multiply the first equation by 12 and the second equation by 12:

$$3x - 6y = 4$$
$$4x - 3y = 8$$

Multiply the second equation by -2:

$$3x - 6y = 4$$
$$-8x + 6y = -16$$

Adding yields:

$$-5x = -12$$
$$x = \frac{12}{5}$$

Substituting into the first equation:

$$3\left(\frac{12}{5}\right) - 6y = 4$$

$$\frac{36}{5} - 6y = 4$$

$$-6y = -\frac{16}{5}$$

$$y = \frac{8}{15}$$

The solution is $\left(\frac{12}{5}, \frac{8}{15}\right)$.

63. To clear each equation of fractions, multiply the first equation by 12 and the second equation by 2:

$$9x + 4y = 24$$
$$x - 2y = 0$$

Multiply the second equation by 2:

$$9x + 4y = 24$$
$$2x - 4y = 0$$

Adding yields:

$$11x = 24$$

$$x = \frac{24}{11}$$

Substituting into the second equation:

$$\frac{24}{11} - 2y = 0$$

$$2y = \frac{24}{11}$$

$$y = \frac{12}{11}$$

The solution is $\left(\frac{24}{11}, \frac{12}{11}\right)$.

65. To clear each equation of fractions, multiply the first equation by 10 and the second equation by 6:

$$6x - 5y = 7$$
$$x - 4y = -3$$

Multiply the second equation by –6:

$$6x - 5y = 7$$
$$-6x + 24y = 18$$

Adding yields:

$$19y = 25$$

$$y = \frac{25}{19}$$

Substituting into the second equation:

$$x - 4\left(\frac{25}{19}\right) = -3$$

$$x - \frac{100}{19} = -3$$

$$x = \frac{43}{19}$$

The solution is $\left(\frac{43}{19}, \frac{25}{19}\right)$.

67. **a.** Simplifying: $(3x - 4y) - 3(x - y) = 3x - 4y - 3x + 3y = -y$

b. Substituting $x = 0$:

$$3(0) - 4y = 8$$
$$-4y = 8$$
$$y = -2$$

c. From part **b**, the y-intercept is -2.

d. Graphing the line:

e. Multiply the second equation by -3:

$$3x - 4y = 8$$
$$-3x + 3y = -6$$

Adding yields:

$$-y = 2$$
$$y = -2$$

Substituting into the first equation:

$$3x - 4(-2) = 8$$
$$3x + 8 = 8$$
$$3x = 0$$
$$x = 0$$

The lines intersect at the point $(0, -2)$.

69. Multiply the second equation by 100:

$$x + y = 10000$$
$$6x + 5y = 56000$$

Multiply the first equation by -5:

$$-5x - 5y = -50000$$
$$6x + 5y = 56000$$

Adding yields $x = 6000$. The solution is $(6000, 4000)$.

71. Multiplying the first equation by $\frac{2}{3}$ yields the equation $4x - 6y = 2$. For the lines to coincide, the value is $c = 2$.

$c = -3$.

73. Substituting $x = 4 - y$ into the second equation:

$$(4 - y) - 2y = 4$$
$$4 - 3y = 4$$
$$-3y = 0$$
$$y = 0$$

The solution is $(4, 0)$.

75. Finding the slope: $m = \dfrac{5 - (-1)}{-2 - (-4)} = \dfrac{5 + 1}{-2 + 4} = \dfrac{6}{2} = 3$

77. Solving for y:
$$2x - 3y = 6$$
$$-3y = -2x + 6$$
$$y = \frac{2}{3}x - 2$$

The slope is $m = \frac{2}{3}$ and the y-intercept is $b = -2$.

79. Using the point-slope formula:
$$y - 2 = \frac{2}{3}(x + 6)$$
$$y - 2 = \frac{2}{3}x + 4$$
$$y = \frac{2}{3}x + 6$$

81. First find the slope: $m = \dfrac{-2-0}{0-3} = \dfrac{2}{3}$. The slope-intercept form is $y = \dfrac{2}{3}x - 2$.

83. First find the slope of $4x - 3y = 6$ by solving for y:
$$4x - 3y = 6$$
$$-3y = -4x + 6$$
$$y = \frac{4}{3}x - 2$$

So the slope of the parallel line is $\dfrac{4}{3}$. Using the point-slope formula:
$$y - 3 = \frac{4}{3}(x + 2)$$
$$y - 3 = \frac{4}{3}x + \frac{8}{3}$$
$$y = \frac{4}{3}x + \frac{17}{3}$$

85. Simplifying: $2 - 2(6) = 2 - 12 = -10$

87. Simplifying: $(x + 3y) - 1(x - 2z) = x + 3y - x + 2z = 3y + 2z$

89. Solving the equation:
$$-9y = -9$$
$$y = 1$$

91. Solving the equation:
$$3(1) + 2z = 9$$
$$3 + 2z = 9$$
$$2z = 6$$
$$z = 3$$

93. Applying the distributive property: $2(5x - z) = 10x - 2z$

95. Applying the distributive property: $3(3x + y - 2z) = 9x + 3y - 6z$

97. Substituting the points $(1, -2)$ and $(3, 1)$ results in the two equations:
$$a - 2b = 7$$
$$3a + b = 7$$
Multiply the second equation by 2:
$$a - 2b = 7$$
$$6a + 2b = 14$$
Adding the two equations:
$$7a = 21$$
$$a = 3$$
Substituting into the second equation:
$$9 + b = 7$$
$$b = -2$$
So $a = 3$ and $b = -2$.

99.   Substituting the points $(-6,0)$ and $(3,6)$ results in the two equations:
$$0 = -6m + b$$
$$6 = 3m + b$$
Multiply the first equation by -1:
$$0 = 6m - b$$
$$6 = 3m + b$$
Adding yields:
$$9m = 6$$
$$m = \frac{2}{3}$$
Substituting into the second equation:
$$6 = 3\left(\frac{2}{3}\right) + b$$
$$6 = 2 + b$$
$$b = 4$$
So $m = \frac{2}{3}$ and $b = 4$.

4.2 Systems of Linear Equations in Three Variables

1.   Adding the first two equations and the first and third equations results in the system:
$$2x + 3z = 5$$
$$2x - 2z = 0$$
Solving the second equation yields $x = z$, now substituting:
$$2z + 3z = 5$$
$$5z = 5$$
$$z = 1$$
So $x = 1$, now substituting into the original first equation:
$$1 + y + 1 = 4$$
$$y + 2 = 4$$
$$y = 2$$
The solution is $(1,2,1)$.

3.   Adding the first two equations and the first and third equations results in the system:
$$2x + 3z = 13$$
$$3x - 3z = -3$$
Adding yields:
$$5x = 10$$
$$x = 2$$
Substituting to find z:
$$2(2) + 3z = 13$$
$$4 + 3z = 13$$
$$3z = 9$$
$$z = 3$$
Substituting into the original first equation:
$$2 + y + 3 = 6$$
$$y + 5 = 6$$
$$y = 1$$
The solution is $(2,1,3)$.

5.   Adding the second and third equations:
$$5x + z = 11$$
Multiplying the second equation by 2:
$$x + 2y + z = 3$$
$$4x - 2y + 4z = 12$$
Adding yields:
$$5x + 5z = 15$$
$$x + z = 3$$
So the system becomes:
$$5x + z = 11$$
$$x + z = 3$$
Multiply the second equation by -1:
$$5x + z = 11$$
$$-x - z = -3$$
Adding yields:
$$4x = 8$$
$$x = 2$$
Substituting to find z:
$$5(2) + z = 11$$
$$z + 10 = 11$$
$$z = 1$$
Substituting into the original first equation:
$$2 + 2y + 1 = 3$$
$$2y + 3 = 3$$
$$2y = 0$$
$$y = 0$$
The solution is $(2,0,1)$.

7.   Multiply the second equation by -1 and add it to the first equation:
$$2x + 3y - 2z = 4$$
$$-x - 3y + 3z = -4$$
Adding results in the equation $x + z = 0$. Multiply the second equation by 2 and add it to the third equation:
$$2x + 6y - 6z = 8$$
$$3x - 6y + z = -3$$
Adding results in the equation:
$$5x - 5z = 5$$
$$x - z = 1$$
So the system becomes:
$$x - z = 1$$
$$x + z = 0$$
Adding yields:
$$2x = 1$$
$$x = \frac{1}{2}$$
Substituting to find z:
$$\frac{1}{2} + z = 0$$
$$z = -\frac{1}{2}$$

Substituting into the original first equation:

$$2\left(\frac{1}{2}\right) + 3y - 2\left(-\frac{1}{2}\right) = 4$$
$$1 + 3y + 1 = 4$$
$$3y + 2 = 4$$
$$3y = 2$$
$$y = \frac{2}{3}$$

The solution is $\left(\frac{1}{2}, \frac{2}{3}, -\frac{1}{2}\right)$.

9. Multiply the first equation by 2 and add it to the second equation:
$$-2x + 8y - 6z = 4$$
$$2x - 8y + 6z = 1$$

Adding yields $0 = 5$, which is false. There is no solution (inconsistent system).

11. To clear the system of fractions, multiply the first equation by 2 and the second equation by 3:
$$x - 2y + 2z = 0$$
$$6x + y + 3z = 6$$
$$x + y + z = -4$$

Multiply the third equation by 2 and add it to the first equation:
$$x - 2y + 2z = 0$$
$$2x + 2y + 2z = -8$$

Adding yields the equation $3x + 4z = -8$. Multiply the third equation by -1 and add it to the second equation:
$$6x + y + 3z = 6$$
$$-x - y - z = 4$$

Adding yields the equation $5x + 2z = 10$. So the system becomes:
$$3x + 4z = -8$$
$$5x + 2z = 10$$

Multiply the second equation by -2:
$$3x + 4z = -8$$
$$-10x - 4z = -20$$

Adding yields:
$$-7x = -28$$
$$x = 4$$

Substituting to find z:
$$3(4) + 4z = -8$$
$$12 + 4z = -8$$
$$4z = -20$$
$$z = -5$$

Substituting into the original third equation:
$$4 + y - 5 = -4$$
$$y - 1 = -4$$
$$y = -3$$

The solution is $(4, -3, -5)$.

13. Multiply the first equation by -2 and add it to the third equation:
$$-4x + 2y + 6z = -2$$
$$4x - 2y - 6z = 2$$

Adding yields $0 = 0$, which is true. Since there are now less equations than unknowns, there is no unique solution (dependent system).

15.   Multiply the second equation by 3 and add it to the first equation:
$$2x - y + 3z = 4$$
$$3x + 6y - 3z = -9$$
Adding yields the equation $5x + 5y = -5$, or $x + y = -1$.

Multiply the second equation by 2 and add it to the third equation:
$$2x + 4y - 2z = -6$$
$$4x + 3y + 2z = -5$$
Adding yields the equation $6x + 7y = -11$. So the system becomes:
$$6x + 7y = -11$$
$$x + y = -1$$
Multiply the second equation by -6:
$$6x + 7y = -11$$
$$-6x - 6y = 6$$
Adding yields $y = -5$. Substituting to find x:
$$6x + 7(-5) = -11$$
$$6x - 35 = -11$$
$$6x = 24$$
$$x = 4$$
Substituting into the original first equation:
$$2(4) - (-5) + 3z = 4$$
$$13 + 3z = 4$$
$$3z = -9$$
$$z = -3$$
The solution is $(4, -5, -3)$.

17.   Adding the second and third equations results in the equation $x + y = 9$. Since this is the same as the first equation, there are less equations than unknowns. There is no unique solution (dependent system).

19.   Adding the second and third equations results in the equation $4x + y = 3$. So the system becomes:
$$4x + y = 3$$
$$2x + y = 2$$
Multiplying the second equation by -1:
$$4x + y = 3$$
$$-2x - y = -2$$
Adding yields:
$$2x = 1$$
$$x = \frac{1}{2}$$
Substituting to find y:
$$2\left(\frac{1}{2}\right) + y = 2$$
$$1 + y = 2$$
$$y = 1$$
Substituting into the original second equation:
$$1 + z = 3$$
$$z = 2$$
The solution is $\left(\frac{1}{2}, 1, 2\right)$.

21. Multiply the third equation by 2 and adding it to the second equation:

$$6y - 4z = 1$$
$$2x + 4z = 2$$

Adding yields the equation $2x + 6y = 3$. So the system becomes:

$$2x - 3y = 0$$
$$2x + 6y = 3$$

Multiply the first equation by 2:

$$4x - 6y = 0$$
$$2x + 6y = 3$$

Adding yields:

$$6x = 3$$
$$x = \frac{1}{2}$$

Substituting to find y:

$$2\left(\frac{1}{2}\right) + 6y = 3$$
$$1 + 6y = 3$$
$$6y = 2$$
$$y = \frac{1}{3}$$

Substituting into the original third equation to find z:

$$\frac{1}{2} + 2z = 1$$
$$2z = \frac{1}{2}$$
$$z = \frac{1}{4}$$

The solution is $\left(\frac{1}{2}, \frac{1}{3}, \frac{1}{4}\right)$.

23. Multiply the first equation by –2 and add it to the second equation:

$$-2x - 2y + 2z = -4$$
$$2x + y + 3z = 4$$

Adding yields $-y + 5z = 0$. Multiply the first equation by –1 and add it to the third equation:

$$-x - y + z = -2$$
$$x - 2y + 2z = 6$$

Adding yields $-3y + 3z = 4$. So the system becomes:

$$-y + 5z = 0$$
$$-3y + 3z = 4$$

Multiply the first equation by –3:

$$3y - 15z = 0$$
$$-3y + 3z = 4$$

Adding yields:

$$-12z = 4$$
$$z = -\frac{1}{3}$$

Substituting to find y:

$$-3y + 3\left(-\frac{1}{3}\right) = 4$$

$$-3y - 1 = 4$$

$$-3y = 5$$

$$y = -\frac{5}{3}$$

Substituting into the original first equation:

$$x - \frac{5}{3} + \frac{1}{3} = 2$$

$$x - \frac{4}{3} = 2$$

$$x = \frac{10}{3}$$

The solution is $\left(\frac{10}{3}, -\frac{5}{3}, -\frac{1}{3}\right)$.

25.   Multiply the third equation by -1 and add it to the first equation:

$$2x + 3y = -\frac{1}{2}$$

$$-3y - 2z = \frac{3}{4}$$

Adding yields the equation $2x - 2z = \frac{1}{4}$. So the system becomes:

$$2x - 2z = \frac{1}{4}$$

$$4x + 8z = 2$$

Multiply the first equation by 4:

$$8x - 8z = 1$$

$$4x + 8z = 2$$

Adding yields:

$$12x = 3$$

$$x = \frac{1}{4}$$

Substituting to find z:

$$4\left(\frac{1}{4}\right) + 8z = 2$$

$$1 + 8z = 2$$

$$8z = 1$$

$$z = \frac{1}{8}$$

Substituting to find y:

$$2\left(\frac{1}{4}\right) + 3y = -\frac{1}{2}$$

$$\frac{1}{2} + 3y = -\frac{1}{2}$$

$$3y = -1$$

$$y = -\frac{1}{3}$$

The solution is $\left(\frac{1}{4}, -\frac{1}{3}, \frac{1}{8}\right)$.

27. To clear each equation of fractions, multiply the first equation by 6, the second equation by 4, and the third equation by 12:

$$2x + 3y - z = 24$$
$$x - 3y + 2z = 6$$
$$6x - 8y - 3z = -64$$

Multiply the first equation by 2 and add it to the second equation:

$$4x + 6y - 2z = 48$$
$$x - 3y + 2z = 6$$

Adding yields the equation $5x + 3y = 54$. Multiply the first equation by -3 and add it to the third equation:

$$-6x - 9y + 3z = -72$$
$$6x - 8y - 3z = -64$$

Adding yields:

$$-17y = -136$$
$$y = 8$$

Substituting to find x:

$$5x + 3(8) = 54$$
$$5x + 24 = 54$$
$$5x = 30$$
$$x = 6$$

Substituting to find z:

$$6 - 3(8) + 2z = 6$$
$$-18 + 2z = 6$$
$$2z = 24$$
$$z = 12$$

The solution is $(6, 8, 12)$.

29. To clear each equation of fractions, multiply the first equation by 6, the second equation by 6, and the third equation by 12:

$$6x - 3y - 2z = -8$$
$$2x + 6y - 3z = 30$$
$$-3x + 8y - 12z = -9$$

Multiply the first equation by 2 and add it to the second equation:

$$12x - 6y - 4z = -16$$
$$2x + 6y - 3z = 30$$

Adding yields the equation:

$$14x - 7z = 14$$
$$2x - z = 2$$

Multiply the first equation by 8 and the third equation by 3:

$$48x - 24y - 16z = -64$$
$$-9x + 24y - 36z = -27$$

Adding yields the equation $39x - 52z = -91$. So the system becomes:

$$2x - z = 2$$
$$39x - 52z = -91$$

Multiply the first equation by -52:

$$-104x + 52z = -104$$
$$39x - 52z = -91$$

Adding yields:

$$-65x = -195$$
$$x = 3$$

Substituting to find z:
$$6 - z = 2$$
$$z = 4$$

Substituting to find y:
$$6 + 6y - 12 = 30$$
$$6y - 6 = 30$$
$$6y = 36$$
$$y = 6$$

The solution is $(3,6,4)$.

31. To clear each equation of fractions, multiply the first equation by 6, the second equation by 10, and the third equation by 12:
$$3x + 4y = 15$$
$$2x - 5z = -3$$
$$4y - 3z = 9$$

Multiply the third equation by -1 and add it to the first equation:
$$3x + 4y = 15$$
$$-4y + 3z = -9$$

Adding yields:
$$3x + 3z = 6$$
$$x + z = 2$$

So the system becomes:
$$x + z = 2$$
$$2x - 5z = -3$$

Multiply the first equation by 5:
$$5x + 5z = 10$$
$$2x - 5z = -3$$

Adding yields:
$$7x = 7$$
$$x = 1$$

Substituting yields $z = 1$. Substituting to find y:
$$3 + 4y = 15$$
$$4y = 12$$
$$y = 3$$

The solution is $(1,3,1)$.

33. To clear each equation of fractions, multiply the first equation by 4, the second equation by 12, and the third equation by 6:
$$2x - y + 2z = -8$$
$$3x - y - 4z = 3$$
$$x + 2y - 3z = 9$$

Multiply the first equation by -1 and add it to the second equation:
$$-2x + y - 2z = 8$$
$$3x - y - 4z = 3$$

Adding yields the equation $x - 6z = 11$. Multiply the first equation by 2 and add it to the third equation:
$$4x - 2y + 4z = -16$$
$$x + 2y - 3z = 9$$

Adding yields the equation $5x + z = -7$. So the system becomes:
$$5x + z = -7$$
$$x - 6z = 11$$

Multiply the first equation by 6:
$$30x + 6z = -42$$
$$x - 6z = 11$$

Adding yields:
$$31x = -31$$
$$x = -1$$
Substituting to find z:
$$-5 + z = -7$$
$$z = -2$$
Substituting to find y:
$$-1 + 2y + 6 = 9$$
$$2y + 5 = 9$$
$$2y = 4$$
$$y = 2$$
The solution is $(-1, 2, -2)$.

35.   Divide the second equation by 5 and the third equation by 10 to produce the system:
$$x - y - z = 0$$
$$x + 4y = 16$$
$$2y - z = 5$$

Multiply the third equation by –1 and add it to the first equation:
$$x - y - z = 0$$
$$-2y + z = -5$$

Adding yields the equation $x - 3y = -5$. So the system becomes:
$$x + 4y = 16$$
$$x - 3y = -5$$

Multiply the second equation by –1:
$$x + 4y = 16$$
$$-x + 3y = 5$$

Adding yields:
$$7y = 21$$
$$y = 3$$
Substituting to find x:
$$x + 12 = 16$$
$$x = 4$$
Substituting to find z:
$$6 - z = 5$$
$$z = 1$$
The currents are 4 amps, 3 amps, and 1 amp.

37.   Simplifying: $1(4) - 3(2) = 4 - 6 = -2$

39.   Simplifying: $1(1) - 3(-2) + (-2)(-2) = 1 + 6 + 4 = 11$

41.   Simplifying: $-3(-1-1) + 4(-2+2) - 5\left[2 - (-2)\right] = -3(-2) + 4(0) - 5(4) = 6 + 0 - 20 = -14$

43.   Solving the equation:
$$-5x = 20$$
$$x = -4$$

45.   The variation equation is $y = Kx^2$. Substituting $x = 5$ and $y = 75$:
$$75 = K \cdot 5^2$$
$$75 = 25K$$
$$K = 3$$
So $y = 3x^2$. Substituting $x = 7$: $y = 3 \cdot 7^2 = 3 \cdot 49 = 147$

47. The variation equation is $y = \dfrac{K}{x}$. Substituting $x = 25$ and $y = 10$:

$$10 = \frac{K}{25}$$
$$K = 250$$

So $y = \dfrac{250}{x}$. Substituting $y = 5$:

$$5 = \frac{250}{x}$$
$$5x = 250$$
$$x = 50$$

49. The variation equation is $z = Kxy^2$. Substituting $z = 40$, $x = 5$, and $y = 2$:

$$40 = K \bullet 5 \bullet 2^2$$
$$40 = 20K$$
$$K = 2$$

So $z = 2xy^2$. Substituting $x = 2$ and $y = 5$: $z = 2 \bullet 2 \bullet 5^2 = 100$

51. Adding the first and second, the first and third, and third and fourth equations results in the system:
$$2x + 3y + 2w = 16$$
$$2x + 3w = 14$$
$$2x - 3y - w = -8$$

Adding the first and third equations results in the system:
$$4x + w = 8$$
$$2x + 3w = 14$$

Multiply the second equation by -2:
$$4x + w = 8$$
$$-4x - 6w = -28$$

Adding yields:
$$-5w = -20$$
$$w = 4$$

Substituting to find x:
$$4x + 4 = 8$$
$$4x = 4$$
$$x = 1$$

Substituting to find y:
$$2 + 3y + 8 = 16$$
$$3y + 10 = 16$$
$$3y = 6$$
$$y = 2$$

Substituting to find z:
$$1 + 2 + z + 4 = 10$$
$$z + 7 = 10$$
$$z = 3$$

The solution is $(1,2,3,4)$.

4.3 Applications of Linear Systems

1. Let x and y represent the two numbers. The system of equations is:
$$y = 2x + 3$$
$$x + y = 18$$
Substituting into the second equation:
$$x + 2x + 3 = 18$$
$$3x = 15$$
$$x = 5$$
$$y = 2(5) + 3 = 13$$
The two numbers are 5 and 13.

3. Let x and y represent the two numbers. The system of equations is:
$$y - x = 6$$
$$2x = 4 + y$$
The second equation is $y = 2x - 4$. Substituting into the first equation:
$$2x - 4 - x = 6$$
$$x = 10$$
$$y = 2(10) - 4 = 16$$
The two numbers are 10 and 16.

5. Let x, y, and z represent the three numbers. The system of equations is:
$$x + y + z = 8$$
$$2x = z - 2$$
$$x + z = 5$$
The third equation is $z = 5 - x$. Substituting into the second equation:
$$2x = 5 - x - 2$$
$$3x = 3$$
$$x = 1$$
$$z = 5 - 1 = 4$$
Substituting into the first equation:
$$1 + y + 4 = 8$$
$$y = 3$$
The three numbers are 1, 3, and 4.

7. Let a represent the number of adult tickets and c represent the number of children's tickets. The system of equations is:
$$a + c = 925$$
$$2a + c = 1150$$
Multiply the first equation by -1:
$$a \quad c = 925$$
$$2a + c = 1150$$
Adding yields:
$$a = 225$$
$$c = 700$$
There were 225 adult tickets and 700 children's tickets sold.

9. Let x represent the amount invested at 6% and y represent the amount invested at 7%. The system of equations is:

$$x + y = 20,000$$
$$0.06x + 0.07y = 1,280$$

Multiplying the first equation by –0.06:

$$-0.06x - 0.06y = -1,200$$
$$0.06x + 0.07y = 1,280$$

Adding yields:

$$0.01y = 80$$
$$y = 8,000$$
$$x = 12,000$$

Mr. Jones invested $12,000 at 6% and $8,000 at 7%.

11. Let x represent the amount invested at 6% and $2x$ represent the amount invested at 7.5%. The equation is:

$$0.075(2x) + 0.06(x) = 840$$
$$0.21x = 840$$
$$x = 4,000$$
$$2x = 8,000$$

Susan invested $4,000 at 6% and $8,000 at 7.5%.

13. Let x, y and z represent the amounts invested in the three accounts. The system of equations is:

$$x + y + z = 2,200$$
$$z = 3x$$
$$0.06x + 0.08y + 0.09z = 178$$

Substituting into the first equation:

$$x + y + 3x = 2,200$$
$$4x + y = 2,200$$

Substituting into the third equation:

$$0.06x + 0.08y + 0.09(3x) = 178$$
$$0.33x + 0.08y = 178$$

The system of equations becomes:

$$4x + y = 2,200$$
$$0.33x + 0.08y = 178$$

Multiply the first equation by –0.08:

$$-0.32x - 0.08y = -176$$
$$0.33x + 0.08y = 178$$

Adding yields:

$$0.01x = 2$$
$$x = 200$$
$$z = 3(200) = 600$$
$$y = 2,200 - 4(200) = 1,400$$

He invested $200 at 6%, $1,400 at 8%, and $600 at 9%.

15. Let x, y and z represent the amounts invested in the three accounts. The system of equations is:

$$y = 2x$$
$$x + y + z = 1,300$$
$$0.06x + 0.08y + 0.12z = 114$$

Substituting into the second equation:

$$x + 2x + z = 1,300$$
$$3x + z = 1,300$$

Substituting into the third equation:

$$0.06x + 0.08(2x) + 0.12z = 114$$
$$0.22x + 0.12z = 114$$

The system of equations becomes:

$$3x + z = 1,300$$
$$0.22x + 0.12z = 114$$

Multiply the first equation by –0.12:
$$-0.36x - 0.12z = -156$$
$$0.22x + 0.12z = 114$$
Adding yields:
$$-0.14x = -42$$
$$x = 300$$
$$y = 2(300) = 600$$
$$z = 1,300 - 300 - 600 = 400$$
Martin has invested \$300 at 6%, \$600 at 8%, and \$400 at 12%.

17.   Let x represent the amount of 20% alcohol and y represent the amount of 50% alcohol. The system of equations is:
$$x + y = 9$$
$$0.20x + 0.50y = 0.30(9)$$
Multiplying the first equation by –0.2:
$$-0.20x - 0.20y = -1.8$$
$$0.20x + 0.50y = 2.7$$
Adding yields:
$$0.30y = 0.9$$
$$y = 3$$
$$x = 6$$
The mixture contains 3 gallons of 50% alcohol and 6 gallons of 20% alcohol.

19.   Let x represent the amount of 20% disinfectant and y represent the amount of 14% disinfectant. The system of equations is:
$$x + y = 15$$
$$0.20x + 0.14y = 0.16(15)$$
Multiplying the first equation by –0.14:
$$-0.14x - 0.14y = -2.1$$
$$0.20x + 0.14y = 2.4$$
Adding yields:
$$0.06x = 0.3$$
$$x = 5$$
$$y = 10$$
The mixture contains 5 gallons of 20% disinfectant and 10 gallons of 14% disinfectant.

21.   Let x represent the amount 40% copper alloy and y represent the amount of 60% copper alloy. The system of equations is:
$$x + y = 50$$
$$0.4x + 0.6y = 0.55(50)$$
Multiplying the first equation by 0.4:
$$-0.4x - 0.4y = -20$$
$$0.4x + 0.6y = 27.5$$
Adding yields:
$$0.2y = 7.5$$
$$y = 37.5$$
$$x = 12.5$$
The mixture contains 12.5 pounds of 40% copper alloy and 37.5 pounds of 60% copper alloy.

23.   Let b represent the rate of the boat and c represent the rate of the current. The system of equations is:

$$2(b+c) = 24$$
$$3(b-c) = 18$$

The system of equations simplifies to:

$$b+c = 12$$
$$b-c = 6$$

Adding yields:

$$2b = 18$$
$$b = 9$$
$$c = 3$$

The rate of the boat is 9 mph and the rate of the current is 3 mph.

25.   Let a represent the rate of the airplane and w represent the rate of the wind. The system of equations is:

$$2(a+w) = 600$$
$$\frac{5}{2}(a-w) = 600$$

The system of equations simplifies to:

$$a+w = 300$$
$$a-w = 240$$

Adding yields:

$$2a = 540$$
$$a = 270$$
$$w = 30$$

The rate of the airplane is 270 mph and the rate of the wind is 30 mph.

27.   Let n represent the number of nickels and d represent the number of dimes. The system of equations is:

$$n+d = 20$$
$$0.05n + 0.10d = 1.40$$

Multiplying the first equation by –0.05:

$$-0.05n - 0.05d = -1$$
$$0.05n + 0.10d = 1.40$$

Adding yields:

$$0.05d = 0.40$$
$$d = 8$$
$$n = 12$$

Bob has 12 nickels and 8 dimes.

29.   Let n, d, and q represent the number of nickels, dimes, and quarters. The system of equations is:

$$n+d+q = 9$$
$$0.05n + 0.10d + 0.25q = 1.20$$
$$d = n$$

Substituting into the first equation:

$$n+n+q = 9$$
$$2n+q = 9$$

Substituting into the second equation:

$$0.05n + 0.10n + 0.25q = 1.20$$
$$0.15n + 0.25q = 1.20$$

The system of equations becomes:

$$2n+q = 9$$
$$0.15n + 0.25q = 1.20$$

Multiplying the first equation by –0.25:

$$-0.50n - 0.25q = -2.25$$
$$0.15n + 0.25q = 1.20$$

Adding yields:
$$-0.35n = -1.05$$
$$n = 3$$
$$d = 3$$
$$q = 9 - 2(3) = 3$$
The collection contains 3 nickels, 3 dimes, and 3 quarters.

31. Let n, d, and q represent the number of nickels, dimes, and quarters. The system of equations is:
$$n + d + q = 140$$
$$0.05n + 0.10d + 0.25q = 10.00$$
$$d = 2q$$

Substituting into the first equation:
$$n + 2q + q = 140$$
$$n + 3q = 140$$

Substituting into the second equation:
$$0.05n + 0.10(2q) + 0.25q = 10.00$$
$$0.05n + 0.45q = 10.00$$

The system of equations becomes:
$$n + 3q = 140$$
$$0.05n + 0.45q = 10.00$$

Multiplying the first equation by -0.05:
$$-0.05n - 0.15q = -7$$
$$0.05n + 0.45q = 10$$

Adding yields:
$$0.30q = 3$$
$$q = 10$$
$$d = 2(10) = 20$$
$$n = 140 - 3(10) = 110$$

There are 110 nickels in the collection.

33. Let n, d, and q represent the number of nickels, dimes, and quarters. The system of equations is:
$$n + d + q = 31$$
$$d = 2n$$
$$0.05n + 0.10d + 0.25q = 5.25$$

Substituting into the first equation:
$$n + 2n + q = 31$$
$$3n + q = 31$$

Substituting into the third equation:
$$0.05n + 0.10(2n) + 0.25q = 5.25$$
$$0.25n + 0.25q = 5.25$$
$$n + q = 21$$

The system of equations becomes:
$$3n + q = 31$$
$$n + q = 21$$

Multiplying the second equation by -1:
$$3n + q = 31$$
$$-n - q = -21$$

Adding yields:
$$2n = 10$$
$$n = 5$$
$$d = 2(5) = 10$$
$$q = 31 - 5 - 10 = 16$$

Justin has 5 nickels, 10 dimes, and 16 quarters in the collection.

35. Let $x = mp + b$ represent the relationship. Using the points (2,300) and (1.5,400) results in the system:

$$300 = 2m + b$$
$$400 = 1.5m + b$$

Multiplying the second equation by –1:

$$300 = 2m + b$$
$$-400 = -1.5m - b$$

Adding yields:

$$-100 = 0.5m$$
$$m = -200$$
$$b = 300 - 2(-200) = 700$$

The equation is $x = -200p + 700$. Substituting $p = 3$: $x = -200(3) + 700 = 100$ items

37. The system of equations is:

$$a + b + c = 128$$
$$9a + 3b + c = 128$$
$$25a + 5b + c = 0$$

Multiply the first equation by –1 and add it to the second equation:

$$-a - b - c = -128$$
$$9a + 3b + c = 128$$

Adding yields:

$$8a + 2b = 0$$
$$4a + b = 0$$

Multiply the first equation by –1 and add it to the third equation:

$$-a - b - c = -128$$
$$25a + 5b + c = 0$$

Adding yields:

$$24a + 4b = -128$$
$$6a + b = -32$$

The system simplifies to:

$$4a + b = 0$$
$$6a + b = -32$$

Multiplying the first equation by –1:

$$-4a - b = 0$$
$$6a + b = -32$$

Adding yields:

$$2a = -32$$
$$a = -16$$

Substituting to find b:

$$4(-16) + b = 0$$
$$b = 64$$

Substituting to find c:

$$-16 + 64 + c = 128$$
$$c = 80$$

The equation for the height is $h = -16t^2 + 64t + 80$.

39. No, the graph does not include the boundary line.

41. Substituting $x = 4 - y$ into the second equation:

$$(4 - y) - 2y = 4$$
$$4 - 3y = 4$$
$$-3y = 0$$
$$y = 0$$

The solution is (4,0).

43. Solving the inequality:

$$20x + 9,300 > 18,000$$
$$20x > 8,700$$
$$x > 435$$

45.   Graphing the inequality:

2x + 3y = 6

47.   Graphing the inequality:

y = -3x - 4

49.   Graphing the inequality:

x = 3

51.   Graphing the data:

53.   a.   Using the points $(1975, 13)$ and $(1980, 16)$ in the model $M = mx + b$ results in the system:

$$1975m + b = 13$$
$$1980m + b = 16$$

Multiply the first equation by -1:

$$-1975m - b = -13$$
$$1980m + b = 16$$

Adding yields:

$$5m = 3$$
$$m = 0.6$$
$$b = 16 - 1980(0.6) = -1172$$

The equation is $M = 0.6x - 1172$.

b.   Using the points (1975,15) and (1980,13) in the model $F = mx + b$ results in the system:

$$1975m + b = 15$$
$$1980m + b = 13$$

Multiply the first equation by –1:

$$-1975m - b = -15$$
$$1980m + b = 13$$

Adding yields:

$$5m = -2$$
$$m = -0.4$$
$$b = 13 - 1980(-0.4) = 805$$

The equation is $F = -0.4x + 805$.

c.   Finding where the two rates are equal:

$$0.6x - 1172 = -0.4x + 805$$
$$x = 1977$$

The dropout rates are equal in the year 1977.

4.4 Matrix Solutions to Linear Systems

1.   Form the augmented matrix: $\begin{bmatrix} 1 & 1 & | & 5 \\ 3 & -1 & | & 3 \end{bmatrix}$

Add –3 times row 1 to row 2: $\begin{bmatrix} 1 & 1 & | & 5 \\ 0 & -4 & | & -12 \end{bmatrix}$

Dividing row 2 by –4: $\begin{bmatrix} 1 & 1 & | & 5 \\ 0 & 1 & | & 3 \end{bmatrix}$

So $y = 3$. Substituting to find x:

$$x + 3 = 5$$
$$x = 2$$

The solution is (2,3).

3.   Using the second equation as row 1, form the augmented matrix: $\begin{bmatrix} -1 & 1 & | & -1 \\ 3 & -5 & | & 7 \end{bmatrix}$

Add 3 times row 1 to row 2: $\begin{bmatrix} -1 & 1 & | & -1 \\ 0 & -2 & | & 4 \end{bmatrix}$

Dividing row 2 by –2 and row 1 by –1: $\begin{bmatrix} 1 & -1 & | & 1 \\ 0 & 1 & | & -2 \end{bmatrix}$

So $y = -2$. Substituting to find x:

$$x + 2 = 1$$
$$x = -1$$

The solution is (–1,–2).

5.   Form the augmented matrix: $\begin{bmatrix} 2 & -8 & | & 6 \\ 3 & -8 & | & 13 \end{bmatrix}$

Dividing row 1 by 2: $\begin{bmatrix} 1 & -4 & | & 3 \\ 3 & -8 & | & 13 \end{bmatrix}$

Add -3 times row 1 to row 2: $\begin{bmatrix} 1 & -4 & | & 3 \\ 0 & 4 & | & 4 \end{bmatrix}$

Dividing row 2 by 4: $\begin{bmatrix} 1 & -4 & | & 3 \\ 0 & 1 & | & 1 \end{bmatrix}$

So $y = 1$. Substituting to find x:
$$x - 4 = 3$$
$$x = 7$$
The solution is $(7,1)$.

7.   Form the augmented matrix: $\begin{bmatrix} 2 & -1 & | & -10 \\ 4 & 3 & | & 0 \end{bmatrix}$

Dividing row 1 by 2: $\begin{bmatrix} 1 & -\dfrac{1}{2} & | & -5 \\ 4 & 3 & | & 0 \end{bmatrix}$

Add -4 times row 1 to row 2: $\begin{bmatrix} 1 & -\dfrac{1}{2} & | & -5 \\ 0 & 5 & | & 20 \end{bmatrix}$

Divide row 2 by 5: $\begin{bmatrix} 1 & -\dfrac{1}{2} & | & -5 \\ 0 & 1 & | & 4 \end{bmatrix}$

So $y = 4$. Substituting to find x:
$$x - \frac{1}{2}(4) = -5$$
$$x - 2 = -5$$
$$x = -3$$
The solution is $(-3,4)$.

9.   Form the augmented matrix: $\begin{bmatrix} 5 & -3 & | & 27 \\ 6 & 2 & | & -18 \end{bmatrix}$

Dividing row 1 by 5: $\begin{bmatrix} 1 & -\dfrac{3}{5} & | & \dfrac{27}{5} \\ 6 & 2 & | & -18 \end{bmatrix}$

Add -6 times row 1 to row 2: $\begin{bmatrix} 1 & -\dfrac{3}{5} & | & \dfrac{27}{5} \\ 0 & \dfrac{28}{5} & | & -\dfrac{252}{5} \end{bmatrix}$

Multiply row 2 by $\dfrac{5}{28}$: $\begin{bmatrix} 1 & -\dfrac{3}{5} & | & \dfrac{27}{5} \\ 0 & 1 & | & -9 \end{bmatrix}$

So $y = -9$. Substituting to find x:

$$x - \frac{3}{5}(-9) = \frac{27}{5}$$

$$x + \frac{27}{5} = \frac{27}{5}$$

$$x = 0$$

The solution is $(0, -9)$.

11.   First rewrite the system as:

$$5x + 2y = -14$$

$$-2x + y = 11$$

Form the augmented matrix: $\begin{bmatrix} 5 & 2 & | & -14 \\ -2 & 1 & | & 11 \end{bmatrix}$

Dividing row 1 by 5: $\begin{bmatrix} 1 & \dfrac{2}{5} & | & -\dfrac{14}{5} \\ -2 & 1 & | & 11 \end{bmatrix}$

Add 2 times row 1 to row 2: $\begin{bmatrix} 1 & \dfrac{2}{5} & | & -\dfrac{14}{5} \\ 0 & \dfrac{9}{5} & | & \dfrac{27}{5} \end{bmatrix}$

Multiply row 2 by $\dfrac{5}{9}$: $\begin{bmatrix} 1 & \dfrac{2}{5} & | & -\dfrac{14}{5} \\ 0 & 1 & | & 3 \end{bmatrix}$

So $y = 3$. Substituting to find x:

$$x + \frac{2}{5}(3) = -\frac{14}{5}$$

$$x + \frac{6}{5} = -\frac{14}{5}$$

$$x = -4$$

The solution is $(-4, 3)$.

13.   Form the augmented matrix: $\begin{bmatrix} 1 & 1 & 1 & | & 4 \\ 1 & -1 & 2 & | & 1 \\ 1 & -1 & -1 & | & -2 \end{bmatrix}$

Add -1 times row 1 to both row 2 and row 3: $\begin{bmatrix} 1 & 1 & 1 & | & 4 \\ 0 & -2 & 1 & | & -3 \\ 0 & -2 & -2 & | & -6 \end{bmatrix}$

Multiply row 2 by -1 and add it to row 3: $\begin{bmatrix} 1 & 1 & 1 & | & 4 \\ 0 & 2 & -1 & | & 3 \\ 0 & 0 & -3 & | & -3 \end{bmatrix}$

Divide row 3 by -3: $\begin{bmatrix} 1 & 1 & 1 & | & 4 \\ 0 & 2 & -1 & | & 3 \\ 0 & 0 & 1 & | & 1 \end{bmatrix}$

So $z = 1$. Substituting to find y:

$$2y - 1 = 3$$

$$2y = 4$$

$$y = 2$$

Substituting to find x:
$$x+2+1=4$$
$$x=1$$
The solution is $(1,2,1)$.

15. Form the augmented matrix: $\begin{bmatrix} 1 & 2 & 1 & | & 3 \\ 2 & -1 & 2 & | & 6 \\ 3 & 1 & -1 & | & 5 \end{bmatrix}$

Add -2 times row 1 to row 2 and -3 times row 1 to row 3: $\begin{bmatrix} 1 & 2 & 1 & | & 3 \\ 0 & -5 & 0 & | & 0 \\ 0 & -5 & -4 & | & -4 \end{bmatrix}$

Dividing row 2 by -5: $\begin{bmatrix} 1 & 2 & 1 & | & 3 \\ 0 & 1 & 0 & | & 0 \\ 0 & -5 & -4 & | & -4 \end{bmatrix}$

Adding 5 times row 2 to row 3: $\begin{bmatrix} 1 & 2 & 1 & | & 3 \\ 0 & 1 & 0 & | & 0 \\ 0 & 0 & -4 & | & -4 \end{bmatrix}$

Dividing row 3 by -4: $\begin{bmatrix} 1 & 2 & 1 & | & 3 \\ 0 & 1 & 0 & | & 0 \\ 0 & 0 & 1 & | & 1 \end{bmatrix}$

So $y=0$ and $z=1$. Substituting to find x:
$$x+0+1=3$$
$$x=2$$
The solution is $(2,0,1)$.

17. Form the augmented matrix: $\begin{bmatrix} 1 & -2 & 1 & | & -4 \\ 2 & 1 & -3 & | & 7 \\ 5 & -3 & 1 & | & -5 \end{bmatrix}$

Add -2 times row 1 to row 2 and -5 times row 1 to row 3: $\begin{bmatrix} 1 & -2 & 1 & | & -4 \\ 0 & 5 & -5 & | & 15 \\ 0 & 7 & -4 & | & 15 \end{bmatrix}$

Dividing row 2 by 5: $\begin{bmatrix} 1 & -2 & 1 & | & -4 \\ 0 & 1 & -1 & | & 3 \\ 0 & 7 & -4 & | & 15 \end{bmatrix}$

Add -7 times row 2 to row 3: $\begin{bmatrix} 1 & -2 & 1 & | & -4 \\ 0 & 1 & -1 & | & 3 \\ 0 & 0 & 3 & | & -6 \end{bmatrix}$

Dividing row 3 by 3: $\begin{bmatrix} 1 & -2 & 1 & | & -4 \\ 0 & 1 & -1 & | & 3 \\ 0 & 0 & 1 & | & -2 \end{bmatrix}$

So $z=-2$. Substituting to find y:
$$y-(-2)=3$$
$$y+2=3$$
$$y=1$$

Substituting to find x:

$$x - 2(1) + (-2) = -4$$
$$x - 4 = -4$$
$$x = 0$$

The solution is $(0, 1, -2)$.

19.   Form the augmented matrix (using the second equation as row 1):
$$\begin{bmatrix} 1 & -2 & -1 & | & 0 \\ 5 & -3 & 1 & | & 10 \\ 3 & -1 & 2 & | & 10 \end{bmatrix}$$

Add -5 times row 1 to row 2 and -3 times row 1 to row 3:
$$\begin{bmatrix} 1 & -2 & -1 & | & 0 \\ 0 & 7 & 6 & | & 10 \\ 0 & 5 & 5 & | & 10 \end{bmatrix}$$

Dividing row 5 by 5, and switching it with row 2:
$$\begin{bmatrix} 1 & -2 & -1 & | & 0 \\ 0 & 1 & 1 & | & 2 \\ 0 & 7 & 6 & | & 10 \end{bmatrix}$$

Add -7 times row 2 to row 3:
$$\begin{bmatrix} 1 & -2 & -1 & | & 0 \\ 0 & 1 & 1 & | & 2 \\ 0 & 0 & -1 & | & -4 \end{bmatrix}$$

Multiplying row 3 by -1:
$$\begin{bmatrix} 1 & -2 & -1 & | & 0 \\ 0 & 1 & 1 & | & 2 \\ 0 & 0 & 1 & | & 4 \end{bmatrix}$$

So $z = 4$. Substituting to find y:

$$y + 4 = 2$$
$$y = -2$$

Substituting to find x:

$$x - 2(-2) - 4 = 0$$
$$x = 0$$

The solution is $(0, -2, 4)$.

21.   Form the augmented matrix (using the third equation as row 1):
$$\begin{bmatrix} 1 & 1 & 2 & | & 5 \\ 2 & -5 & 3 & | & 2 \\ 3 & -7 & 1 & | & 0 \end{bmatrix}$$

Add -2 times row 1 to row 2 and -3 times row 1 to row 3:
$$\begin{bmatrix} 1 & 1 & 2 & | & 5 \\ 0 & -7 & -1 & | & -8 \\ 0 & -10 & -5 & | & -15 \end{bmatrix}$$

Dividing row 3 by -10 and switching it with row 2:
$$\begin{bmatrix} 1 & 1 & 2 & | & 5 \\ 0 & 1 & \frac{1}{2} & | & \frac{3}{2} \\ 0 & -7 & -1 & | & -8 \end{bmatrix}$$

Add 7 times row 2 to row 3:
$$\begin{bmatrix} 1 & 1 & 2 & | & 5 \\ 0 & 1 & \frac{1}{2} & | & \frac{3}{2} \\ 0 & 0 & \frac{5}{2} & | & \frac{5}{2} \end{bmatrix}$$

Multiplying row 3 by $\dfrac{2}{5}$: $\begin{bmatrix} 1 & 1 & 2 & | & 5 \\ 0 & 1 & \frac{1}{2} & | & \frac{3}{2} \\ 0 & 0 & 1 & | & 1 \end{bmatrix}$

So $z = 1$. Substituting to find y:

$$y + \frac{1}{2}(1) = \frac{3}{2}$$

$$y + \frac{1}{2} = \frac{3}{2}$$

$$y = 1$$

Substituting to find x:

$$x + 1(1) + 2(1) = 5$$

$$x + 3 = 5$$

$$x = 2$$

The solution is $(2, 1, 1)$.

23.   Form the augmented matrix: $\begin{bmatrix} 1 & 2 & 0 & | & 3 \\ 0 & 1 & 1 & | & 3 \\ 4 & 0 & -1 & | & 2 \end{bmatrix}$

Add −4 times row 1 to row 3: $\begin{bmatrix} 1 & 2 & 0 & | & 3 \\ 0 & 1 & 1 & | & 3 \\ 0 & -8 & -1 & | & -10 \end{bmatrix}$

Add 8 times row 2 to row 3: $\begin{bmatrix} 1 & 2 & 0 & | & 3 \\ 0 & 1 & 1 & | & 3 \\ 0 & 0 & 7 & | & 14 \end{bmatrix}$

Dividing row 3 by 7: $\begin{bmatrix} 1 & 2 & 0 & | & 3 \\ 0 & 1 & 1 & | & 3 \\ 0 & 0 & 1 & | & 2 \end{bmatrix}$

So $z = 2$. Substituting to find y:

$$y + 2 = 3$$

$$y = 1$$

Substituting to find x:

$$x + 2 = 3$$

$$x = 1$$

The solution is $(1, 1, 2)$.

25.   Form the augmented matrix: $\begin{bmatrix} 1 & 3 & 0 & | & 7 \\ 3 & 0 & -4 & | & -8 \\ 0 & 5 & -2 & | & -5 \end{bmatrix}$

Add -3 times row 1 to row 2: $\begin{bmatrix} 1 & 3 & 0 & | & 7 \\ 0 & -9 & -4 & | & -29 \\ 0 & 5 & -2 & | & -5 \end{bmatrix}$

Add 2 times row 3 to row 2: $\begin{bmatrix} 1 & 3 & 0 & | & 7 \\ 0 & 1 & -8 & | & -39 \\ 0 & 5 & -2 & | & -5 \end{bmatrix}$

Add -5 times row 2 to row 3: $\begin{bmatrix} 1 & 3 & 0 & | & 7 \\ 0 & 1 & -8 & | & -39 \\ 0 & 0 & 38 & | & 190 \end{bmatrix}$

Dividing row 3 by 38: $\begin{bmatrix} 1 & 3 & 0 & | & 7 \\ 0 & 1 & -8 & | & -39 \\ 0 & 0 & 1 & | & 5 \end{bmatrix}$

So $z = 5$. Substituting to find y:
$$y - 40 = -39$$
$$y = 1$$
Substituting to find x:
$$x + 3 = 7$$
$$x = 4$$
The solution is $(4,1,5)$.

27.   Form the augmented matrix: $\begin{bmatrix} 1 & 2 & 0 & | & 13 \\ 1 & 0 & -3 & | & 11 \\ 0 & 3 & 4 & | & 4 \end{bmatrix}$

Add -1 times row 1 to row 2: $\begin{bmatrix} 1 & 2 & 0 & | & 13 \\ 0 & -2 & -3 & | & -2 \\ 0 & 3 & 4 & | & 4 \end{bmatrix}$

Dividing row 2 by -2: $\begin{bmatrix} 1 & 2 & 0 & | & 13 \\ 0 & 1 & \frac{3}{2} & | & 1 \\ 0 & 3 & 4 & | & 4 \end{bmatrix}$

Add -3 times row 2 to row 3: $\begin{bmatrix} 1 & 2 & 0 & | & 13 \\ 0 & 1 & \frac{3}{2} & | & 1 \\ 0 & 0 & -\frac{1}{2} & | & 1 \end{bmatrix}$

Multiplying row 3 by -2: $\begin{bmatrix} 1 & 2 & 0 & | & 13 \\ 0 & 1 & \frac{3}{2} & | & 1 \\ 0 & 0 & 1 & | & -2 \end{bmatrix}$

So $z = -2$. Substituting to find y:

$$y + \frac{3}{2}(-2) = 1$$

$$y - 3 = 1$$

$$y = 4$$

Substituting to find x:

$$x + 2(4) = 13$$

$$x + 8 = 13$$

$$x = 5$$

The solution is $(5, 4, -2)$.

29. Form the augmented matrix:
$\begin{bmatrix} \frac{1}{3} & \frac{1}{5} & \bigg| & 2 \\ \frac{1}{3} & -\frac{1}{2} & \bigg| & -\frac{1}{3} \end{bmatrix}$

Multiplying row 1 by 15 and row 2 by 6:
$\begin{bmatrix} 5 & 3 & \big| & 30 \\ 2 & -3 & \big| & -2 \end{bmatrix}$

Dividing row 2 by 2:
$\begin{bmatrix} 5 & 3 & \big| & 30 \\ 1 & -\frac{3}{2} & \big| & -1 \end{bmatrix}$

Add -5 times row 2 to row 1:
$\begin{bmatrix} 0 & \frac{21}{2} & \big| & 35 \\ 1 & -\frac{3}{2} & \big| & -1 \end{bmatrix}$

Multiplying row 1 by $\frac{2}{21}$:
$\begin{bmatrix} 0 & 1 & \big| & \frac{10}{3} \\ 1 & -\frac{3}{2} & \big| & -1 \end{bmatrix}$

So $y = \frac{10}{3}$. Substituting to find x:

$$x - \frac{3}{2}\left(\frac{10}{3}\right) = -1$$

$$x - 5 = -1$$

$$x = 4$$

The solution is $\left(4, \frac{10}{3}\right)$.

31. Form the augmented matrix:
$\begin{bmatrix} 2 & -3 & \big| & 4 \\ 4 & -6 & \big| & 4 \end{bmatrix}$

Add -2 times row 1 to row 2:
$\begin{bmatrix} 2 & -3 & \big| & 4 \\ 0 & 0 & \big| & -4 \end{bmatrix}$

The second row states that $0 = -4$, which is false. There is no solution.

33.   Form the augmented matrix: $\begin{bmatrix} -6 & 4 & | & 8 \\ -3 & 2 & | & 4 \end{bmatrix}$

Divide row 1 by -2: $\begin{bmatrix} 3 & -2 & | & -4 \\ -3 & 2 & | & 4 \end{bmatrix}$

Adding row 1 to row 2: $\begin{bmatrix} 3 & -2 & | & -4 \\ 0 & 0 & | & 0 \end{bmatrix}$

The second row states that $0 = 0$, which is true. The system is dependent.

35.   Graphing the line:

37.   Graphing the line:

39.   Graphing the line:

4.5 Systems of Linear Inequalities

1.   Graphing the solution set:

3.   Graphing the solution set:

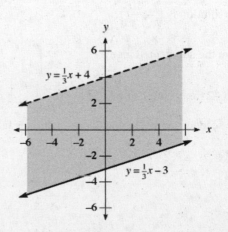

5. Graphing the solution set:

7. Graphing the solution set:

9. Graphing the solution set:

11. Graphing the solution set:

13. Graphing the solution set:

15. Graphing the solution set:

17. Graphing the solution set:

19. The system of inequalities is:
$$x + y \leq 4$$
$$-x + y < 4$$

21. The system of inequalities is:
$$x + y \geq 4$$
$$-x + y < 4$$

23. **a.** The system of inequalities is:
$$0.55x + 0.65y \leq 40$$
$$x \geq 2y$$
$$x > 15$$
$$y \geq 0$$

Graphing the solution set:

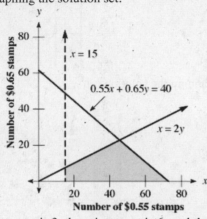

b. Substitute $x = 20$:
$$2y \leq 20$$
$$y \leq 10$$

The most he can purchase is ten 65-cent stamps.

25. The x-intercept is 3, the y-intercept is 6, and the slope is –2.

27. The x-intercept is –2, there is no y-intercept, and there is no slope.

29. First find the slope: $m = \dfrac{-1-2}{4-(-3)} = \dfrac{-3}{4+3} = -\dfrac{3}{7}$. Using the point-slope formula:

$$y - 2 = -\frac{3}{7}(x+3)$$
$$y - 2 = -\frac{3}{7}x - \frac{9}{7}$$
$$y = -\frac{3}{7}x + \frac{5}{7}$$

31. Since the line is vertical, its equation is $x = 4$.

33. The domain is all real numbers and the range is $\{y \mid y \geq -9\}$. This is a function.

35. Evaluating the function: $h(0) + g(0) = \left[3 \cdot 0^2 - 2 \cdot 0 - 8\right] + \left[3 \cdot 0 + 4\right] = -8 + 4 = -4$

37. Evaluating the function: $g(f(2)) = g(2-2) = g(0) = 3 \cdot 0 + 4 = 4$

39. The variation equation is $z = Kxy^3$. Substituting $x = 5$, $y = 2$, and $z = 15$:

$$15 = K(5)(2)^3$$
$$15 = 40K$$
$$K = \frac{3}{8}$$

The equation is $z = \dfrac{3}{8}xy^3$. Substituting $x = 2$ and $y = 3$: $z = \dfrac{3}{8}(2)(3)^3 = \dfrac{3}{8} \cdot 54 = \dfrac{81}{4}$

Chapter 4 Review

1. Adding the two equations yields:
$$3x = 18$$
$$x = 6$$
Substituting to find y:
$$6 + y = 4$$
$$y = -2$$
The solution is $(6, -2)$.

3. Multiply the second equation by 2:
$$2x - 4y = 5$$
$$-2x + 4y = 6$$
Adding yields $0 = 11$, which is false. There is no solution (parallel lines).

5. Multiply the second equation by -2:
$$6x - 5y = -5$$
$$-6x - 2y = -2$$
Adding yields:
$$-7y = -7$$
$$y = 1$$
Substituting to find x:
$$3x + 1 = 1$$
$$3x = 0$$
$$x = 0$$
The solution is $(0, 1)$.

7.   Multiply the first equation by 4 and the second equation by 3:
$$12x - 28y = 8$$
$$-12x + 18y = -18$$
Adding yields:
$$-10y = -10$$
$$y = 1$$
Substitute to find x:
$$3x - 7 = 2$$
$$3x = 9$$
$$x = 3$$
The solution is $(3,1)$.

9.   Multiply the first equation by 3 and the second equation by 4:
$$-21x + 12y = -3$$
$$20x - 12y = 0$$
Adding yields $x = 3$. Substitute to find y:
$$-21 + 4y = -1$$
$$4y = 20$$
$$y = 5$$
The solution is $(3,5)$.

11. To clear each equation of fractions, multiply the first equation by 6 and the second equation by 6:
$$4x - y = 0$$
$$8x + 5y = 84$$
Multiply the first equation by -2:
$$-8x + 2y = 0$$
$$8x + 5y = 84$$
Adding yields:
$$7y = 84$$
$$y = 12$$
Substitute to find x:
$$4x - 12 = 0$$
$$4x = 12$$
$$x = 3$$
The solution is $(3,12)$.

13. Substitute into the first equation:
$$x + x - 1 = 2$$
$$2x - 1 = 2$$
$$2x = 3$$
$$x = \frac{3}{2}$$
$$y = \frac{3}{2} - 1 = \frac{1}{2}$$
The solution is $\left(\frac{3}{2}, \frac{1}{2}\right)$.

15. Write the first equation as $y = 4 - x$. Substitute into the second equation:
$$2x + 5(4 - x) = 2$$
$$2x + 20 - 5x = 2$$
$$-3x = -18$$
$$x = 6$$
$$y = 4 - 6 = -2$$
The solution is $(6,-2)$.

17. Substitute into the first equation:
$$3(-3y+4)+7y=6$$
$$-9y+12+7y=6$$
$$-2y=-6$$
$$y=3$$
$$x=-9+4=-5$$
The solution is $(-5,3)$.

19. Adding the first and second equations yields:
$$2x-2z=-2$$
$$x-z=-1$$
Adding the second and third equations yields $2x-5z=-14$. So the system becomes:
$$x-z=-1$$
$$2x-5z=-14$$
Multiply the first equation by -2:
$$-2x+2z=2$$
$$2x-5z=-14$$
Adding yields:
$$-3z=-12$$
$$z=4$$
Substitute to find x:
$$x-4=-1$$
$$x=3$$
Substitute to find y:
$$3+y+4=6$$
$$y=-1$$
The solution is $(3,-1,4)$.

21. Multiply the second equation by 2 and add it to the first equation:
$$5x+8y-4z=-7$$
$$14x+8y+4z=-4$$
Adding yields the equation $19x+16y=-11$. Multiply the first equation by 2 and add it to the third equation:
$$10x+16y-8z=-14$$
$$3x-2y+8z=8$$
Adding yields the equation $13x+14y=-6$. So the system of equations becomes:
$$19x+16y=-11$$
$$13x+14y=-6$$
Multiply the first equation by 7 and the second equation by -8:
$$133x+112y=-77$$
$$-104x-112y=48$$
Adding yields:
$$29x=-29$$
$$x=-1$$
Substituting to find y:
$$-13+14y=-6$$
$$14y=7$$
$$y=\frac{1}{2}$$

Substituting to find z:

$$7(-1) + 4\left(\frac{1}{2}\right) + 2z = -2$$
$$-5 + 2z = -2$$
$$2z = 3$$
$$z = \frac{3}{2}$$

The solution is $\left(-1, \frac{1}{2}, \frac{3}{2}\right)$.

23.   Multiply the second equation by 2 and add it to the third equation:

$$-6x + 8y - 2z = 4$$
$$6x - 8y + 2z = -4$$

Adding yields $0 = 0$. Since this is a true statement, there is no unique solution (dependent system).

25.   Multiply the third equation by 2 and add it to the second equation:

$$3x - 2z = -2$$
$$10y + 2z = -2$$

Adding yields the equation $3x + 10y = -4$. So the system becomes:

$$2x - y = 5$$
$$3x + 10y = -4$$

Multiplying the first equation by 10:

$$20x - 10y = 50$$
$$3x + 10y = -4$$

Adding yields:

$$23x = 46$$
$$x = 2$$

Substituting to find y:

$$4 - y = 5$$
$$y = -1$$

Substituting to find z:

$$-5 + z = -1$$
$$z = 4$$

The solution is $(2, -1, 4)$.

27.   Let a represent the adult tickets and c represent the children's tickets. The system of equations is:

$$a + c = 127$$
$$2a + 1.5c = 214$$

Multiply the first equation by -2:

$$-2a - 2c = -254$$
$$2a + 1.5c = 214$$

Adding yields:

$$-0.5c = -40$$
$$c = 80$$
$$a = 47$$

There were 47 adult tickets and 80 children's tickets sold.

29. Let x represent the amount invested at 12% and y represent the amount invested at 15%. The system of equations is:

$$x + y = 12000$$
$$0.12x + 0.15y = 1650$$

Multiplying the first equation by -0.12:

$$-0.12x - 0.12y = -1440$$
$$0.12x + 0.15y = 1650$$

Adding yields:

$$0.03y = 210$$
$$y = 7000$$
$$x = 5000$$

Ms. Jones invested \$5,000 at 12% and \$7,000 at 15%.

31. Form the augmented matrix: $\begin{bmatrix} 5 & -3 & | & 13 \\ 2 & 6 & | & -2 \end{bmatrix}$

Dividing row 2 by 2 and switching it with row 1: $\begin{bmatrix} 1 & 3 & | & -1 \\ 5 & -3 & | & 13 \end{bmatrix}$

Add -5 times row 1 to row 2: $\begin{bmatrix} 1 & 3 & | & -1 \\ 0 & -18 & | & 18 \end{bmatrix}$

Dividing row 2 by -18: $\begin{bmatrix} 1 & 3 & | & -1 \\ 0 & 1 & | & -1 \end{bmatrix}$

So $y = -1$. Substituting to find x:

$$x + 3(-1) = -1$$
$$x - 3 = -1$$
$$x = 2$$

The solution is $(2, -1)$.

33. Form the augmented matrix using equation 2 as row 1: $\begin{bmatrix} 1 & -4 & -3 & | & 4 \\ 5 & 3 & 1 & | & 6 \\ 3 & 1 & -5 & | & -6 \end{bmatrix}$

Add -5 times row 1 to row 2 and -3 times row 1 to row 3: $\begin{bmatrix} 1 & -4 & -3 & | & 4 \\ 0 & 23 & 16 & | & -14 \\ 0 & 13 & 4 & | & -18 \end{bmatrix}$

Dividing row 2 by 23: $\begin{bmatrix} 1 & -4 & -3 & | & 4 \\ 0 & 1 & \frac{16}{23} & | & -\frac{14}{23} \\ 0 & 13 & 4 & | & -18 \end{bmatrix}$

Add -13 times row 2 to row 3: $\begin{bmatrix} 1 & -4 & -3 & | & 4 \\ 0 & 1 & \frac{16}{23} & | & -\frac{14}{23} \\ 0 & 0 & -\frac{116}{23} & | & -\frac{232}{23} \end{bmatrix}$

Multiplying row 3 by $-\frac{23}{116}$: $\begin{bmatrix} 1 & -4 & -3 & | & 4 \\ 0 & 1 & \frac{16}{23} & | & -\frac{14}{23} \\ 0 & 0 & 1 & | & 2 \end{bmatrix}$

So $z = 2$. Substituting to find y:

$$y + \frac{16}{23}(2) = -\frac{14}{23}$$

$$y + \frac{32}{23} = -\frac{14}{23}$$

$$y = -2$$

Substituting to find x:

$$x - 4(-2) - 3(2) = 4$$

$$x + 2 = 4$$

$$x = 2$$

The solution is $(2, -2, 2)$.

35. Graphing the solution set:

37. Graphing the solution set:

Chapter 4 Cumulative Review

1. Simplifying: $4^2 - 8^2 = 16 - 64 = -48$

3. Simplifying: $15 - 12 \div 4 - 3 \cdot 2 = 15 - 3 - 6 = 6$

5. Simplifying: $8\left(\frac{3}{2}\right) - 24\left(\frac{3}{2}\right)^2 = 8\left(\frac{3}{2}\right) - 24\left(\frac{9}{4}\right) = 12 - 54 = -42$

7. Evaluating when $x = -3$: $x^2 + 12x - 36 = (-3)^2 + 12(-3) - 36 = 9 - 36 - 36 = -63$

9. Simplifying: $4(3x - 2) + 3(2x + 5) = 12x - 8 + 6x + 15 = 18x + 7$

11. Simplifying: $-4\left(\frac{3}{4}x - \frac{3}{2}y\right) = -4\left(\frac{3}{4}x\right) + 4\left(\frac{3}{2}y\right) = -3x + 6y$

13. Solving the equation:
$$-4y - 2 = 6y + 8$$
$$-10y = 10$$
$$y = -1$$

15. Solving the equation:
$$|x + 5| + 3 = 2$$
$$|x + 5| = -1$$

Since this statement is false, there is no solution, or $\varnothing$.

16. Solving the equation:
$$|2x - 3| + 7 = 1$$
$$|2x - 3| = -6$$

Since this statement is false, there is no solution, or $\varnothing$.

17. Solving for y:

$$y - 3 = -5(x - 2)$$
$$y - 3 = -5x + 10$$
$$y = -5x + 13$$

19. Solving for x:

$$ax - 5 = bx - 3$$
$$ax - bx = 2$$
$$x(a - b) = 2$$
$$x = \frac{2}{a - b}$$

21. Solving the inequality:

$$-3t \geq 12$$
$$t \leq -4$$

The solution is $(-\infty, -4]$.

23. The set is: $A \cup B = \{0, 2, 3, 4, 6, 9\}$

25. The rational numbers are: $-1, 0, 2.35, 4$

27. Multiply the first equation by 3 and the second equation by 2:

$$6x - 15y = -21$$
$$-6x + 8y = 0$$

Adding yields:

$$-7y = -21$$
$$y = 3$$

Substituting into the first equation:

$$2x - 5(3) = -7$$
$$2x - 15 = -7$$
$$2x = 8$$
$$x = 4$$

The solution is $(4, 3)$.

29. Substituting into the first equation:

$$2x + (-2x + 3) = 3$$
$$3 = 3$$

Since the result is true, the two lines coincide. Any point on the line is a solution to the system.

31. Graphing the line:

$$0.03x + 0.04y = 0.04$$

33. Graphing the solution set:

35. Solving for y:

$$3x - 5y = 15$$
$$-5y = -3x + 15$$
$$y = \frac{3}{5}x - 3$$

So $m = \frac{3}{5}, b = -3$.

37.   First solve for y to find the slope:

$$2x - 5y = 10$$
$$-5y = -2x + 10$$
$$y = \frac{2}{5}x - 2$$

So the perpendicular slope is $-\frac{5}{2}$. Since the y-intercept is 1, the equation is $y = -\frac{5}{2}x + 1$.

39.   Evaluating: $f(-2) = 2(-2) + 3 = -4 + 3 = -1$

41.   Simplifying: $(f - g)(x) = f(x) - g(x) = (2x + 3) - (x^2 - 5) = 2x + 3 - x^2 + 5 = -x^2 + 2x + 8$

43.   Evaluating: $f(3) = 3$

45.   Evaluating: $(g \circ f)(6) = g[f(6)] = g(1) = -1$

Chapter 4 Test

1.   Multiply the second equation by 5:

$$2x - 5y = -8$$
$$15x + 5y = 25$$

Adding yields:

$$17x = 17$$
$$x = 1$$

Substituting to find y:

$$3 + y = 5$$
$$y = 2$$

The solution is $(1, 2)$.

2.   To clear each equation of fractions, multiply the first equation by 6 and the second equation by 20:

$$2x - y = 18$$
$$-4x + 5y = 0$$

Multiply the first equation by 2:

$$4x - 2y = 36$$
$$-4x + 5y = 0$$

Adding yields:

$$3y = 36$$
$$y = 12$$

Substituting to find x:

$$2x - 12 = 18$$
$$2x = 30$$
$$x = 15$$

The solution is $(15, 12)$.

3.   Substituting into the first equation:

$$2x - 5(3x + 8) = 14$$
$$2x - 15x - 40 = 14$$
$$-13x = 54$$
$$x = -\frac{54}{13}$$

Substituting to find y: $y = 3\left(-\frac{54}{13}\right) + 8 = -\frac{162}{13} + \frac{104}{13} = -\frac{58}{13}$. The solution is $\left(-\frac{54}{13}, -\frac{58}{13}\right)$.

4. The first equation is equivalent to $y = 2x$. Substituting into the second equation:

$$x + 2(2x) = 5$$
$$5x = 5$$
$$x = 1$$
$$y = 2$$

The solution is (1,2).

5. Adding the first and third equations:

$$5x = 15$$
$$x = 3$$

Adding the first and second equations:

$$3x - 2z = 7$$
$$9 - 2z = 7$$
$$-2z = -2$$
$$z = 1$$

Substituting to find y:

$$3 + y - 3 = -2$$
$$y = -2$$

The solution is (3,–2,1).

6. Divide the second equation by 5 and the third equation by 10 to produce the system:

$$x - y - z = 0$$
$$x + 4y = 12$$
$$2y - z = 4$$

Multiply the third equation by –1 and add it to the first equation:

$$x - y - z = 0$$
$$-2y + z = -4$$

Adding yields the equation $x - 3y = -4$. So the system becomes:

$$x + 4y = 12$$
$$x - 3y = -4$$

Multiply the second equation by –1:

$$x + 4y = 12$$
$$-x + 3y = 4$$

Adding yields:

$$7y = 16$$
$$y = \frac{16}{7}$$

Substituting to find x:

$$x + 4\left(\frac{16}{7}\right) = 12$$
$$x + \frac{64}{7} = 12$$
$$x = \frac{20}{7}$$

Substituting to find z:

$$2\left(\frac{16}{7}\right) - z = 4$$
$$\frac{32}{7} - z = 4$$
$$-z = -\frac{4}{7}$$
$$z = \frac{4}{7}$$

The currents are $\frac{20}{7}$ amps, $\frac{16}{7}$ amps, and $\frac{4}{7}$ amp.

7. Let x and $2x - 1$ represent the two numbers. The equation is:

$$x + 2x - 1 = 14$$
$$3x = 15$$
$$x = 5$$
$$2x - 1 = 9$$

The two numbers are 5 and 9.

8. Let x and $2x$ represent the two investments. The equation is:

$$0.05(x) + 0.06(2x) = 680$$
$$0.17x = 680$$
$$x = 4000$$
$$2x = 8000$$

John invested \$4,000 at 5% and \$8,000 at 6%.

9. Let a represent the adult ticket sales and c represent the children's ticket sales. The system of equations is:

$$a + c = 750$$
$$2a + 1c = 1090$$

Multiplying the first equation by -1:

$$-a - c = -750$$
$$2a + c = 1090$$

Adding yields $a = 340$ and $c = 410$. There were 340 adult tickets and 410 children's tickets sold.

10. Let b and c represent the rate of the boat and current. The system of equations is:

$$2(b + c) = 20$$
$$3(b - c) = 18$$

Simplifying the system:

$$b + c = 10$$
$$b - c = 6$$

Adding yields:

$$2b = 16$$
$$b = 8$$
$$c = 2$$

The boat's rate is 8 mph and the current's rate is 2 mph.

11. Let n, d, and q represent the number of nickels, dimes, and quarters. The system of equations is:

$$n + d + q = 15$$
$$0.05n + 0.10d + 0.25q = 1.10$$
$$n = 4d - 1$$

Substituting into the first equation:

$$4d - 1 + d + q = 15$$
$$5d + q = 16$$

Substituting into the second equation:

$$0.05(4d - 1) + 0.10d + 0.25q = 1.10$$
$$0.3d + 0.25q = 1.15$$

The system of equations becomes:

$$5d + q = 16$$
$$0.3d + 0.25q = 1.15$$

Multiply the first equation by -0.25:

$$-1.25d - 0.25q = -4$$
$$0.3d + 0.25q = 1.15$$

Adding yields:

$$-0.95d = -2.85$$
$$d = 3$$

Substituting to find q:

$$15 + q = 16$$
$$q = 1$$

Substituting to find n:

$$n + 3 + 1 = 15$$
$$n = 11$$

The collection contains 11 nickels, 3 dimes, and 1 quarter.

12. First rewrite the system as:

$$2x - y = -13$$
$$-4x + y = 23$$

Form the augmented matrix: $\begin{bmatrix} 2 & -1 & | & -13 \\ -4 & 1 & | & 23 \end{bmatrix}$

Dividing row 1 by 2: $\begin{bmatrix} 1 & -\dfrac{1}{2} & | & -\dfrac{13}{2} \\ -4 & 1 & | & 23 \end{bmatrix}$

Add 4 times row 1 to row 2: $\begin{bmatrix} 1 & -\dfrac{1}{2} & | & -\dfrac{13}{2} \\ 0 & -1 & | & -3 \end{bmatrix}$

Dividing row 2 by -1: $\begin{bmatrix} 1 & -\dfrac{1}{2} & | & -\dfrac{13}{2} \\ 0 & 1 & | & 3 \end{bmatrix}$

So $y = 3$. Substituting to find x:

$$x - \frac{1}{2}(3) = -\frac{13}{2}$$
$$x - \frac{3}{2} = -\frac{13}{2}$$
$$x = -5$$

The solution is $(-5, 3)$.

13. Form the augmented matrix using equation 2 as row 1: $\begin{bmatrix} 1 & 3 & 4 & | & 19 \\ 3 & -4 & -3 & | & -3 \\ 0 & 2 & -5 & | & -20 \end{bmatrix}$

Add -3 times row 1 to row 2: $\begin{bmatrix} 1 & 3 & 4 & | & 19 \\ 0 & -13 & -15 & | & -60 \\ 0 & 2 & -5 & | & -20 \end{bmatrix}$

Dividing row 3 by 2 and switching it with row 2: $\begin{bmatrix} 1 & 3 & 4 & | & 19 \\ 0 & 1 & -\dfrac{5}{2} & | & -10 \\ 0 & -13 & -15 & | & -60 \end{bmatrix}$

Add 13 times row 2 to row 3: $\begin{bmatrix} 1 & 3 & 4 & | & 19 \\ 0 & 1 & -\dfrac{5}{2} & | & -10 \\ 0 & 0 & -\dfrac{95}{2} & | & -190 \end{bmatrix}$

Multiplying row 3 by $-\dfrac{2}{95}$: $\begin{bmatrix} 1 & 3 & 4 & | & 19 \\ 0 & 1 & -\dfrac{5}{2} & | & -10 \\ 0 & 0 & 1 & | & 4 \end{bmatrix}$

So $z = 4$. Substituting to find y:

$$y - \frac{5}{2}(4) = -10$$

$$y - 10 = -10$$

$$y = 0$$

Substituting to find x:

$$x + 3(0) + 4(4) = 19$$

$$x + 16 = 19$$

$$x = 3$$

The solution is $(3,0,4)$.

14. Graphing the solution set:

15. Graphing the solution set:

16. The system is:

$$y \geq \frac{4}{3}x - 4$$

$$y > -\frac{4}{3}x - 4$$

17. The system is:

$$y \leq \frac{4}{3}x - 4$$

$$y > -\frac{4}{3}x - 4$$

Chapter 5
Exponents and Polynomials

Getting Ready for Chapter 5

1. Simplifying: $-6 - (-9) = -6 + 9 = 3$
2. Simplifying: $-5 - (-7) = -5 + 7 = 2$
3. Simplifying: $2 - 18 = -16$
4. Simplifying: $3 - 10 = -7$
5. Simplifying: $-8(4) = -32$
6. Simplifying: $-3(2) = -6$
7. Simplifying: $\dfrac{6}{12} = \dfrac{1}{2}$
8. Simplifying: $\dfrac{18}{36} = \dfrac{1}{2}$
9. Simplifying: $3^4 - 3 \cdot 3 \cdot 3 \cdot 3 = 81$
10. Simplifying: $2^3 = 2 \cdot 2 \cdot 2 = 8$
11. Simplifying: $10^3 = 10 \cdot 10 \cdot 10 = 1{,}000$
12. Simplifying: $10^4 = 10 \cdot 10 \cdot 10 \cdot 10 = 10{,}000$
13. Simplifying: $(-2)^3 = (-2)(-2)(-2) = -8$
14. Simplifying: $(-3)^3 = (-3)(-3)(-3) = -27$
15. Simplifying: $\dfrac{1}{5^2} = \dfrac{1}{5 \cdot 5} = \dfrac{1}{25}$
16. Simplifying: $\dfrac{1}{3^2} = \dfrac{1}{3 \cdot 3} = \dfrac{1}{9}$
17. Simplifying: $\dfrac{1}{(-2)^3} = \dfrac{1}{(-2)(-2)(-2)} = -\dfrac{1}{8}$
18. Simplifying: $\dfrac{1}{(-5)^3} = \dfrac{1}{(-5)(-5)(-5)} = -\dfrac{1}{125}$
19. Simplifying: $\dfrac{1}{\left(\dfrac{2}{3}\right)^2} = \dfrac{1}{\dfrac{2}{3} \cdot \dfrac{2}{3}} = \dfrac{1}{\dfrac{4}{9}} = \dfrac{9}{4}$
20. Simplifying: $\dfrac{1}{\left(\dfrac{3}{4}\right)^2} = \dfrac{1}{\dfrac{3}{4} \cdot \dfrac{3}{4}} = \dfrac{1}{\dfrac{9}{16}} = \dfrac{16}{9}$
21. Simplifying: $4.52 \times 1{,}000 = 4{,}520$
22. Simplifying: $3.91 \times 10{,}000 = 39{,}100$
23. Simplifying: $376{,}000 \div 100{,}000 = 3.76$
24. Simplifying: $27{,}400 \div 10{,}000 = 2.74$

5.1 Properties of Exponents

1. Evaluating: $4^2 = 4 \cdot 4 = 16$
3. Evaluating: $-4^2 = -4 \cdot 4 = -16$
5. Evaluating: $-0.3^3 = -0.3 \cdot 0.3 \cdot 0.3 = -0.027$
7. Evaluating: $2^5 = 2 \cdot 2 \cdot 2 \cdot 2 \cdot 2 = 32$
9. Evaluating: $\left(\dfrac{1}{2}\right)^3 = \dfrac{1}{2} \cdot \dfrac{1}{2} \cdot \dfrac{1}{2} = \dfrac{1}{8}$
11. Evaluating: $-\left(\dfrac{5}{6}\right)^2 = -\dfrac{5}{6} \cdot \dfrac{5}{6} = -\dfrac{25}{36}$
13. Using properties of exponents: $x^5 \cdot x^4 = x^{5+4} = x^9$
15. Using properties of exponents: $\left(2^3\right)^2 = 2^{3 \cdot 2} = 2^6 = 64$

141

17. Using properties of exponents: $\left(-\frac{2}{3}x^2\right)^3 = \left(-\frac{2}{3}x^2\right)\left(-\frac{2}{3}x^2\right)\left(-\frac{2}{3}x^2\right) = -\frac{8}{27}x^6$

19. Using properties of exponents: $-3a^2\left(2a^4\right) = -6a^{2+4} = -6a^6$

21. Writing with positive exponents: $3^{-2} = \frac{1}{3^2} = \frac{1}{9}$

23. Writing with positive exponents: $(-2)^{-5} = \frac{1}{(-2)^5} = -\frac{1}{32}$

25. Writing with positive exponents: $\left(\frac{3}{4}\right)^{-2} = \left(\frac{4}{3}\right)^2 = \frac{16}{9}$

27. Writing with positive exponents: $\left(\frac{1}{3}\right)^{-2} + \left(\frac{1}{2}\right)^{-3} = 3^2 + 2^3 = 9 + 8 = 17$

29. Simplifying: $x^{-4}x^7 = x^{-4+7} = x^3$

31. Simplifying: $\left(a^2b^{-5}\right)^3 = a^6b^{-15} = \frac{a^6}{b^{15}}$

33. Simplifying: $\left(5y^4\right)^{-3}\left(2y^{-2}\right)^3 = 5^{-3}y^{-12}2^3y^{-6} = \frac{2^3}{5^3}y^{-18} = \frac{8}{125y^{18}}$

35. Simplifying: $\left(\frac{1}{2}x^3\right)\left(\frac{2}{3}x^4\right)\left(\frac{3}{5}x^{-7}\right) = \frac{1}{2} \cdot \frac{2}{3} \cdot \frac{3}{5}x^0 = \frac{1}{5}$

37. Simplifying: $\left(4a^5b^2\right)\left(2b^{-5}c^2\right)\left(3a^7c^4\right) = 24a^{5+7}b^{2-5}c^{2+4} = 24a^{12}b^{-3}c^6 = \frac{24a^{12}c^6}{b^3}$

39. Simplifying: $\left(2x^2y^{-5}\right)^3\left(3x^{-4}y^2\right)^{-4} = 2^3x^6y^{-15} \cdot 3^{-4}x^{16}y^{-8} = \frac{2^3}{3^4}x^{22}y^{-23} = \frac{8x^{22}}{81y^{23}}$

41. Simplifying: $\frac{x^{-1}}{x^9} = x^{-1-9} = x^{-10} = \frac{1}{x^{10}}$

43. Simplifying: $\frac{a^4}{a^{-6}} = a^{4-(-6)} = a^{4+6} = a^{10}$

45. Simplifying: $\frac{t^{-10}}{t^{-4}} = t^{-10-(-4)} = t^{-10+4} = t^{-6} = \frac{1}{t^6}$

47. Simplifying: $\left(\frac{x^5}{x^3}\right)^6 = \left(x^{5-3}\right)^6 = \left(x^2\right)^6 = x^{12}$

49. Simplifying: $\frac{\left(x^5\right)^6}{\left(x^3\right)^4} = \frac{x^{30}}{x^{12}} = x^{30-12} = x^{18}$

51. Simplifying: $\frac{\left(x^{-2}\right)^3\left(x^3\right)^{-2}}{x^{10}} = \frac{x^{-6}x^{-6}}{x^{10}} = \frac{x^{-12}}{x^{10}} = x^{-12-10} = x^{-22} = \frac{1}{x^{22}}$

53. Simplifying: $\frac{5a^8b^3}{20a^5b^{-4}} = \frac{5}{20}a^{8-5}b^{3-(-4)} = \frac{1}{4}a^3b^7 = \frac{a^3b^7}{4}$

55. Simplifying: $\frac{\left(3x^{-2}y^8\right)^4}{\left(9x^4y^{-3}\right)^2} = \frac{81x^{-8}y^{32}}{81x^8y^{-6}} = x^{-8-8}y^{32+6} = x^{-16}y^{38} = \frac{y^{38}}{x^{16}}$

57. Simplifying: $\left(\frac{8x^2y}{4x^4y^{-3}}\right)^4 = \left(2x^{2-4}y^{1+3}\right)^4 = \left(2x^{-2}y^4\right)^4 = 16x^{-8}y^{16} = \frac{16y^{16}}{x^8}$

59. Simplifying: $\left(\frac{x^{-5}y^2}{x^{-3}y^5}\right)^{-2} = \left(x^{-5+3}y^{2-5}\right)^{-2} = \left(x^{-2}y^{-3}\right)^{-2} = x^4y^6$

61. Writing as a perfect square: $x^4 y^2 = \left(x^2 y\right)^2$

63. Writing as a perfect square: $9a^2 b^4 = \left(3ab^2\right)^2$

65. Writing as a perfect cube: $8a^3 = \left(2a\right)^3$

67. Writing as a perfect cube: $64x^3 y^{12} = \left(4xy^4\right)^3$

69. **a.** Substituting $x = 2$: $x^3 x^2 = 2^3 2^2 = 8 \cdot 4 = 32$

    b. Substituting $x = 2$: $\left(x^3\right)^2 = \left(2^3\right)^2 = (8)^2 = 64$

    c. Substituting $x = 2$: $x^5 = 2^5 = 32$

    d. Substituting $x = 2$: $x^6 = 2^6 = 64$

71. **a.** Substituting $x = 2$: $\dfrac{x^5}{x^2} = \dfrac{2^5}{2^2} = \dfrac{32}{4} = 8$

    b. Substituting $x = 2$: $x^3 = 2^3 = 8$

    c. Substituting $x = 2$: $\dfrac{x^2}{x^6} = \dfrac{2^2}{2^6} = \dfrac{4}{64} = \dfrac{1}{16}$

    d. Substituting $x = 2$: $x^{-4} = 2^{-4} = \dfrac{1}{2^4} = \dfrac{1}{16}$

73. **a.** Writing as a perfect square: $\dfrac{1}{49} = \left(\dfrac{1}{7}\right)^2$

    b. Writing as a perfect square: $\dfrac{1}{121} = \left(\dfrac{1}{11}\right)^2$

    c. Writing as a perfect square: $\dfrac{1}{4x^2} = \left(\dfrac{1}{2x}\right)^2$

    d. Writing as a perfect square: $\dfrac{1}{64x^4} = \left(\dfrac{1}{8x^2}\right)^2$

75. Simplifying: $2 \cdot 2^{n-1} = 2^{1+n-1} = 2^n$

77. Simplifying: $\dfrac{ar^6}{ar^3} = r^{6-3} = r^3$

79. Writing in scientific notation: $378{,}000 = 3.78 \times 10^5$

81. Writing in scientific notation: $4{,}900 = 4.9 \times 10^3$

83. Writing in scientific notation: $0.00037 = 3.7 \times 10^{-4}$

85. Writing in scientific notation: $0.00495 = 4.95 \times 10^{-3}$

87. Writing in expanded form: $5.34 \times 10^3 = 5{,}340$

89. Writing in expanded form: $7.8 \times 10^6 = 7{,}800{,}000$

91. Writing in expanded form: $3.44 \times 10^{-3} = 0.00344$

93. Writing in expanded form: $4.9 \times 10^{-1} = 0.49$

95. Simplifying: $\left(4 \times 10^{10}\right)\left(2 \times 10^{-6}\right) = 8 \times 10^{10-6} = 8 \times 10^4$

97. Simplifying: $\dfrac{8 \times 10^{14}}{4 \times 10^5} = 2 \times 10^{14-5} = 2 \times 10^9$

99. Simplifying: $\dfrac{\left(5 \times 10^6\right)\left(4 \times 10^{-8}\right)}{8 \times 10^4} = \dfrac{20 \times 10^{-2}}{8 \times 10^4} = \dfrac{20}{8} \times 10^{-2-4} = 2.5 \times 10^{-6}$

101. Multiplying: $8x^3 \cdot 10y^6 = 80x^3 y^6$

103. Multiplying: $8x^3 \cdot 9y^3 = 72x^3 y^3$

105. Multiplying: $3x \cdot 5y = 15xy$

107. Multiplying: $4x^6 y^6 \cdot 3x = 12x^7 y^6$

109. Multiplying: $27a^6 e^3 \cdot 2b^2 e = 54a^6 b^2 e^4$

111. Dividing: $\dfrac{10x^5}{5x^2} = 2x^{5-2} = 2x^3$

113. Dividing: $\dfrac{20x^3}{5x^2} = 4x^{3-2} = 4x$

115. Dividing: $\dfrac{8x^3 y^5}{-2x^2 y} = -4x^{3-2} y^{5-1} = -4xy^4$

117. Dividing: $\dfrac{4x^4 y^3}{-2x^2 y} = -2x^{4-2} y^{3-1} = -2x^2 y^2$

119. Dividing: $\dfrac{7x^4 y^3}{21xy^4} = \dfrac{1}{3} x^{4-1} y^{3-4} = \dfrac{1}{3} x^3 y^{-1} = \dfrac{x^3}{3y}$

121. Dividing: $\dfrac{25x^5 y^3}{10x^4 y^2} = \dfrac{5}{2} x^{5-4} y^{3-2} = \dfrac{5}{2} xy = \dfrac{5xy}{2}$

123. The area of the base is: $A = \left(525 \text{ ft}\right)^2 = 275{,}625 \text{ feet}^2$

125. Writing in scientific notation: $200{,}000{,}000{,}000 = 2.0 \times 10^{11}$ stars

In words, this is two hundred billion stars.

127. Multiplying to find the distance: $\left(1.7 \times 10^6 \text{ light-years}\right)\left(5.9 \times 10^{12} \text{ miles/light-year}\right) \approx 1.003 \times 10^{19}$ miles

129. Writing in scientific notation: $15,000,000,000,000 = 1.5 \times 10^{13}$

131. Simplifying: $-4x + 9x = 5x$ **133.** Simplifying: $5x^2 + 3x^2 = 8x^2$

135. Simplifying: $-8x^3 + 10x^3 = 2x^3$ **137.** Simplifying: $2x + 3 - 2x - 8 = -5$

139. Simplifying: $-1(2x - 3) = -2x + 3$ **141.** Simplifying: $-3(-3x - 2) = 9x + 6$

143. Simplifying: $-500 + 27(100) - 0.1(100)^2 = -500 + 2,700 - 1,000 = 1,200$

145. Multiply the second equation by -2:

$$4x + 3y = 10$$
$$-4x - 2y = -8$$

Adding yields $y = 2$. Substituting into the first equation:

$$4x + 3(2) = 10$$
$$4x + 6 = 10$$
$$4x = 4$$
$$x = 1$$

The solution is $(1,2)$.

147. Multiply the second equation by -5:

$$4x + 5y = 5$$
$$-6x - 5y = -10$$

Adding yields:

$$-2x = -5$$
$$x = \frac{5}{2}$$

Substituting into the first equation:

$$4\left(\frac{5}{2}\right) + 5y = 5$$
$$10 + 5y = 5$$
$$5y = -5$$
$$y = -1$$

The solution is $\left(\frac{5}{2}, -1\right)$.

149. Substituting into the first equation:

$$x + (x + 3) = 3$$
$$2x + 3 = 3$$
$$2x = 0$$
$$x = 0$$

The solution is $(0,3)$.

151. Substituting into the first equation:

$$2x - 3(3x - 5) = -6$$
$$2x - 9x + 15 = -6$$
$$-7x + 15 = -6$$
$$-7x = -21$$
$$x = 3$$

The solution is $(3,4)$.

153. First rewrite the system as:
$$4x - 3y = 11$$
$$x - 2y = 4$$

Form the augmented matrix using equation 2 as row 1: $\begin{bmatrix} 1 & -2 & | & 4 \\ 4 & -3 & | & 11 \end{bmatrix}$

Add -4 times row 1 to row 2: $\begin{bmatrix} 1 & -2 & | & 4 \\ 0 & 5 & | & -5 \end{bmatrix}$

Dividing row 2 by 5: $\begin{bmatrix} 1 & -2 & | & 4 \\ 0 & 1 & | & -1 \end{bmatrix}$

So $y = -1$. Substituting to find x:
$$x - 2(-1) = 4$$
$$x + 2 = 4$$
$$x = 2$$
The solution is $(2, -1)$.

155. Form the augmented matrix: $\begin{bmatrix} 1 & -1 & 1 & | & 2 \\ -1 & 2 & -1 & | & 0 \\ 2 & -1 & 2 & | & 6 \end{bmatrix}$

Add row 1 to row 2 and -2 times row 1 to row 3: $\begin{bmatrix} 1 & -1 & 1 & | & 2 \\ 0 & 1 & 0 & | & 2 \\ 0 & 1 & 0 & | & 2 \end{bmatrix}$

Add -1 times row 2 to row 3: $\begin{bmatrix} 1 & -1 & 1 & | & 2 \\ 0 & 1 & 0 & | & 2 \\ 0 & 0 & 0 & | & 0 \end{bmatrix}$

Since the last row corresponds to $0 = 0$, which is true, the system is dependent.

157. Simplifying: $x^{m+2} \bullet x^{-2m} \bullet x^{m-5} = x^{m+2-2m+m-5} = x^{-3} = \dfrac{1}{x^3}$

159. Simplifying: $\left(y^m\right)^2 \left(y^{-3}\right)^m \left(y^{m+3}\right) = y^{2m} \bullet y^{-3m} \bullet y^{m+3} = y^{2m-3m+m+3} = y^3$

161. Simplifying: $\dfrac{x^{n+2}}{x^{n-3}} = x^{n+2-(n-3)} = x^{n+2-n+3} = x^5$

5.2 Polynomials, Sums, and Differences

1. This is a trinomial. The degree is 2 and the leading coefficient is 5.

3. This is a binomial. The degree is 1 and the leading coefficient is 3.

5. This is a trinomial. The degree is 2 and the leading coefficient is 8.

7. This is a polynomial. The degree is 3 and the leading coefficient is 4.

9. This is a monomial. The degree is 0 and the leading coefficient is $-\dfrac{3}{4}$.

11. This is a trinomial. The degree is 3 and the leading coefficient is 6.

13. Simplifying: $(4x + 2) + (3x - 1) = 7x + 1$

15. Simplifying: $2x^2 - 3x + 10x - 15 = 2x^2 + 7x - 15$

17. Simplifying: $12a^2 + 8ab - 15ab - 10b^2 = 12a^2 - 7ab - 10b^2$

19. Simplifying: $\left(5x^2 - 6x + 1\right) - \left(4x^2 + 7x - 2\right) = 5x^2 - 6x + 1 - 4x^2 - 7x + 2 = x^2 - 13x + 3$

21. Simplifying:
$$\left(\frac{1}{2}x^2 - \frac{1}{3}x - \frac{1}{6}\right) - \left(\frac{1}{4}x^2 + \frac{7}{12}x\right) + \left(\frac{1}{3}x - \frac{1}{12}\right) = \frac{1}{2}x^2 - \frac{1}{3}x - \frac{1}{6} - \frac{1}{4}x^2 - \frac{7}{12}x + \frac{1}{3}x - \frac{1}{12} = \frac{1}{4}x^2 - \frac{7}{12}x - \frac{1}{4}$$

23. Simplifying: $\left(y^3 - 2y^2 - 3y + 4\right) - \left(2y^3 - y^2 + y - 3\right) = y^3 - 2y^2 - 3y + 4 - 2y^3 + y^2 - y + 3 = -y^3 - y^2 - 4y + 7$

25. Simplifying:
$$\left(5x^3 - 4x^2\right) - (3x + 4) + \left(5x^2 - 7\right) - \left(3x^3 + 6\right) = 5x^3 - 4x^2 - 3x - 4 + 5x^2 - 7 - 3x^3 - 6 = 2x^3 + x^2 - 3x - 17$$

27. Simplifying:
$$\left(\frac{4}{7}x^2 - \frac{1}{7}xy + \frac{1}{14}y^2\right) - \left(\frac{1}{2}x^2 - \frac{2}{7}xy - \frac{9}{14}y^2\right) = \frac{4}{7}x^2 - \frac{1}{7}xy + \frac{1}{14}y^2 - \frac{1}{2}x^2 + \frac{2}{7}xy + \frac{9}{14}y^2$$
$$= \frac{8}{14}x^2 - \frac{7}{14}x^2 - \frac{1}{7}xy + \frac{2}{7}xy + \frac{1}{14}y^2 + \frac{9}{14}y^2$$
$$= \frac{1}{14}x^2 + \frac{1}{7}xy + \frac{5}{7}y^2$$

29. Simplifying:
$$\left(3a^3 + 2a^2b + ab^2 - b^3\right) - \left(6a^3 - 4a^2b + 6ab^2 - b^3\right) = 3a^3 + 2a^2b + ab^2 - b^3 - 6a^3 + 4a^2b - 6ab^2 + b^3$$
$$= -3a^3 + 6a^2b - 5ab^2$$

31. Subtracting: $\left(2x^2 - 7x\right) - \left(2x^2 - 4x\right) = 2x^2 - 7x - 2x^2 + 4x = -3x$

33. Adding: $\left(x^2 - 6xy + y^2\right) + \left(2x^2 - 6xy - y^2\right) = x^2 - 6xy + y^2 + 2x^2 - 6xy - y^2 = 3x^2 - 12xy$

35. Subtracting: $\left(9x^5 - 4x^3 - 6\right) - \left(-8x^5 - 4x^3 + 6\right) = 9x^5 - 4x^3 - 6 + 8x^5 + 4x^3 - 6 = 17x^5 - 12$

37. Adding:
$$\left(11a^2 + 3ab + 2b^2\right) + \left(9a^2 - 2ab + b^2\right) + \left(-6a^2 - 3ab + 5b^2\right)$$
$$= 11a^2 + 3ab + 2b^2 + 9a^2 - 2ab + b^2 - 6a^2 - 3ab + 5b^2$$
$$= 14a^2 - 2ab + 8b^2$$

39. Simplifying: $-[2 - (4 - x)] = -(2 - 4 + x) = -(-2 + x) = 2 - x$

41. Simplifying: $-5[-(x - 3) - (x + 2)] = -5(-x + 3 - x - 2) = -5(-2x + 1) = 10x - 5$

43. Simplifying: $4x - 5[3 - (x - 4)] = 4x - 5(3 - x + 4) = 4x - 5(7 - x) = 4x - 35 + 5x = 9x - 35$

45. Simplifying:
$$-(3x - 4y) - [(4x + 2y) - (3x + 7y)] = -(3x - 4y) - (4x + 2y - 3x - 7y)$$
$$= -(3x - 4y) - (x - 5y)$$
$$= -3x + 4y - x + 5y$$
$$= -4x + 9y$$

47. Simplifying:
$$4a - \{3a + 2[a - 5(a + 1) + 4]\} = 4a - [3a + 2(a - 5a - 5 + 4)]$$
$$= 4a - [3a + 2(-4a - 1)]$$
$$= 4a - (3a - 8a - 2)$$
$$= 4a - (-5a - 2)$$
$$= 4a + 5a + 2$$
$$= 9a + 2$$

49. Evaluating when $x = 2$: $2(2)^2 - 3(2) - 4 = 2(4) - 3(2) - 4 = 8 - 6 - 4 = -2$

51. a. Evaluating when $x = 12$: $P(12) = \frac{3}{2}(12)^2 - \frac{3}{4}(12) + 1 = \frac{3}{2}(144) - \frac{3}{4}(12) + 1 = 216 - 9 + 1 = 208$

   b. Evaluating when $x = -8$: $P(-8) = \frac{3}{2}(-8)^2 - \frac{3}{4}(-8) + 1 = \frac{3}{2}(64) - \frac{3}{4}(-8) + 1 = 96 + 6 + 1 = 103$

53. **a.** Evaluating when $x = 4$: $Q(4) = (4)^3 - (4)^2 + (4) - 1 = 64 - 16 + 4 - 1 = 51$

b. Evaluating when $x = -2$: $Q(-2) = (-2)^3 - (-2)^2 + (-2) - 1 = -8 - 4 - 2 - 1 = -15$

55. **a.** Evaluating when $x = 10$: $R(10) = 11.5(10) - 0.05(10)^2 = 115 - 5 = 110$

b. Evaluating when $x = -10$: $R(-10) = 11.5(-10) - 0.05(-10)^2 = -115 - 5 = -120$

57. **a.** Evaluating when $x = -4$: $P(-4) = 600 + 1,000(-4) - 100(-4)^2 = 600 - 4,000 - 1,600 = -5,000$

b. Evaluating when $x = 4$: $P(4) = 600 + 1,000(4) - 100(4)^2 = 600 + 4,000 - 1,600 = 3,000$

59. The total income for the first year is: $53,160x + 63,935y$

The total income for the second year is: $53,160(2x) + 63,935y = 106,320x + 63,935y$

Combining the results: $(53,160x + 63,935y) + (106,320x + 63,935y) = 159,480x + 127,870y$

61. Substituting $t = 3$: $h = -16(3)^2 + 128(3) = 240$ feet   Substituting $t = 5$: $h = -16(5)^2 + 128(5) = 240$ feet

63. The weekly profit is given by:

$$P(x) = R(x) - C(x) = \left(100x - 0.5x^2\right) - (60x + 300) = 100x - 0.5x^2 - 60x - 300 = -0.5x^2 + 40x - 300$$

Substituting $x = 60$: $P(60) = -0.5(60)^2 + 40(60) - 300 = \$300$

65. The weekly profit is given by:

$$P(x) = R(x) \quad C(x) = \left(10x - 0.002x^2\right) - (800 + 6.5x) = 10x - 0.002x^2 - 800 - 6.5x = -0.002x^2 + 3.5x - 800$$

Substituting $x = 1000$: $P(1000) = -0.002(1000)^2 + 3.5(1000) - 800 = \$700$

67. Simplifying: $2x^2 - 3x + 10x - 15 = 2x^2 + 7x - 15$

69. Simplifying: $\left(6x^3 - 2x^2y + 8xy^2\right) + \left(-9x^2y + 3xy^2 - 12y^3\right) = 6x^3 - 11x^2y + 11xy^2 - 12y^3$

71. Simplifying: $4x^3(-3x) = -12x^{3+1} = -12x^4$ **73.** Simplifying: $4x^3\left(5x^2\right) = 20x^{3+2} = 20x^5$

75. Simplifying: $\left(a^3\right)^2 = a^{3 \bullet 2} = a^6$

77. Simplifying: $11.5(130) - 0.05(130)^2 = 1,495 - 845 = 650$

79. Simplifying: $-1(5 - x) = -5 + x = x - 5$ **81.** Simplifying: $-1(7 - x) = -7 + x = x - 7$

83. Simplifying: $5\left(x - \dfrac{1}{5}\right) = 5 \bullet x - 5 \bullet \dfrac{1}{5} = 5x - 1$ **85.** Simplifying: $x\left(1 - \dfrac{1}{x}\right) = x \bullet 1 - x \bullet \dfrac{1}{x} = x - 1$

87. Simplifying: $12\left(\dfrac{1}{4}x + \dfrac{2}{3}y\right) = 12 \bullet \dfrac{1}{4}x + 12 \bullet \dfrac{2}{3}y = 3x + 8y$

89. **a.** From the graph: $f(-3) = 5$ **b.** From the graph: $f(0) = -4$

c. From the graph: $f(1) = -3$ **d.** From the graph: $g(-1) = 3$

e. From the graph: $g(0) = 4$ **f.** From the graph: $g(2) = 0$

g. From the graph: $f(g(2)) = f(0) = -4$ **h.** From the graph: $g(f(2)) = g(0) = 4$

5.3 Multiplication of Polynomials

1. Multiplying: $2x\left(6x^2 - 5x + 4\right) = 2x \cdot 6x^2 - 2x \cdot 5x + 2x \cdot 4 = 12x^3 - 10x^2 + 8x$

3. Multiplying: $-3a^2\left(a^3 - 6a^2 + 7\right) = -3a^2 \cdot a^3 - \left(-3a^2\right) \cdot 6a^2 + \left(-3a^2\right) \cdot 7 = -3a^5 + 18a^4 - 21a^2$

5. Multiplying: $2a^2b\left(a^3 - ab + b^3\right) = 2a^2b \cdot a^3 - 2a^2b \cdot ab + 2a^2b \cdot b^3 = 2a^5b - 2a^3b^2 + 2a^2b^4$

7. Multiplying using the vertical format:

$$
\begin{array}{rrr}
x & -5 & \\
x & +3 & \\
\hline
x^2 & -5x & \\
& +3x & -15 \\
\hline
x^2 & -2x & -15
\end{array}
$$

The product is $x^2 - 2x - 15$.

9. Multiplying using the vertical format:

$$
\begin{array}{rrr}
2x^2 & -3 & \\
3x^2 & -5 & \\
\hline
6x^4 & -9x^2 & \\
& -10x^2 & +15 \\
\hline
6x^4 & -19x^2 & +15
\end{array}
$$

The product is $6x^4 - 19x^2 + 15$.

11. Multiplying using the vertical format:

$$
\begin{array}{rrr}
x^2 & +6x & +5 \\
& x & +3 \\
\hline
x^3 & +6x^2 & +5x \\
& +3x^2 & +18x & +15 \\
\hline
x^3 & +9x^2 & +23x & +15
\end{array}
$$

The product is $x^3 + 9x^2 + 23x + 15$.

13. Multiplying using the vertical format:

$$
\begin{array}{rrr}
a^2 & +ab & +b^2 \\
& a & -b \\
\hline
a^3 & +a^2b & +ab^2 \\
& -a^2b & -ab^2 & -b^3 \\
\hline
a^3 & & & -b^3
\end{array}
$$

The product is $a^3 - b^3$.

15. Multiplying using the vertical format:

$$
\begin{array}{rrr}
4x^2 & -2xy & +y^2 \\
& 2x & +y \\
\hline
8x^3 & -4x^2y & +2xy^2 \\
& +4x^2y & -2xy^2 & +y^3 \\
\hline
8x^3 & & & +y^3
\end{array}
$$

The product is $8x^3 + y^3$.

17. Multiplying using the vertical format:

$$
\begin{array}{rrr}
a^2 & +ab & +b^2 \\
& 2a & -3b \\
\hline
2a^3 & +2a^2b & +2ab^2 \\
& -3a^2b & -3ab^2 & -3b^3 \\
\hline
2a^3 & -a^2b & -ab^2 & -3b^3
\end{array}
$$

The product is $2a^3 - a^2b - ab^2 - 3b^3$.

19. Multiplying using FOIL: $(x-2)(x+3) = x^2 + 3x - 2x - 6 = x^2 + x - 6$

21. Multiplying using FOIL: $(2a+3)(3a+2) = 6a^2 + 4a + 9a + 6 = 6a^2 + 13a + 6$

23. Multiplying using FOIL: $(5-3t)(4+2t) = 20 + 10t - 12t - 6t^2 = 20 - 2t - 6t^2$

25. Multiplying using FOIL: $\left(x^3+3\right)\left(x^3-5\right) = x^6 - 5x^3 + 3x^3 - 15 = x^6 - 2x^3 - 15$

27. Multiplying using FOIL: $(5x-6y)(4x+3y) = 20x^2 + 15xy - 24xy - 18y^2 = 20x^2 - 9xy - 18y^2$

29. Multiplying using FOIL: $\left(3t+\dfrac{1}{3}\right)\left(6t-\dfrac{2}{3}\right) = 18t^2 - 2t + 2t - \dfrac{2}{9} = 18t^2 - \dfrac{2}{9}$

31. Finding the product: $(5x+2y)^2 = (5x)^2 + 2(5x)(2y) + (2y)^2 = 25x^2 + 20xy + 4y^2$

33. Finding the product: $\left(5-3t^3\right)^2 = (5)^2 - 2(5)\left(3t^3\right) + \left(3t^3\right)^2 = 25 - 30t^3 + 9t^6$

35. Finding the product: $(2a+3b)(2a-3b) = (2a)^2 - (3b)^2 = 4a^2 - 9b^2$

37. Finding the product: $\left(3r^2+7s\right)\left(3r^2-7s\right) = \left(3r^2\right)^2 - (7s)^2 = 9r^4 - 49s^2$

39. Finding the product: $\left(y+\frac{3}{2}\right)^2 = (y)^2 + 2(y)\left(\frac{3}{2}\right) + \left(\frac{3}{2}\right)^2 = y^2 + 3y + \frac{9}{4}$

41. Finding the product: $\left(a-\frac{1}{2}\right)^2 = (a)^2 - 2(a)\left(\frac{1}{2}\right) + \left(\frac{1}{2}\right)^2 = a^2 - a + \frac{1}{4}$

43. Finding the product: $\left(x+\frac{1}{4}\right)^2 = (x)^2 + 2(x)\left(\frac{1}{4}\right) + \left(\frac{1}{4}\right)^2 = x^2 + \frac{1}{2}x + \frac{1}{16}$

45. Finding the product: $\left(t+\frac{1}{3}\right)^2 = (t)^2 + 2(t)\left(\frac{1}{3}\right) + \left(\frac{1}{3}\right)^2 = t^2 + \frac{2}{3}t + \frac{1}{9}$

47. Finding the product: $\left(\frac{1}{3}x-\frac{2}{5}\right)\left(\frac{1}{3}x+\frac{2}{5}\right) = \left(\frac{1}{3}x\right)^2 - \left(\frac{2}{5}\right)^2 = \frac{1}{9}x^2 - \frac{4}{25}$

49. Expanding and simplifying:

$$
\begin{aligned}
(x-2)^3 &= (x-2)(x-2)^2 \\
&= (x-2)\left(x^2 - 4x + 4\right) \\
&= x^3 - 4x^2 + 4x - 2x^2 + 8x - 8 \\
&= x^3 \;\; 6x^2 \;\text{I}\; 12x \;\; 8
\end{aligned}
$$

51. Expanding and simplifying:

$$
\begin{aligned}
\left(x-\frac{1}{2}\right)^3 &= \left(x-\frac{1}{2}\right)\left(x-\frac{1}{2}\right)^2 \\
&= \left(x-\frac{1}{2}\right)\left(x^2 - x + \frac{1}{4}\right) \\
&= x^3 - x^2 + \frac{1}{4}x - \frac{1}{2}x^2 + \frac{1}{2}x - \frac{1}{8} \\
&= x^3 - \frac{3}{2}x^2 + \frac{3}{4}x - \frac{1}{8}
\end{aligned}
$$

53. Expanding and simplifying:

$$
\begin{aligned}
3(x-1)(x-2)(x-3) &= 3(x-1)\left(x^2 - 5x + 6\right) \\
&= 3\left(x^3 - 5x^2 + 6x - x^2 + 5x - 6\right) \\
&= 3\left(x^3 - 6x^2 + 11x - 6\right) \\
&= 3x^3 - 18x^2 + 33x - 18
\end{aligned}
$$

55. Expanding and simplifying: $\left(b^2 + 8\right)\left(a^2 + 1\right) = a^2 b^2 + b^2 + 8a^2 + 8$

57. Expanding and simplifying:
$$(x+1)^2 + (x+2)^2 + (x+3)^2 = x^2 + 2x + 1 + x^2 + 4x + 4 + x^2 + 6x + 9 = 3x^2 + 12x + 14$$

59. Expanding and simplifying:
$$(2x+3)^2 - (2x-3)^2 = \left(4x^2 + 12x + 9\right) - \left(4x^2 - 12x + 9\right) = 4x^2 + 12x + 9 - 4x^2 + 12x - 9 = 24x$$

61. Simplifying: $(x+3)^2 - 2(x+3) - 8 = x^2 + 6x + 9 - 2x - 6 - 8 = x^2 + 4x - 5$

63. Simplifying: $(2a-3)^2 - 9(2a-3) + 20 = 4a^2 - 12a + 9 - 18a + 27 + 20 = 4a^2 - 30a + 56$

65. Simplifying:
$$
\begin{aligned}
2(4a+2)^2 - 3(4a+2) - 20 &= 2\left(16a^2 + 16a + 4\right) - 3(4a+2) - 20 \\
&= 32a^2 + 32a + 8 - 12a - 6 - 20 \\
&= 32a^2 + 20a - 18
\end{aligned}
$$

67. Evaluating when $a = 2$ and $b = 3$:
$$
\begin{aligned}
a^4 - b^4 &= 2^4 - 3^4 = 16 - 81 = -65 \\
(a-b)^4 &= (2-3)^4 = (-1)^4 = 1 \\
\left(a^2 + b^2\right)(a+b)(a-b) &= \left(2^2 + 3^2\right)(2+3)(2-3) = (13)(5)(-1) = -65
\end{aligned}
$$

69. The expression is: $x(6,200 - 25x) = 6,200x - 25x^2$

Substituting $x = 3$: $6,200(3) - 25(3)^2 = 18,600 - 225 = \$18,375$

71. Since $R = xp$, substitute $x = 900 - 300p$ to obtain: $R(p) = (900 - 300p)p = 900p - 300p^2$

To find $R(x)$, we first solve for p:

$$x = 900 - 300p$$
$$300p = 900 - x$$
$$p = 3 - \frac{x}{300}$$

Now substitute to obtain: $R(x) = x\left(3 - \frac{x}{300}\right) = 3x - \frac{x^2}{300}$

Substituting $p = \$1.60$: $R(1.60) = 900(1.60) - 300(1.60)^2 = \$672$. The revenue is \$672.

73. Since $R = xp$, substitute $x = 350 - 10p$ to obtain: $R(p) = (350 - 10p)p = 350p - 10p^2$

To find $R(x)$, we first solve for p:

$$x = 350 - 10p$$
$$10p = 350 - x$$
$$p = 35 - \frac{x}{10}$$

Now substitute to obtain: $R(x) = x\left(35 - \frac{x}{10}\right) = 35x - \frac{x^2}{10}$

Substituting $x = 65$: $R(65) = 35(65) - \frac{(65)^2}{10} = \$1,852.50$. The revenue is \$1,852.50.

75. Since $R(x) = 35x - \frac{x^2}{10}$ and $C(x) = 5x + 500$, then:

$$P(x) = R(x) - C(x) = 35x - \frac{x^2}{10} - (5x + 500) = -\frac{x^2}{10} + 30x - 500$$

$$P(60) = -\frac{(60)^2}{10} + 30(60) - 500 = \$940$$

77. Expanding the formula:

$$A = 100(1+r)^4$$
$$= 100(1+r)^2(1+r)^2$$
$$= 100(1 + 2r + r^2)(1 + 2r + r^2)$$
$$= 100(1 + 2r + r^2 + 2r + 4r^2 + 2r^3 + r^2 + 2r^3 + r^4)$$
$$= 100(1 + 4r + 6r^2 + 4r^3 + r^4)$$
$$= 100 + 400r + 600r^2 + 400r^3 + 100r^4$$

79. Simplifying: $\dfrac{8a^3}{a} = 8a^{3-1} = 8a^2$

81. Simplifying: $\dfrac{-48a}{a} = -48$

83. Simplifying: $\dfrac{16a^5b^4}{8a^2b^3} = 2a^{5-2}b^{4-3} = 2a^3b$

85. Simplifying: $\dfrac{-24a^5b^5}{8a^5b^3} = -3a^{5-5}b^{5-3} = -3b^2$

87. Simplifying: $\dfrac{x^3y^4}{-x^3} = -x^{3-3}y^4 = -y^4$

89. Multiply the first equation by –1 and add it to the second equation:

$$-x - y - z = -6$$
$$2x - y + z = 3$$

Adding yields the equation $x - 2y = -3$. Multiply the first equation by 3 and add it to the third equation:

$$3x + 3y + 3z = 18$$
$$x + 2y - 3z = -4$$

Adding yields the equation $4x + 5y = 14$. So the system becomes:

$$x - 2y = -3$$
$$4x + 5y = 14$$

Multiply the first equation by –4 and add it to the second equation:

$$-4x + 8y = 12$$
$$4x + 5y = 14$$

Adding yields:

$$13y = 26$$
$$y = 2$$

Substituting to find x:

$$4x + 5(2) = 14$$
$$4x + 10 = 14$$
$$4x = 4$$
$$x = 1$$

Substituting into the original first equation:

$$1 + 2 + z = 6$$
$$z + 3 = 6$$
$$z = 3$$

The solution is (1,2,3).

91. Multiply the third equation by –1 and add it to the first equation:

$$3x + 4y = 15$$
$$-4y + 3z = -9$$

Adding yields the equation $3x + 3z = 6$, or $x + z = 2$. So the system becomes:

$$2x - 5z = -3$$
$$x + z = 2$$

Multiply the second equation by –2:

$$2x - 5z = -3$$
$$-2x - 2z = -4$$

Adding yields:

$$-7z = -7$$
$$z = 1$$

Substituting to find x:

$$2x - 5(1) = -3$$
$$2x - 5 = -3$$
$$2x = 2$$
$$x = 1$$

Substituting into the original first equation:

$$3(1) + 4y = 15$$
$$4y + 3 = 15$$
$$4y = 12$$
$$y = 3$$

The solution is (1,3,1).

93. Multiplying using FOIL: $[(x+y) - 4][(x+y) + 5] = (x+y)^2 + (x+y) - 20 = x^2 + 2xy + y^2 + x + y - 20$

95. Multiplying: $(x^n - 2)(x^n - 3) = x^{2n} - 3x^n - 2x^n + 6 = x^{2n} - 5x^n + 6$

97. Multiplying: $(2x^n + 3)(5x^n - 1) = 10x^{2n} - 2x^n + 15x^n - 3 = 10x^{2n} + 13x^n - 3$

99. Multiplying: $(x^n + 5)^2 = (x^n)^2 + 2(x^n)(5) + (5)^2 = x^{2n} + 10x^n + 25$

101. Multiplying: $(x^n + 1)(x^{2n} - x^n + 1) = x^{3n} - x^{2n} + x^n + x^{2n} - x^n + 1 = x^{3n} + 1$

5.4 The Greatest Common Factor and Factoring by Grouping

1. Factoring the expression: $10x^3 - 15x^2 = 5x^2(2x - 3)$

3. Factoring the expression: $9y^6 + 18y^3 = 9y^3(y^3 + 2)$

5. Factoring the expression: $9a^2b - 6ab^2 = 3ab(3a - 2b)$

7. Factoring the expression: $21xy^4 + 7x^2y^2 = 7xy^2(3y^2 + x)$

9. Factoring the expression: $3a^2 - 21a + 30 = 3(a^2 - 7a + 10)$

11. Factoring the expression: $4x^3 - 16x^2 - 20x = 4x(x^2 - 4x - 5)$

13. Factoring the expression: $10x^4y^2 + 20x^3y^3 - 30x^2y^4 = 10x^2y^2(x^2 + 2xy - 3y^2)$

15. Factoring the expression: $-x^2y + xy^2 - x^2y^2 = xy(-x + y - xy)$

17. Factoring the expression: $4x^3y^2z - 8x^2y^2z^2 + 6xy^2z^3 = 2xy^2z(2x^2 - 4xz + 3z^2)$

19. Factoring the expression: $20a^2b^2c^2 - 30ab^2c + 25a^2bc^2 = 5abc(4abc - 6b + 5ac)$

21. Factoring the expression: $5x(a - 2b) - 3y(a - 2b) = (a - 2b)(5x - 3y)$

23. Factoring the expression: $3x^2(x + y)^2 - 6y^2(x + y)^2 = 3(x + y)^2(x^2 - 2y^2)$

25. Factoring the expression: $2x^2(x + 5) + 7x(x + 5) + 6(x + 5) = (x + 5)(2x^2 + 7x + 6)$

27. Factoring by grouping: $3xy + 3y + 2ax + 2a = 3y(x + 1) + 2a(x + 1) = (x + 1)(3y + 2a)$

29. Factoring by grouping: $x^2y + x + 3xy + 3 = x(xy + 1) + 3(xy + 1) = (x + 3)(xy + 1)$

31. Factoring by grouping: $3xy^2 - 6y^2 + 4x - 8 = 3y^2(x - 2) + 4(x - 2) = (x - 2)(3y^2 + 4)$

33. Factoring by grouping: $x^2 - ax - bx + ab = x(x - a) - b(x - a) = (x - a)(x - b)$

35. Factoring by grouping: $ab + 5a - b - 5 = a(b + 5) - 1(b + 5) = (b + 5)(a - 1)$

37. Factoring by grouping: $a^4b^2 + a^4 - 5b^2 - 5 = a^4(b^2 + 1) - 5(b^2 + 1) = (b^2 + 1)(a^4 - 5)$

39. Factoring by grouping: $x^3 + 3x^2 - 4x - 12 = x^2(x + 3) - 4(x + 3) = (x + 3)(x^2 - 4)$

41. Factoring by grouping: $x^3 + 2x^2 - 25x - 50 = x^2(x + 2) - 25(x + 2) = (x + 2)(x^2 - 25)$

43. Factoring by grouping: $2x^3 + 3x^2 - 8x - 12 = x^2(2x + 3) - 4(2x + 3) = (2x + 3)(x^2 - 4)$

45. Factoring by grouping: $4x^3 + 12x^2 - 9x - 27 = 4x^2(x + 3) - 9(x + 3) = (x + 3)(4x^2 - 9)$

47. It will be $3 \cdot 2 = 6$.

49. Using factoring by grouping:

$$P + Pr + (P + Pr)r = (P + Pr) + (P + Pr)r = (P + Pr)(1 + r) = P(1 + r)(1 + r) = P(1 + r)^2$$

51. Factoring the revenue: $R(x) = 11.5x - 0.05x^2 = x(11.5 - 0.05x)$. So the price is $p = 11.5 - 0.05x$.

Substituting $x = 125$: $p = 11.5 - 0.05(125) = \$5.25$. The price is $5.25.

53.   Factoring the revenue: $R(x) = 35x - 0.1x^2 = x(35 - 0.1x)$. So the price is $p = 35 - 0.1x$.

Substituting $x = 65$: $p = 35 - 0.1(65) = \$28.50$. The price is $\$28.50$.

55.   Factoring out the greatest common factor: $3x^4 - 9x^3 y - 18x^2 y^2 = 3x^2 \left(x^2 - 3xy - 6y^2 \right)$

57.   Factoring out the greatest common factor: $2x^2 (x - 3) - 4x(x - 3) - 3(x - 3) = (x - 3)\left(2x^2 - 4x - 3 \right)$

59.   Multiplying using FOIL: $(x + 2)(3x - 1) = 3x^2 - x + 6x - 2 = 3x^2 + 5x - 2$

61.   Multiplying using FOIL: $(x - 1)(3x - 2) = 3x^2 - 2x - 3x + 2 = 3x^2 - 5x + 2$

63.   Multiplying using FOIL: $(x + 2)(x + 3) = x^2 + 3x + 2x + 6 = x^2 + 5x + 6$

65.   Multiplying using FOIL: $(2y + 5)(3y - 7) = 6y^2 - 14y + 15y - 35 = 6y^2 + y - 35$

67.   Multiplying using FOIL: $(4 - 3a)(5 - a) = 20 - 4a - 15a + 3a^2 = 20 - 19a + 3a^2$

69.   Completing the table:

Two Numbers a and b	Their Product ab	Their Sum $a + b$
1, −24	−24	−23
−1, 24	−24	23
2, −12	−24	−10
−2, 12	−24	10
3, −8	−24	−5
−3, 8	−24	5
4, −6	−24	−2
−4, 6	−24	2

71.   Multiply the first equation by 2 and the second equation by 5:

$$4x + 10y = 8$$
$$15x - 10y = -65$$

Adding:

$$19x = -57$$
$$x = -3$$

Substituting into the first equation:

$$2(-3) + 5y = 4$$
$$-6 + 5y = 4$$
$$5y = 10$$
$$y = 2$$

The solution is $(-3, 2)$.

73.   Substituting into the first equation:

$$2x + 6(x - 14) = -12$$
$$2x + 6x - 84 = -12$$
$$8x - 84 = -12$$
$$8x = 72$$
$$x = 9$$
$$y = 9 - 14 = -5$$

The solution is $(9, -5)$.

5.5 Factoring Trinomials

1. Factoring the trinomial: $x^2 + 7x + 12 = (x+3)(x+4)$ **3.** Factoring the trinomial: $x^2 - x - 12 = (x+3)(x-4)$

5. Factoring the trinomial: $y^2 + y - 6 = (y+3)(y-2)$ **7.** Factoring the trinomial: $16 - 6x - x^2 = (2-x)(8+x)$

9. Factoring the trinomial: $12 + 8x + x^2 = (2+x)(6+x)$

11. Factoring the trinomial: $3a^2 - 21a + 30 = 3(a^2 - 7a + 10) = 3(a-5)(a-2)$

13. Factoring the trinomial: $4x^3 - 16x^2 - 20x = 4x(x^2 - 4x - 5) = 4x(x-5)(x+1)$

15. Factoring the trinomial: $x^2 + 3xy + 2y^2 = (x+2y)(x+y)$

17. Factoring the trinomial: $a^2 + 3ab - 18b^2 = (a+6b)(a-3b)$

19. Factoring the trinomial: $x^2 - 2xa - 48a^2 = (x-8a)(x+6a)$

21. Factoring the trinomial: $x^2 - 12xb + 36b^2 = (x-6b)^2$

23. Factoring the trinomial: $3x^2 - 6xy - 9y^2 = 3(x^2 - 2xy - 3y^2) = 3(x-3y)(x+y)$

25. Factoring the trinomial: $2a^5 + 4a^4b + 4a^3b^2 = 2a^3(a^2 + 2ab + 2b^2)$

27. Factoring the trinomial: $10x^4y^2 + 20x^3y^3 - 30x^2y^4 = 10x^2y^2(x^2 + 2xy - 3y^2) = 10x^2y^2(x+3y)(x-y)$

29. Factoring completely: $2x^2 + 7x - 15 = (2x-3)(x+5)$

31. Factoring completely: $2x^2 + x - 15 = (2x-5)(x+3)$

33. Factoring completely: $2x^2 - 13x + 15 = (2x-3)(x-5)$

35. Factoring completely: $2x^2 - 11x + 15 = (2x-5)(x-3)$

37. The trinomial $2x^2 + 7x + 15$ does not factor (prime).

39. Factoring completely: $2 + 7a + 6a^2 = (2+3a)(1+2a)$

41. Factoring completely: $60y^2 - 15y - 45 = 15(4y^2 - y - 3) = 15(4y+3)(y-1)$

43. Factoring completely: $6x^4 - x^3 - 2x^2 = x^2(6x^2 - x - 2) = x^2(3x-2)(2x+1)$

45. Factoring completely: $40r^3 - 120r^2 + 90r = 10r(4r^2 - 12r + 9) = 10r(2r-3)^2$

47. Factoring completely: $4x^2 - 11xy - 3y^2 = (4x+y)(x-3y)$

49. Factoring completely: $10x^2 - 3xa - 18a^2 = (2x-3a)(5x+6a)$

51. Factoring completely: $18a^2 + 3ab - 28b^2 = (3a+4b)(6a-7b)$

53. Factoring completely: $600 + 800t - 800t^2 = 200(3 + 4t - 4t^2) = 200(1+2t)(3-2t)$

55. Factoring completely: $9y^4 + 9y^3 - 10y^2 = y^2(9y^2 + 9y - 10) = y^2(3y-2)(3y+5)$

57. Factoring completely: $24a^2 - 2a^3 - 12a^4 = 2a^2(12 - a - 6a^2) = 2a^2(3+2a)(4-3a)$

59. Factoring completely: $8x^4y^2 - 2x^3y^3 - 6x^2y^4 = 2x^2y^2(4x^2 - xy - 3y^2) = 2x^2y^2(4x+3y)(x-y)$

61. Factoring completely: $300x^4 + 1000x^2 + 300 = 100(3x^4 + 10x^2 + 3) = 100(3x^2+1)(x^2+3)$

63. Factoring completely: $20a^4 + 37a^2 + 15 = (5a^2+3)(4a^2+5)$

65. Factoring completely: $9 + 3r^2 - 12r^4 = 3(3 + r^2 - 4r^4) = 3(3+4r^2)(1-r^2)$

67. Factoring completely: $2x^2(x+5)+7x(x+5)+6(x+5)=(x+5)(2x^2+7x+6)=(x+5)(2x+3)(x+2)$

69. Factoring completely: $x^2(2x+3)+7x(2x+3)+10(2x+3)=(2x+3)(x^2+7x+10)=(2x+3)(x+5)(x+2)$

71. Factoring completely: $3x^2(x-3)+7x(x-3)-20(x-3)=(x-3)(3x^2+7x-20)=(x-3)(3x-5)(x+4)$

73. Factoring completely: $6x^2(x-2)-17x(x-2)+12(x-2)=(x-2)(6x^2-17x+12)=(x-2)(3x-4)(2x-3)$

75. Factoring completely: $12x^2(x+3)+7x(x+3)-45(x+3)=(x+3)(12x^2+7x-45)=(x+3)(4x+9)(3x-5)$

77. Factoring completely:
$$6x^2(5x-2)-11x(5x-2)-10(5x-2)=(5x-2)(6x^2-11x-10)=(5x-2)(3x+2)(2x-5)$$

79. Factoring completely:
$$20x^2(2x+3)+47x(2x+3)+21(2x+3)=(2x+3)(20x^2+47x+21)=(2x+3)(5x+3)(4x+7)$$

81. Multiplying out, the polynomial is: $(3x+5y)(3x-5y)=9x^2-25y^2$

83. The polynomial factors as $a^2+260a+2500=(a+10)(a+250)$, so the other factor is $a+250$.

85. The polynomial factors as $12x^2-107x+210=(x-6)(12x-35)$, so the other factor is $12x-35$.

87. The polynomial factors as $54x^2+111x+56=(6x+7)(9x+8)$, so the other factor is $9x+8$.

89. The polynomial factors as $35x^2+19x-24=(5x-3)(7x+8)$, so the other factor is $7x+8$.

91. Factoring the right side: $y=4x^2+18x-10=2(2x^2+9x-5)=2(2x-1)(x+5)$

Evaluating when $x=\dfrac{1}{2}$: $y=2\left(2\cdot\dfrac{1}{2}-1\right)\left(\dfrac{1}{2}+5\right)=2(0)\left(\dfrac{11}{2}\right)=0$

Evaluating when $x=-5$: $y=2(2\cdot(-5)-1)(-5+5)=2(-11)(0)=0$

Evaluating when $x=2$: $y=2(2\cdot2-1)(2+5)=2(3)(7)=42$

93. Multiplying: $(2x-3)(2x+3)=(2x)^2-(3)^2=4x^2-9$

95. Multiplying: $(2x-3)^2=(2x-3)(2x-3)=4x^2-12x+9$

97. Multiplying: $(2x-3)(4x^2+6x+9)=8x^3+12x^2+18x-12x^2-18x-27=8x^3-27$

99. Writing as a square: $\dfrac{25}{64}=\left(\dfrac{5}{8}\right)^2$

101. Writing as a square: $x^6=\left(x^3\right)^2$

103. Writing as a square: $16x^4=\left(4x^2\right)^2$

105. Writing as a cube: $\dfrac{1}{8}=\left(\dfrac{1}{2}\right)^3$

107. Writing as a cube: $x^6=\left(x^2\right)^3$

109. Writing as a cube: $27x^3=(3x)^3$

111. Writing as a cube: $8y^3=(2y)^3$

113. Factoring completely: $8x^6+26x^3y^2+15y^4=\left(2x^3+5y^2\right)\left(4x^3+3y^2\right)$

115. Factoring completely: $3x^2+295x-500=(3x-5)(x+100)$

117. Factoring completely: $\dfrac{1}{8}x^2+x+2=\left(\dfrac{1}{4}x+1\right)\left(\dfrac{1}{2}x+2\right)$

119. Factoring completely: $2x^2+1.5x+0.25=(2x+0.5)(x+0.5)$

5.6 Special Factoring

1. Factoring the trinomial: $x^2 - 6x + 9 = (x-3)^2$

3. Factoring the trinomial: $a^2 - 12a + 36 = (a-6)^2$

5. Factoring the trinomial: $25 - 10t + t^2 = (5-t)^2$

7. Factoring the trinomial: $\frac{1}{9}x^2 + 2x + 9 = \left(\frac{1}{3}x + 3\right)^2$

9. Factoring the trinomial: $4y^4 - 12y^2 + 9 = \left(2y^2 - 3\right)^2$

11. Factoring the trinomial: $16a^2 + 40ab + 25b^2 = (4a + 5b)^2$

13. Factoring the trinomial: $\frac{1}{25} + \frac{1}{10}t^2 + \frac{1}{16}t^4 = \left(\frac{1}{5} + \frac{1}{4}t^2\right)^2$

15. Factoring the trinomial: $y^2 + 3y + \frac{9}{4} = \left(y + \frac{3}{2}\right)^2$

17. Factoring the trinomial: $a^2 - a + \frac{1}{4} = \left(a - \frac{1}{2}\right)^2$

19. Factoring the trinomial: $x^2 - \frac{1}{2}x + \frac{1}{16} = \left(x - \frac{1}{4}\right)^2$

21. Factoring the trinomial: $t^2 + \frac{2}{3}t + \frac{1}{9} = \left(t + \frac{1}{3}\right)^2$

23. Factoring the trinomial: $16x^2 - 48x + 36 = 4\left(4x^2 - 12x + 9\right) = 4(2x - 3)^2$

25. Factoring the trinomial: $75a^3 + 30a^2 + 3a = 3a\left(25a^2 + 10a + 1\right) = 3a(5a + 1)^2$

27. Factoring the trinomial: $(x+2)^2 + 6(x+2) + 9 = (x+2+3)^2 = (x+5)^2$

29. Factoring: $x^2 - 9 = (x+3)(x-3)$

31. Factoring: $49x^2 - 64y^2 = (7x + 8y)(7x - 8y)$

33. Factoring: $4a^2 - \frac{1}{4} = \left(2a + \frac{1}{2}\right)\left(2a - \frac{1}{2}\right)$

35. Factoring: $x^2 - \frac{9}{25} = \left(x + \frac{3}{5}\right)\left(x - \frac{3}{5}\right)$

37. Factoring: $9x^2 - 16y^2 = (3x + 4y)(3x - 4y)$

39. Factoring: $250 - 10t^2 = 10\left(25 - t^2\right) = 10(5 + t)(5 - t)$

41. Factoring: $x^4 - 81 = \left(x^2 + 9\right)\left(x^2 - 9\right) = \left(x^2 + 9\right)(x+3)(x-3)$

43. Factoring: $9x^6 - 1 = \left(3x^3 + 1\right)\left(3x^3 - 1\right)$

45. Factoring: $16a^4 - 81 = \left(4a^2 + 9\right)\left(4a^2 - 9\right) = \left(4a^2 + 9\right)(2a + 3)(2a - 3)$

47. Factoring: $\frac{1}{81} - \frac{y^4}{16} = \left(\frac{1}{9} + \frac{y^2}{4}\right)\left(\frac{1}{9} - \frac{y^2}{4}\right) = \left(\frac{1}{9} + \frac{y^2}{4}\right)\left(\frac{1}{3} + \frac{y}{2}\right)\left(\frac{1}{3} - \frac{y}{2}\right)$

49. Factoring: $x^6 - y^6 = \left(x^3 + y^3\right)\left(x^3 - y^3\right) = (x+y)(x-y)\left(x^2 - xy + y^2\right)\left(x^2 + xy + y^2\right)$

51. Factoring: $2a^7 - 128a = 2a\left(a^6 - 64\right) = 2a\left(a^3 + 8\right)\left(a^3 - 8\right) = 2a(a+2)(a-2)\left(a^2 - 2a + 4\right)\left(a^2 + 2a + 4\right)$

53. Factoring: $(x-2)^2 - 9 = (x-2+3)(x-2-3) = (x+1)(x-5)$

55. Factoring: $(y+4)^2 - 16 = (y+4+4)(y+4-4) = y(y+8)$

57. Factoring: $x^2 - 10x + 25 - y^2 = (x-5)^2 - y^2 = (x-5+y)(x-5-y)$

59. Factoring: $a^2 + 8a + 16 - b^2 = (a+4)^2 - b^2 = (a+4+b)(a+4-b)$

61. Factoring: $x^2 + 2xy + y^2 - a^2 = (x+y)^2 - a^2 = (x+y+a)(x+y-a)$

63. Factoring: $x^3 + 3x^2 - 4x - 12 = x^2(x+3) - 4(x+3) = (x+3)\left(x^2 - 4\right) = (x+3)(x+2)(x-2)$

65. Factoring: $x^3 + 2x^2 - 25x - 50 = x^2(x+2) - 25(x+2) = (x+2)\left(x^2 - 25\right) = (x+2)(x+5)(x-5)$

67. Factoring: $2x^3 + 3x^2 - 8x - 12 = x^2(2x+3) - 4(2x+3) = (2x+3)(x^2-4) = (2x+3)(x+2)(x-2)$

69. Factoring: $4x^3 + 12x^2 - 9x - 27 = 4x^2(x+3) - 9(x+3) = (x+3)(4x^2-9) = (x+3)(2x+3)(2x-3)$

71. Factoring: $(2x-5)^2 - 100 = (2x-5-10)(2x-5+10) = (2x-15)(2x+5)$

73. Factoring: $(a-3)^2 - (4b)^2 = (a-3-4b)(a-3+4b)$

75. Factoring: $a^2 - 6a + 9 - 16b^2 = (a-3)^2 - (4b)^2 = (a-3-4b)(a-3+4b)$

77. Factoring: $x^2(x+4) - 6x(x+4) + 9(x+4) = (x+4)(x^2-6x+9) = (x+4)(x-3)^2$

79. Factoring: $x^3 - y^3 = (x-y)(x^2+xy+y^2)$

81. Factoring: $a^3 + 8 = (a+2)(a^2-2a+4)$

83. Factoring: $27 + x^3 = (3+x)(9-3x+x^2)$

85. Factoring: $y^3 - 1 = (y-1)(y^2+y+1)$

87. Factoring: $10r^3 - 1{,}250 = 10(r^3-125) = 10(r-5)(r^2+5r+25)$

89. Factoring: $64 + 27a^3 = (4+3a)(16-12a+9a^2)$

91. Factoring: $8x^3 - 27y^3 = (2x-3y)(4x^2+6xy+9y^2)$

93. Factoring: $t^3 + \frac{1}{27} = \left(t+\frac{1}{3}\right)\left(t^2-\frac{1}{3}t+\frac{1}{9}\right)$

95. Factoring: $27x^3 - \frac{1}{27} = \left(3x-\frac{1}{3}\right)\left(9x^2+x+\frac{1}{9}\right)$

97. Factoring: $64a^3 + 125b^3 = (4a+5b)(16a^2-20ab+25b^2)$

99. Since $9x^2 + 30x + 25 = (3x+5)^2$ and $9x^2 - 30x + 25 = (3x-5)^2$, two values of b are $b = 30$ and $b = -30$.

101. Factoring the greatest common factor: $y^3 + 25y = y(y^2+25)$

103. Factoring the greatest common factor: $2ab^5 + 8ab^4 + 2ab^3 = 2ab^3(b^2+4b+1)$

105. Factoring by grouping: $4x^2 - 6x + 2ax - 3a = 2x(2x-3) + a(2x-3) = (2x-3)(2x+a)$

107. Factoring the difference of squares: $x^2 - 4 = (x+2)(x-2)$

109. Factoring the perfect square trinomial: $x^2 - 6x + 9 = (x-3)^2$

111. Factoring: $6a^2 - 11a + 4 = (3a-4)(2a-1)$

113. Factoring the sum of cubes: $x^3 + 8 = (x+2)(x^2-2x+4)$

115. Form the augmented matrix: $\begin{bmatrix} 2 & -4 & | & -2 \\ 3 & -2 & | & 9 \end{bmatrix}$

Dividing row 1 by 2: $\begin{bmatrix} 1 & -2 & | & -1 \\ 3 & -2 & | & 9 \end{bmatrix}$

Add -3 times row 1 to row 2: $\begin{bmatrix} 1 & -2 & | & -1 \\ 0 & 4 & | & 12 \end{bmatrix}$

Dividing row 2 by 4: $\begin{bmatrix} 1 & -2 & | & -1 \\ 0 & 1 & | & 3 \end{bmatrix}$

So $y = 3$. Substituting to find x:

$$x - 2(3) = -1$$
$$x - 6 = -1$$
$$x = 5$$

The solution is $(5,3)$.

117. Form the augmented matrix: $\begin{bmatrix} 3 & 4 & 0 & | & 15 \\ 2 & 0 & -5 & | & -3 \\ 0 & 4 & -3 & | & 9 \end{bmatrix}$

Divide row 1 by 3: $\begin{bmatrix} 1 & \frac{4}{3} & 0 & | & 5 \\ 2 & 0 & -5 & | & -3 \\ 0 & 4 & -3 & | & 9 \end{bmatrix}$

Add -2 times row 1 to row 2: $\begin{bmatrix} 1 & \frac{4}{3} & 0 & | & 5 \\ 0 & -\frac{8}{3} & -5 & | & -13 \\ 0 & 4 & -3 & | & 9 \end{bmatrix}$

Multiply row 2 by $-\frac{3}{8}$: $\begin{bmatrix} 1 & \frac{4}{3} & 0 & | & 5 \\ 0 & 1 & \frac{15}{8} & | & \frac{39}{8} \\ 0 & 4 & -3 & | & 9 \end{bmatrix}$

Multiply row 2 by -4 and add it to row 3: $\begin{bmatrix} 1 & \frac{4}{3} & 0 & | & 5 \\ 0 & 1 & \frac{15}{8} & | & \frac{39}{8} \\ 0 & 0 & -\frac{21}{2} & | & -\frac{21}{2} \end{bmatrix}$

Multiply row 3 by $-\frac{2}{21}$: $\begin{bmatrix} 1 & \frac{4}{3} & 0 & | & 5 \\ 0 & 1 & \frac{15}{8} & | & \frac{39}{8} \\ 0 & 0 & 1 & | & 1 \end{bmatrix}$

So $z = 1$. Substituting to find y:

$$y + \frac{15}{8}(1) = \frac{39}{8}$$

$$y + \frac{15}{8} = \frac{39}{8}$$

$$y = 3$$

Substituting to find x:

$$x + \frac{4}{3}(3) = 5$$

$$x + 4 = 5$$

$$x = 1$$

The solution is $(1,3,1)$.

119. Factoring completely: $a^2 - b^2 + 6b - 9 = a^2 - \left(b^2 - 6b + 9\right) = a^2 - (b-3)^2 = (a - b + 3)(a + b - 3)$

121. Factoring completely: $(x-3)^2 - (y+5)^2 = ((x-3) - (y+5))((x-3) + (y+5)) = (x - y - 8)(x + y + 2)$

123. Since $144x^2 - 168xy + 49y^2 = (12x - 7y)^2$, a value of k is $k = 144$.

125. Since $49x^2 + 126x + 81 = (7x + 9)^2$ and $49x^2 - 126x + 81 = (7x - 9)^2$, two values of k are $k = \pm 126$.

5.7 Factoring: A General Review

1. Factoring: $x^2 - 81 = (x + 9)(x - 9)$

3. Factoring: $x^2 + 2x - 15 = (x - 3)(x + 5)$

5. Factoring: $x^2(x+2) + 6x(x+2) + 9(x+2) = (x+2)\left(x^2 + 6x + 9\right) = (x+2)(x+3)^2$

7. Factoring: $x^2 y^2 + 2y^2 + x^2 + 2 = y^2\left(x^2 + 2\right) + 1\left(x^2 + 2\right) = \left(x^2 + 2\right)\left(y^2 + 1\right)$

9. Factoring: $2a^3 b + 6a^2 b + 2ab = 2ab\left(a^2 + 3a + 1\right)$

11. The polynomial $x^2 + x + 1$ does not factor (prime).

13. Factoring: $12a^2 - 75 = 3\left(4a^2 - 25\right) = 3(2a + 5)(2a - 5)$

15. Factoring: $9x^2 - 12xy + 4y^2 = (3x - 2y)^2$

17. Factoring: $25 - 10t + t^2 = (5 - t)^2$

19. Factoring: $4x^3 + 16xy^2 = 4x\left(x^2 + 4y^2\right)$

21. Factoring: $2y^3 + 20y^2 + 50y = 2y\left(y^2 + 10y + 25\right) = 2y(y + 5)^2$

23. Factoring: $a^7 + 8a^4 b^3 = a^4\left(a^3 + 8b^3\right) = a^4(a + 2b)\left(a^2 - 2ab + 4b^2\right)$

25. Factoring: $t^2 + 6t + 9 - x^2 = (t + 3)^2 - x^2 = (t + 3 + x)(t + 3 - x)$

27. Factoring: $x^3 + 5x^2 - 9x - 45 = x^2(x+5) - 9(x+5) = (x+5)\left(x^2 - 9\right) = (x+5)(x+3)(x-3)$

29. Factoring: $5a^2 + 10ab + 5b^2 = 5\left(a^2 + 2ab + b^2\right) = 5(a + b)^2$

31. The polynomial $x^2 + 49$ does not factor (prime).

33. Factoring: $3x^2 + 15xy + 18y^2 = 3\left(x^2 + 5xy + 6y^2\right) = 3(x + 2y)(x + 3y)$

35. Factoring: $9a^2 + 2a + \dfrac{1}{9} = \left(3a + \dfrac{1}{3}\right)^2$

37. Factoring: $x^2(x-3) - 14x(x-3) + 49(x-3) = (x-3)\left(x^2 - 14x + 49\right) = (x-3)(x-7)^2$

39. Factoring: $x^2 - 64 = (x + 8)(x - 8)$

41. Factoring: $8 - 14x - 15x^2 = (2 - 5x)(4 + 3x)$

43. Factoring: $49a^7 - 9a^5 = a^5\left(49a^2 - 9\right) = a^5(7a + 3)(7a - 3)$

45. Factoring: $r^2 - \dfrac{1}{25} = \left(r + \dfrac{1}{5}\right)\left(r - \dfrac{1}{5}\right)$

47. The polynomial $49x^2 + 9y^2$ does not factor (prime).

49. Factoring: $100x^2 - 100x - 600 = 100\left(x^2 - x - 6\right) = 100(x - 3)(x + 2)$

51. Factoring: $25a^3 + 20a^2 + 3a = a\left(25a^2 + 20a + 3\right) = a(5a + 3)(5a + 1)$

53. Factoring: $3x^4 - 14x^2 - 5 = \left(3x^2 + 1\right)\left(x^2 - 5\right)$

55. Factoring: $24a^5 b - 3a^2 b = 3a^2 b\left(8a^3 - 1\right) = 3a^2 b(2a - 1)\left(4a^2 + 2a + 1\right)$

57. Factoring: $64 - r^3 = (4 - r)\left(16 + 4r + r^2\right)$

59. Factoring: $20x^4 - 45x^2 = 5x^2\left(4x^2 - 9\right) = 5x^2\left(2x + 3\right)\left(2x - 3\right)$

61. Factoring: $400t^2 - 900 = 100\left(4t^2 - 9\right) = 100\left(2t + 3\right)\left(2t - 3\right)$

63. Factoring: $16x^5 - 44x^4 + 30x^3 = 2x^3\left(8x^2 - 22x + 15\right) = 2x^3\left(4x - 5\right)\left(2x - 3\right)$

65. Factoring: $y^6 - 1 = \left(y^3 + 1\right)\left(y^3 - 1\right) = \left(y + 1\right)\left(y - 1\right)\left(y^2 - y + 1\right)\left(y^2 + y + 1\right)$

67. Factoring: $50 - 2a^2 = 2\left(25 - a^2\right) = 2\left(5 + a\right)\left(5 - a\right)$

69. Factoring: $12x^4y^2 + 36x^3y^3 + 27x^2y^4 = 3x^2y^2\left(4x^2 + 12xy + 9y^2\right) = 3x^2y^2\left(2x + 3y\right)^2$

71. Factoring: $x^2 - 4x + 4 - y^2 = \left(x - 2\right)^2 - y^2 = \left(x - 2 + y\right)\left(x - 2 - y\right)$

73. Factoring: $a^2 - \dfrac{4}{3}ab + \dfrac{4}{9}b^2 = \left(a - \dfrac{2}{3}b\right)^2$
75. Factoring: $x^2 - \dfrac{4}{5}xy + \dfrac{4}{25}y^2 = \left(x - \dfrac{2}{5}y\right)^2$

77. Factoring: $a^2 - \dfrac{5}{3}ab + \dfrac{25}{36}b^2 = \left(a - \dfrac{5}{6}b\right)^2$
79. Factoring: $x^2 - \dfrac{8}{5}xy + \dfrac{16}{25}y^2 = \left(x - \dfrac{4}{5}y\right)^2$

81. Factoring: $2x^2\left(x + 2\right) - 13x\left(x + 2\right) + 15\left(x + 2\right) = \left(x + 2\right)\left(2x^2 - 13x + 15\right) = \left(x + 2\right)\left(2x - 3\right)\left(x - 5\right)$

83. Factoring: $\left(x - 4\right)^3 + \left(x - 4\right)^4 = \left(x - 4\right)^3\left(1 + x - 4\right) = \left(x - 4\right)^3\left(x - 3\right)$

85. Factoring: $2y^3 - 54 = 2\left(y^3 - 27\right) = 2\left(y - 3\right)\left(y^2 + 3y + 9\right)$

87. Factoring: $2a^3 - 128b^3 = 2\left(a^3 - 64b^3\right) = 2\left(a - 4b\right)\left(a^2 + 4ab + 16b^2\right)$

89. Factoring: $2x^3 + 432y^3 = 2\left(x^3 + 216y^3\right) = 2\left(x + 6y\right)\left(x^2 - 6xy + 36y^2\right)$

91. Let g and d represent the number of geese and ducks. The system of equations is:
$$g + d = 108$$
$$1.4g + 0.6d = 112.80$$
Substituting $d = 108 - g$ into the second equation:
$$1.4g + 0.6\left(108 - g\right) = 112.80$$
$$1.4g + 64.8 - 0.6g = 112.80$$
$$0.8g = 48$$
$$g = 60$$
$$d = 48$$
He bought 60 geese and 48 ducks.

93. Let o represent the number of oranges and a represent the number of apples. The system of equations is:
$$\frac{o}{3}(0.10) + \frac{a}{12}(0.15) = 6.80$$
$$\frac{5o}{3}(0.10) + \frac{a}{48}(0.15) = 25.45$$
Clearing each equation of fractions:
$$12 \cdot \frac{o}{3}(0.10) + 12 \cdot \frac{a}{12}(0.15) = 12 \cdot 6.80 \qquad 48 \cdot \frac{5o}{3}(0.10) + 48 \cdot \frac{a}{48}(0.15) = 48 \cdot 25.45$$
$$0.4o + 0.15a = 81.6 \qquad\qquad 8o + 0.15a = 1221.6$$
So the system becomes:
$$0.4o + 0.15a = 81.6$$
$$8o + 0.15a = 1221.6$$
Multiply the first equation by -20:
$$-8o - 3a = -1632$$
$$8o + 0.15a = 1221.6$$

Adding yields:
$$-2.85a = -410.4$$
$$a = 144$$
Substituting to find o:
$$8o + 0.15(144) = 1221.6$$
$$8o + 21.6 = 1221.6$$
$$8o = 1200$$
$$o = 150$$
So 150 oranges and 144 apples were bought.

95. Simplifying: $x^2 + (x+1)^2 = x^2 + x^2 + 2x + 1 = 2x^2 + 2x + 1$

97. Simplifying: $\dfrac{16t^2 - 64t + 48}{16} = \dfrac{16\left(t^2 - 4t + 3\right)}{16} = t^2 - 4t + 3$

99. Factoring: $x^2 - 2x - 24 = (x-6)(x+4)$

101. Factoring: $2x^3 - 5x^2 - 3x = x\left(2x^2 - 5x - 3\right) = x(2x+1)(x-3)$

103. Factoring: $x^3 + 2x^2 - 9x - 18 = x^2(x+2) - 9(x+2) = (x+2)\left(x^2 - 9\right) = (x+2)(x+3)(x-3)$

105. Solving the equation:
$$x - 6 = 0$$
$$x = 6$$

107. Solving the equation:
$$2x + 1 = 0$$
$$2x = -1$$
$$x = -\frac{1}{2}$$

5.8 Solving Equations by Factoring

1. Solving the equation:
$$x^2 - 5x - 6 = 0$$
$$(x+1)(x-6) = 0$$
$$x = -1, 6$$

3. Solving the equation:
$$x^3 - 5x^2 + 6x = 0$$
$$x\left(x^2 - 5x + 6\right) = 0$$
$$x(x-2)(x-3) = 0$$
$$x = 0, 2, 3$$

5. Solving the equation:
$$3y^2 + 11y - 4 = 0$$
$$(3y-1)(y+4) = 0$$
$$y = -4, \frac{1}{3}$$

7. Solving the equation:
$$60x^2 - 130x + 60 = 0$$
$$10\left(6x^2 - 13x + 6\right) = 0$$
$$10(3x - 2)(2x - 3) = 0$$
$$x = \frac{2}{3}, \frac{3}{2}$$

9. Solving the equation:
$$\frac{1}{10}t^2 - \frac{5}{2} = 0$$
$$10\left(\frac{1}{10}t^2 - \frac{5}{2}\right) = 10(0)$$
$$t^2 - 25 = 0$$
$$(t+5)(t-5) = 0$$
$$t = -5, 5$$

11. Solving the equation:
$$100x^4 = 400x^3 + 2100x^2$$
$$100x^4 - 400x^3 - 2100x^2 = 0$$
$$100x^2\left(x^2 - 4x - 21\right) = 0$$
$$100x^2(x-7)(x+3) = 0$$
$$x = -3, 0, 7$$

13.   Solving the equation:

$$\frac{1}{5}y^2 - 2 = -\frac{3}{10}y$$

$$10\left(\frac{1}{5}y^2 - 2\right) = 10\left(-\frac{3}{10}y\right)$$

$$2y^2 - 20 = -3y$$

$$2y^2 + 3y - 20 = 0$$

$$(y+4)(2y-5) = 0$$

$$y = -4, \frac{5}{2}$$

15.   Solving the equation:

$$9x^2 - 12x = 0$$

$$3x(3x - 4) = 0$$

$$x = 0, \frac{4}{3}$$

17.   Solving the equation:

$$0.02r + 0.01 = 0.15r^2$$

$$2r + 1 = 15r^2$$

$$15r^2 - 2r - 1 = 0$$

$$(5r+1)(3r-1) = 0$$

$$r = -\frac{1}{5}, \frac{1}{3}$$

19.   Solving the equation:

$$9a^3 = 16a$$

$$9a^3 - 16a = 0$$

$$a\left(9a^2 - 16\right) = 0$$

$$a(3a+4)(3a-4) = 0$$

$$a = -\frac{4}{3}, 0, \frac{4}{3}$$

21.   Solving the equation:

$$-100x = 10x^2$$

$$0 = 10x^2 + 100x$$

$$0 = 10x(x + 10)$$

$$x = -10, 0$$

23.   Solving the equation:

$$(x+6)(x-2) = -7$$

$$x^2 + 4x - 12 = -7$$

$$x^2 + 4x - 5 = 0$$

$$(x+5)(x-1) = 0$$

$$x = -5, 1$$

25.   Solving the equation:

$$(y-4)(y+1) = -6$$

$$y^2 - 3y - 4 = -6$$

$$y^2 - 3y + 2 = 0$$

$$(y-2)(y-1) = 0$$

$$y = 1, 2$$

27.   Solving the equation:

$$(x+1)^2 = 3x + 7$$

$$x^2 + 2x + 1 = 3x + 7$$

$$x^2 - x - 6 = 0$$

$$(x+2)(x-3) = 0$$

$$x = -2, 3$$

29.   Solving the equation:

$$(2r+3)(2r-1) = -(3r+1)$$

$$4r^2 + 4r - 3 = -3r - 1$$

$$4r^2 + 7r - 2 = 0$$

$$(r+2)(4r-1) = 0$$

$$r = -2, \frac{1}{4}$$

31.   Solving the equation:

$$x^3 + 3x^2 - 4x - 12 = 0$$

$$x^2(x+3) - 4(x+3) = 0$$

$$(x+3)\left(x^2 - 4\right) = 0$$

$$(x+3)(x+2)(x-2) = 0$$

$$x = -3, -2, 2$$

33.   Solving the equation:

$$x^3 + 2x^2 - 25x - 50 = 0$$

$$x^2(x+2) - 25(x+2) = 0$$

$$(x+2)\left(x^2 - 25\right) = 0$$

$$(x+2)(x+5)(x-5) = 0$$

$$x = -5, -2, 5$$

35.   Solving the equation:

$$2x^3 + 3x^2 - 8x - 12 = 0$$

$$x^2(2x+3) - 4(2x+3) = 0$$

$$(2x+3)\left(x^2 - 4\right) = 0$$

$$(2x+3)(x+2)(x-2) = 0$$

$$x = -2, -\frac{3}{2}, 2$$

37. Solving the equation:

$$4x^3 + 12x^2 - 9x - 27 = 0$$
$$4x^2(x+3) - 9(x+3) = 0$$
$$(x+3)(4x^2 - 9) = 0$$
$$(x+3)(2x+3)(2x-3) = 0$$
$$x = -3, -\frac{3}{2}, \frac{3}{2}$$

39. Solving the equation:

$$3x^2 + x = 10$$
$$3x^2 + x - 10 = 0$$
$$(3x-5)(x+2) = 0$$
$$x = -2, \frac{5}{3}$$

41. Solving the equation:

$$12(x+3) + 12(x-3) = 3(x^2 - 9)$$
$$12x + 36 + 12x - 36 = 3x^2 - 27$$
$$24x = 3x^2 - 27$$
$$3x^2 - 24x - 27 = 0$$
$$3(x^2 - 8x - 9) = 0$$
$$3(x-9)(x+1) = 0$$
$$x = -1, 9$$

43. Solving the equation:

$$(y+3)^2 + y^2 = 9$$
$$y^2 + 6y + 9 + y^2 = 9$$
$$2y^2 + 6y = 0$$
$$2y(y+3) = 0$$
$$y = -3, 0$$

45. Solving the equation:

$$(x+3)^2 + 1^2 = 2$$
$$x^2 + 6x + 9 + 1 = 2$$
$$x^2 + 6x + 8 = 0$$
$$(x+4)(x+2) = 0$$
$$x = -4, -2$$

47. Solving the equation:

$$(x+2)(x) = 2^3$$
$$x^2 + 2x = 8$$
$$x^2 + 2x - 8 = 0$$
$$(x+4)(x-2) = 0$$
$$x = -4, 2$$

49. Solving $f(x) = 0$:

$$\left(x + \frac{3}{2}\right)^2 = 0$$
$$x + \frac{3}{2} = 0$$
$$x = -\frac{3}{2}$$

51. Solving $f(x) = 0$:

$$(x-3)^2 - 25 = 0$$
$$x^2 - 6x + 9 - 25 = 0$$
$$x^2 - 6x - 16 = 0$$
$$(x+2)(x-8) = 0$$
$$x = -2, 8$$

53. Solving $f(x) = g(x)$:

$$x^2 + 6x + 3 = -6$$
$$x^2 + 6x + 9 = 0$$
$$(x+3)^2 = 0$$
$$x + 3 = 0$$
$$x = -3$$

55. Solving $f(x) = g(x)$:

$$x^2 + 6x + 3 = 10$$
$$x^2 + 6x - 7 = 0$$
$$(x+7)(x-1) = 0$$
$$x = -7, 1$$

57. Solving $h(x) = f(x)$:

$$x^2 - 5x = 0$$
$$x(x-5) = 0$$
$$x = 0, 5$$

59. Solving $h(x) = f(x)$:

$$x^2 - 5x = 2x + 8$$
$$x^2 - 7x - 8 = 0$$
$$(x-8)(x+1) = 0$$
$$x = -1, 8$$

61.　**a.**　Solving the equation:
$$9x - 25 = 0$$
$$9x = 25$$
$$x = \frac{25}{9}$$

b.　Solving the equation:
$$9x^2 - 25 = 0$$
$$(3x + 5)(3x - 5) = 0$$
$$x = -\frac{5}{3}, \frac{5}{3}$$

c.　Solving the equation:
$$9x^2 - 25 = 56$$
$$9x^2 - 81 = 0$$
$$9(x + 3)(x - 3) = 0$$
$$x = -3, 3$$

d.　Solving the equation:
$$9x^2 - 25 = 30x - 50$$
$$9x^2 - 30x + 25 = 0$$
$$(3x - 5)^2 = 0$$
$$3x - 5 = 0$$
$$3x = 5$$
$$x = \frac{5}{3}$$

63.　Let t represent the required time. Using the Pythagorean theorem:
$$(15t)^2 + (20t)^2 = 75^2$$
$$225t^2 + 400t^2 = 5625$$
$$625t^2 = 5625$$
$$625t^2 - 5625 = 0$$
$$625(t^2 - 9) = 0$$
$$t = -3, 3$$
Since the time must be positive, the time is 3 hours.

65.　Let x and $x + 1$ represent the two integers. The equation is:
$$(x + x + 1)^2 = 81$$
$$(2x + 1)^2 = 81$$
$$4x^2 + 4x + 1 = 81$$
$$4x^2 + 4x - 80 = 0$$
$$x^2 + x - 20 = 0$$
$$(x + 5)(x - 4) = 0$$
$$x = -5, 4$$
$$x + 1 = -4, 5$$
The integers are either –5 and –4, or 4 and 5.

67.　Let h represent the height the ladder makes with the building. Using the Pythagorean theorem:
$$7^2 + h^2 = 25^2$$
$$49 + h^2 = 625$$
$$h^2 = 576$$
$$h = 24$$
The ladder reaches a height of 24 feet along the building.

69.　Let x, $x + 2$, and $x + 4$ represent the three sides. Using the Pythagorean theorem:
$$x^2 + (x + 2)^2 = (x + 4)^2$$
$$x^2 + x^2 + 4x + 4 = x^2 + 8x + 16$$
$$2x^2 + 4x + 4 = x^2 + 8x + 16$$
$$x^2 - 4x - 12 = 0$$
$$(x - 6)(x + 2) = 0$$
$$x = 6 \quad (x = -2 \text{ is impossible})$$
The lengths of the three sides are 6, 8, and 10.

71. Let w represent the width and $3w + 2$ represent the length. Using the area formula:

$$w(3w + 2) = 16$$
$$3w^2 + 2w = 16$$
$$3w^2 + 2w - 16 = 0$$
$$(3w + 8)(w - 2) = 0$$
$$w = 2 \quad (w = -\frac{8}{3} \text{ is impossible})$$
$$3w + 2 = 8$$

The dimensions are 2 feet by 8 feet.

73. Let h represent the height and $4h + 2$ represent the base. Using the area formula:

$$\frac{1}{2}(4h + 2)(h) = 36$$
$$4h^2 + 2h = 72$$
$$4h^2 + 2h - 72 = 0$$
$$2(2h^2 + h - 36) = 0$$
$$2(2h + 9)(h - 4) = 0$$
$$h = 4 \quad (h = -\frac{9}{2} \text{ is impossible})$$
$$4h + 2 = 18$$

The base is 18 inches and the height is 4 inches.

75. Setting $h = 0$ in the equation:

$$32t - 16t^2 = 0$$
$$16t(2 - t) = 0$$
$$t = 0, 2$$

The object is on the ground at 0 and 2 seconds.

77. Substituting $v = 48$ and $h = 32$:

$$h = vt - 16t^2$$
$$32 = 48t - 16t^2$$
$$0 = -16t^2 + 48t - 32$$
$$0 = -16(t^2 - 3t + 2)$$
$$0 = -16(t - 1)(t - 2)$$
$$t = 1, 2$$

It will reach a height of 32 feet after 1 sec and 2 sec.

79. Substituting $v = 24$ and $h = 0$:

$$h = vt - 16t^2$$
$$0 = 24t - 16t^2$$
$$0 = -16t^2 + 24t$$
$$0 = -8t(2t - 3)$$
$$t = 0, \frac{3}{2}$$

It will be on the ground after 0 sec and $\frac{3}{2}$ sec.

81. Substituting $h = 192$:

$$192 = 96 + 80t - 16t^2$$
$$0 = -16t^2 + 80t - 96$$
$$0 = -16(t^2 - 5t + 6)$$
$$0 = -16(t - 2)(t - 3)$$
$$t = 2, 3$$

The bullet will be 192 feet in the air after 2 sec and 3 sec.

83. Substituting $R = \$3,200$:

$$R = xp$$
$$3200 = (1200 - 100p)p$$
$$3200 = 1200p - 100p^2$$
$$0 = -100p^2 + 1200p - 3200$$
$$0 = -100(p^2 - 12p + 32)$$
$$0 = -100(p - 4)(p - 8)$$
$$p = 4, 8$$

The cartridges should be sold for either $4 or $8.

85. Substituting $R = \$7,000$:

$$R = xp$$
$$7000 = (1700 - 100p)p$$
$$7000 = 1700p - 100p^2$$
$$0 = -100p^2 + 1700p - 7000$$
$$0 = -100(p^2 - 17p + 70)$$
$$0 = -100(p - 7)(p - 10)$$
$$p = 7, 10$$

The calculators should be sold for either $7 or $10.

87.   Multiply the second equation by 5:
$$2x - 5y = -8$$
$$15x + 5y = 25$$

Adding yields:
$$17x = 17$$
$$x = 1$$

Substituting to find y:
$$3 + y = 5$$
$$y = 2$$

The solution is $(1, 2)$.

89.   To clear each equation of fractions, multiply the first equation by 6 and the second equation by 20:
$$2x - y = 18$$
$$-4x + 5y = 0$$

Multiply the first equation by 2:
$$4x - 2y = 36$$
$$-4x + 5y = 0$$

Adding yields:
$$3y = 36$$
$$y = 12$$

Substituting to find x:
$$2x - 12 = 18$$
$$2x = 30$$
$$x = 15$$

The solution is $(15, 12)$.

91.   Adding the first and third equations:
$$5x = 15$$
$$x = 3$$

Adding the first and second equations:
$$3x - 2z = 7$$
$$9 - 2z = 7$$
$$-2z = -2$$
$$z = 1$$

Substituting to find y:
$$3 + y - 3 = -2$$
$$y = -2$$

The solution is $(3, -2, 1)$.

93.   Let x and $2x$ represent the two investments. The equation is:
$$0.05(x) + 0.06(2x) = 680$$
$$0.17x = 680$$
$$x = 4,000$$
$$2x = 8,000$$

John invested \$4,000 at 5% and \$8,000 at 6%.

95. Graphing the solution set:

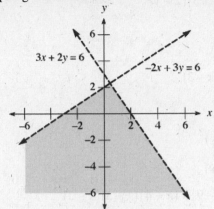

$3x + 2y = 6$

$-2x + 3y = 6$

97. Graphing the solution set:

$y = 2$

$x = 4$

Chapter 5 Review

1. Simplifying: $x^3 \cdot x^7 = x^{3+7} = x^{10}$

3. Simplifying: $\left(2x^3 y\right)^2 \left(-2x^4 y^2\right)^3 = 4x^6 y^2 \cdot \left(-8x^{12} y^6\right) = -32x^{18} y^8$

5. Writing with positive exponents: $\left(\dfrac{2}{3}\right)^{-2} = \left(\dfrac{3}{2}\right)^2 = \dfrac{9}{4}$

7. Writing in scientific notation: $34,500,000 = 3.45 \times 10^7$

9. Writing in expanded form: $4.45 \times 10^4 = 44,500$

11. Simplifying: $\dfrac{a^{-4}}{a^5} = a^{-4-5} = a^{-9} = \dfrac{1}{a^9}$

13. Simplifying: $\dfrac{x^n x^{3n}}{x^{4n-2}} = \dfrac{x^{4n}}{x^{4n-2}} = x^{4n-(4n-2)} = x^{4n-4n+2} = x^2$

15. Simplifying: $\dfrac{(600,000)(0.000008)}{(4,000)(3,000,000)} = \dfrac{\left(6 \times 10^5\right)\left(8 \times 10^{-6}\right)}{\left(4 \times 10^3\right)\left(3 \times 10^6\right)} = \dfrac{48 \times 10^{-1}}{12 \times 10^9} = \dfrac{48}{12} \times 10^{-1-9} = 4 \times 10^{-10}$

17. Simplifying: $\left(x^3 - x\right) - \left(x^2 + x\right) + \left(x^3 - 3\right) - \left(x^2 + 1\right) = x^3 - x - x^2 - x + x^3 - 3 - x^2 - 1 = 2x^3 - 2x^2 - 2x - 4$

19. Simplifying: $-3\left[2x - 4(3x+1)\right] = -3(2x - 12x - 4) = -3(-10x - 4) = 30x + 12$

21. Multiplying: $3x\left(4x^2 - 2x + 1\right) = 12x^3 - 6x^2 + 3x$

23. Multiplying: $(6 - y)(3 - y) = 18 - 6y - 3y + y^2 = 18 - 9y + y^2$

25. Multiplying: $2t(t+1)(t-3) = 2t\left(t^2 - 2t - 3\right) = 2t^3 - 4t^2 - 6t$

27. Multiplying: $(2x - 3)\left(4x^2 + 6x + 9\right) = 8x^3 + 12x^2 + 18x - 12x^2 - 18x - 27 = 8x^3 - 27$

29. Multiplying: $(3x + 5)^2 = (3x)^2 + 2(3x)(5) + 5^2 = 9x^2 + 30x + 25$

31. Multiplying: $\left(x - \dfrac{1}{3}\right)\left(x + \dfrac{1}{3}\right) = (x)^2 - \left(\dfrac{1}{3}\right)^2 = x^2 - \dfrac{1}{9}$

33. Multiplying: $(x-1)^3 = (x-1)(x-1)^2 = (x-1)\left(x^2 - 2x + 1\right) = x^3 - 2x^2 + x - x^2 + 2x - 1 = x^3 - 3x^2 + 3x - 1$

35. Factoring: $6x^4 y - 9xy^4 + 18x^3 y^3 = 3xy\left(2x^3 - 3y^3 + 6x^2 y^2\right)$

37. Factoring: $8x^2 + 10 - 4x^2y - 5y = 2(4x^2 + 5) - y(4x^2 + 5) = (4x^2 + 5)(2 - y)$

39. Factoring: $x^2 - 5x + 6 = (x - 2)(x - 3)$

41. Factoring: $20a^2 - 41ab + 20b^2 = (5a - 4b)(4a - 5b)$

43. Factoring: $24x^2y - 6xy - 45y = 3y(8x^2 - 2x - 15) = 3y(4x + 5)(2x - 3)$

45. Factoring: $3a^4 + 18a^2 + 27 = 3(a^4 + 6a^2 + 9) = 3(a^2 + 3)^2$

47. Factoring: $5x^3 + 30x^2y + 45xy^2 = 5x(x^2 + 6xy + 9y^2) = 5x(x + 3y)^2$

49. Factoring: $x^2 - 10x + 25 - y^2 = (x - 5)^2 - y^2 = (x - 5 + y)(x - 5 - y)$

51. Factoring: $x^3 + 4x^2 - 9x - 36 = x^2(x + 4) - 9(x + 4) = (x + 4)(x^2 - 9) = (x + 4)(x + 3)(x - 3)$

53. Solving the equation:

$$\frac{5}{6}y^2 = \frac{1}{4}y + \frac{1}{3}$$
$$12\left(\frac{5}{6}y^2\right) = 12\left(\frac{1}{4}y + \frac{1}{3}\right)$$
$$10y^2 = 3y + 4$$
$$10y^2 - 3y - 4 = 0$$
$$(5y - 4)(2y + 1) = 0$$
$$y = -\frac{1}{2}, \frac{4}{5}$$

55. Solving the equation:

$$5x^2 = -10x$$
$$5x^2 + 10x = 0$$
$$5x(x + 2) = 0$$
$$x = -2, 0$$

57. Solving the equation:

$$x^3 + 4x^2 - 9x - 36 = 0$$
$$x^2(x + 4) - 9(x + 4) = 0$$
$$(x + 4)(x^2 - 9) = 0$$
$$(x + 4)(x + 3)(x - 3) = 0$$
$$x = -4, -3, 3$$

59. Let x and $x + 1$ represent the two integers. The equation is:

$$x^2 + (x + 1)^2 = 41$$
$$x^2 + x^2 + 2x + 1 = 41$$
$$2x^2 + 2x - 40 = 0$$
$$x^2 + x - 20 = 0$$
$$(x + 5)(x - 4) = 0$$
$$x = -5, 4$$
$$x + 1 = -4, 5$$

The integers are either –5 and –4, or 4 and 5.

61. Let x, $x + 2$, and $x + 4$ represent the three sides. Using the Pythagorean theorem:

$$x^2 + (x + 2)^2 = (x + 4)^2$$
$$x^2 + x^2 + 4x + 4 = x^2 + 8x + 16$$
$$2x^2 + 4x + 4 = x^2 + 8x + 16$$
$$x^2 - 4x - 12 = 0$$
$$(x - 6)(x + 2) = 0$$
$$x = 6 \quad (x = -2 \text{ is impossible})$$

The lengths of the three sides are 6, 8, and 10.

Chapter 5 Cumulative Review

1. Simplifying: $(-5)^3 = (-5)(-5)(-5) = -125$

3. Simplifying: $|-32 - 41| - 13 = |-73| - 13 = 73 - 13 = 60$

5. Simplifying: $96 \div 8 \cdot 4 = 12 \cdot 4 = 48$

7. Simplifying: $\dfrac{3^3 + 5}{-2\left(5^2 - 3^2\right)} = \dfrac{27 + 5}{-2(25 - 9)} = \dfrac{32}{-2(16)} = \dfrac{32}{-32} = -1$

9. Evaluating: $P(1) = 11.5(1) - 0.01(1)^2 = 11.5 - 0.01(1) = 11.5 - 0.01 = 11.49$

11. Evaluating: $Q(4) = \dfrac{3}{2}(4)^2 + \dfrac{3}{4}(4) - 5 = \dfrac{3}{2}(16) + \dfrac{3}{4}(4) - 5 = 24 + 3 - 5 = 22$

13. Simplifying: $6\left(x - \dfrac{1}{6}\right) = 6 \cdot x - 6 \cdot \dfrac{1}{6} = 6x - 1$

15. Simplifying: $(3x - 2) - (5x + 2) = 3x - 2 - 5x - 2 = -2x - 4$

17. Simplifying:
$$\left(3x^3 - 2\right) - \left(5x^2 + 2\right) + \left(5x^3 - 6\right) - \left(2x^2 + 4\right) = \left(3x^3 + 5x^3\right) + \left(-5x^2 - 2x^2\right) + (-2 - 2 - 6 - 4)$$
$$= 8x^3 - 7x^2 - 14$$

19. Multiplying: $\left(3x^2 y\right)^4 = 3^4 \left(x^2\right)^4 y^4 = 81x^8 y^4$

21. Dividing: $\dfrac{32x^2 y^6}{16xy^2} = -2x^{2-1} y^{6-2} = -2xy^4$

23. Multiplying: $(3x - 2)(5x + 2) = 15x^2 + 6x - 10x - 4 = 15x^2 - 4x - 4$

25. Multiplying: $\left(x - \dfrac{2}{3}\right)^2 = x^2 - \dfrac{4}{3}x + \dfrac{4}{9}$

27. Multiplying: $\left(4 \times 10^3\right)\left(7 \times 10^6\right) = 28 \times 10^9 = 2.8 \times 10^{10}$

29. Factoring: $x^2 - xy - ax + ay = x(x - y) - a(x - y) = (x - y)(x - a)$

31. Factoring: $(x + 3)^2 - 25 = (x + 3 - 5)(x + 3 + 5) = (x - 2)(x + 8)$

33. Solving the equation:
$$3x + 2 = 8$$
$$3x = 6$$
$$x = 2$$

35. Solving the equation:
$$\dfrac{3}{4}x - 7 = -34$$
$$\dfrac{3}{4}x = -27$$
$$3x = -108$$
$$x = -36$$

37. Solving the equation:
$$25a^3 = 36a$$
$$25a^3 - 36a = 0$$
$$a\left(25a^2 - 36\right) = 0$$
$$a(5a + 6)(5a - 6) = 0$$
$$a = -\dfrac{6}{5}, 0, \dfrac{6}{5}$$

39.   Multiply the first equation by 3:
$$3x - 3y = -12$$
$$x + 3y = 4$$
Adding yields:
$$4x = -8$$
$$x = -2$$
Substituting into the first equation:
$$-2 - y = -4$$
$$-y = -2$$
$$y = 2$$
The solution is $(-2, 2)$.

41.   Finding the slope: $m = \dfrac{-4 - 1}{5 - (-2)} = -\dfrac{5}{7}$

43.   Evaluating: $f(-7) = \dfrac{3(-7) - 5}{2} = \dfrac{-21 - 5}{2} = \dfrac{-26}{2} = -13$

45.   Evaluating: $(g \circ f)(2) = g\big[f(2)\big] = g\left[\dfrac{3(2) - 5}{2}\right] = g\left(\dfrac{1}{2}\right) = \dfrac{2\left(\frac{1}{2}\right) + 5}{3} = \dfrac{1 + 5}{3} = 2$

47.   The variation equation is $y = \dfrac{K}{x^2}$. Substituting $x = 4$ and $y = 4$:
$$4 = \dfrac{K}{4^2}$$
$$4 = \dfrac{K}{16}$$
$$K = 64$$
So the equation is $y = \dfrac{64}{x^2}$. Substituting $x = 6$: $y = \dfrac{64}{6^2} = \dfrac{64}{36} = \dfrac{16}{9}$

49.   Let b represent the base and $2b - 5$ represent the height. Using the area formula:
$$75 = \dfrac{1}{2}b(2b - 5)$$
$$150 = 2b^2 - 5b$$
$$2b^2 - 5b - 150 = 0$$
$$(2b + 15)(b - 10) = 0$$
$$b = 10 \qquad \left(b = -\dfrac{15}{2} \text{ is impossible}\right)$$
$$2b - 5 = 15$$
The base is 10 feet and the height is 15 feet.

51.   The sum is: $31 + 16 + 27 = 74$ users

Chapter 5 Test

1. Using properties of exponents: $\left(\dfrac{3}{4}\right)^{-2} = \left(\dfrac{4}{3}\right)^2 = \dfrac{16}{9}$

2. Using properties of exponents: $2^{-5} = \dfrac{1}{2^5} = \dfrac{1}{32}$

3. Using properties of exponents: $x^4 \cdot x^7 \cdot x^{-3} = x^{4+7-3} = x^8$

4. Using properties of exponents: $\dfrac{a^{-5}}{a^{-7}} = a^{-5+7} = a^2$

5. Using properties of exponents: $\left(2x^2 y\right)^3 \left(2x^3 y^4\right)^2 = \left(8x^6 y^3\right)\left(4x^6 y^8\right) = 32x^{12} y^{11}$

6. Using properties of exponents:

$$\dfrac{\left(2ab^3\right)^{-2}}{\left(a^{-4}b^3\right)^4} = \dfrac{2^{-2} a^{-2} b^{-6}}{a^{-16} b^{12}} = 2^{-2} a^{-2+16} b^{-6-12} = 2^{-2} a^{14} b^{-18} = \dfrac{a^{14}}{4b^{18}}$$

7. Writing in scientific notation: $6{,}530{,}000 = 6.53 \times 10^6$

8. Writing in scientific notation: $0.00087 = 8.7 \times 10^{-4}$

9. Performing the operations: $\left(2.9 \times 10^{12}\right)\left(3 \times 10^{-5}\right) = 8.7 \times 10^7$

10. Performing the operations: $\dfrac{\left(6 \times 10^{-4}\right)\left(4 \times 10^9\right)}{8 \times 10^{-3}} = \dfrac{24 \times 10^5}{8 \times 10^{-3}} = \dfrac{24}{8} \times 10^{5+3} = 3 \times 10^8$

11. Evaluating: $P(10) = 15(10) - 0.01(10)^2 = 150 - 1 = 149$

12. Evaluating: $P(-10) = 15(-10) - 0.01(-10)^2 = -150 - 1 = -151$

13. Simplifying: $\left(\dfrac{3}{4}x^3 - x^2 - \dfrac{3}{2}\right) - \left(\dfrac{1}{4}x^2 + 2x - \dfrac{1}{2}\right) = \dfrac{3}{4}x^3 - x^2 - \dfrac{3}{2} - \dfrac{1}{4}x^2 - 2x + \dfrac{1}{2} = \dfrac{3}{4}x^3 - \dfrac{5}{4}x^2 - 2x - 1$

14. Simplifying: $3 - 4\left[2x - 3(x+6)\right] = 3 - 4(2x - 3x - 18) = 3 - 4(-x - 18) = 3 + 4x + 72 = 4x + 75$

15. Multiplying: $(3y - 7)(2y + 5) = 6y^2 + 15y - 14y - 35 = 6y^2 + y - 35$

16. Multiplying: $(2x - 5)\left(x^2 + 4x - 3\right) = 2x^3 + 8x^2 - 6x - 5x^2 - 20x + 15 = 2x^3 + 3x^2 - 26x + 15$

17. Multiplying: $\left(8 - 3t^3\right)^2 = (8)^2 - 2(8)\left(3t^3\right) + \left(3t^3\right)^2 = 64 - 48t^3 + 9t^6$

18. Multiplying: $(1 - 6y)(1 + 6y) = (1)^2 - (6y)^2 = 1 - 36y^2$

19. Multiplying: $2x(x - 3)(2x + 5) = 2x\left(2x^2 - x - 15\right) = 4x^3 - 2x^2 - 30x$

20. Multiplying: $\left(5t^2 - \dfrac{1}{2}\right)\left(2t^2 + \dfrac{1}{5}\right) = 10t^4 - t^2 + t^2 - \dfrac{1}{10} = 10t^4 - \dfrac{1}{10}$

21. Factoring: $x^2 + x - 12 = (x + 4)(x - 3)$

22. Factoring: $12x^4 + 26x^2 - 10 = 2\left(6x^4 + 13x^2 - 5\right) = 2\left(3x^2 - 1\right)\left(2x^2 + 5\right)$

23. Factoring: $16a^4 - 81y^4 = \left(4a^2 + 9y^2\right)\left(4a^2 - 9y^2\right) = \left(4a^2 + 9y^2\right)(2a + 3y)(2a - 3y)$

24. Factoring: $7ax^2 - 14ay - b^2 x^2 + 2b^2 y = 7a\left(x^2 - 2y\right) - b^2\left(x^2 - 2y\right) = \left(x^2 - 2y\right)\left(7a - b^2\right)$

25. Factoring: $t^3 + \dfrac{1}{8} = t^3 + \left(\dfrac{1}{2}\right)^3 = \left(t + \dfrac{1}{2}\right)\left(t^2 - \dfrac{1}{2}t + \dfrac{1}{4}\right)$

26. Factoring: $4a^5b - 24a^4b^2 - 64a^3b^3 = 4a^3b\left(a^2 - 6ab - 16b^2\right) = 4a^3b\left(a - 8b\right)\left(a + 2b\right)$

27. Factoring: $x^2 - 10x + 25 - b^2 = \left(x - 5\right)^2 - b^2 = \left(x - 5 + b\right)\left(x - 5 - b\right)$

28. Factoring: $81 - x^4 = \left(9 + x^2\right)\left(9 - x^2\right) = \left(9 + x^2\right)\left(3 + x\right)\left(3 - x\right)$

29. Solving the equation:

$$\frac{1}{5}x^2 = \frac{1}{3}x + \frac{2}{15}$$
$$15\left(\frac{1}{5}x^2\right) = 15\left(\frac{1}{3}x + \frac{2}{15}\right)$$
$$3x^2 = 5x + 2$$
$$3x^2 - 5x - 2 = 0$$
$$\left(3x + 1\right)\left(x - 2\right) = 0$$
$$x = -\frac{1}{3}, 2$$

30. Solving the equation:

$$100x^3 = 500x^2$$
$$100x^3 - 500x^2 = 0$$
$$100x^2\left(x - 5\right) = 0$$
$$x = 0, 5$$

31. Solving the equation:

$$\left(x + 1\right)\left(x + 2\right) = 12$$
$$x^2 + 3x + 2 = 12$$
$$x^2 + 3x - 10 = 0$$
$$\left(x + 5\right)\left(x - 2\right) = 0$$
$$x = -5, 2$$

32. Solving the equation:

$$x^3 + 2x^2 - 16x - 32 = 0$$
$$x^2\left(x + 2\right) - 16\left(x + 2\right) = 0$$
$$\left(x + 2\right)\left(x^2 - 16\right) = 0$$
$$\left(x + 2\right)\left(x + 4\right)\left(x - 4\right) = 0$$
$$x = -4, -2, 4$$

33. Setting $f(x) = g(x)$:

$$x^2 - 5x + 6 = 0$$
$$\left(x - 2\right)\left(x - 3\right) = 0$$
$$x = 2, 3$$

34. Setting $f(x) = g(x)$:

$$x^2 - 5x + 6 = x - 2$$
$$x^2 - 6x + 8 = 0$$
$$\left(x - 4\right)\left(x - 2\right) = 0$$
$$x = 2, 4$$

35. Using the Pythagorean theorem:

$$x^2 + 12^2 = \left(2x - 3\right)^2$$
$$x^2 + 144 = 4x^2 - 12x + 9$$
$$3x^2 - 12x - 135 = 0$$
$$x^2 - 4x - 45 = 0$$
$$\left(x - 9\right)\left(x + 5\right) = 0$$
$$x = 9 \qquad \left(x = -5 \text{ is impossible}\right)$$

36. Using the area formula:

$$\frac{1}{2}\left(2x + 4\right)\left(x\right) = 35$$
$$x^2 + 2x = 35$$
$$x^2 + 2x - 35 = 0$$
$$\left(x + 7\right)\left(x - 5\right) = 0$$
$$x = 5 \quad \left(x = -7 \text{ is impossible}\right)$$

37. Using the area formula:

$$x\left(2x - 4\right) = 48$$
$$2x^2 - 4x = 48$$
$$2x^2 - 4x - 48 = 0$$
$$x^2 - 2x - 24 = 0$$
$$\left(x - 6\right)\left(x + 4\right) = 0$$
$$x = 6 \quad \left(x = -4 \text{ is impossible}\right)$$

38. The revenue is given by: $R(x) = xp = x(25 - 0.2x) = 25x - 0.2x^2$

39. The profit is given by:

$$P(x) = R(x) - C(x) = (25x - 0.2x^2) - (2x + 100) = 25x - 0.2x^2 - 2x - 100 = -0.2x^2 + 23x - 100$$

40. Substituting $x = 100$ into the revenue equation: $R(100) = 25(100) - 0.2(100)^2 = 2500 - 2000 = \$500$.
The revenue is \$500.

41. Substituting $x = 100$ into the cost equation: $C(100) = 2(100) + 100 = 200 + 100 = \$300$. The cost is \$300.

42. Substituting $x = 100$ into the profit equation: $P(100) = -0.2(100)^2 + 23(100) - 100 = -2000 + 2300 - 100 = \$200$
The profit is \$200. Note that this also could have been found by subtracting the answers from problems 40 and 41.

Chapter 6
Rational Expressions and Rational Functions

Getting Ready for Chapter 6

1. Reducing the fraction: $\dfrac{6}{8} = \dfrac{3}{4}$

2. Reducing the fraction: $\dfrac{10}{25} = \dfrac{2}{5}$

3. Reducing the fraction: $\dfrac{200}{5} = 40$

4. Reducing the fraction: $\dfrac{240}{6} = 40$

5. Multiplying: $250(3.14) = 785$

6. Multiplying: $165(3.14) = 518.1$

7. Multiplying: $-1(5 - x) = -5 + x = x - 5$

8. Multiplying: $-1(a - b) - -a + b = b - a$

9. Multiplying: $4x(2x - 4) = 8x^2 - 16x$

10. Simplifying: $\dfrac{0 + 6}{0 - 3} = \dfrac{6}{-3} = -2$

11. Simplifying: $\dfrac{-6 + 6}{-6 - 3} = \dfrac{0}{-9} = 0$

12. Simplifying: $\dfrac{3 + 6}{3 - 3} = \dfrac{9}{0}$, which is undefined

13. Simplifying: $\dfrac{8x^3 y^5}{-2x^2 y} = -4xy^4$

14. a. Evaluating when $t = 15$: $\dfrac{250}{15} \approx 16.7$

   b. Evaluating when $t = 25$: $\dfrac{250}{25} = 10$

   c. Evaluating when $t = 35$: $\dfrac{250}{35} \approx 7.1$

15. Factoring: $x^2 - a^2 = (x + a)(x - a)$

16. Factoring: $x^2 - 6x + 9 = (x - 3)(x - 3) = (x - 3)^2$

17. Factoring: $x^3 - x^2 y = x^2(x - y)$

18. Factoring: $xy - y = y(x - 1)$

19. Factoring: $2y^2 - 2 = 2(y^2 - 1) - 2(y + 1)(y - 1)$

20. Factoring: $8y^3 - 27 = (2y - 3)(4y^2 + 6y + 9)$

175

6.1 Basic Properties and Reducing to Lowest Terms

1. a. Simplifying: $\dfrac{6+1}{36-1} = \dfrac{7}{35} = \dfrac{1}{5}$

    b. Simplifying: $\dfrac{x+3}{x^2-9} = \dfrac{x+3}{(x+3)(x-3)} = \dfrac{1}{x-3}$ if $x \neq -3, 3$

    c. Simplifying: $\dfrac{x^2-3x}{x^2-9} = \dfrac{x(x-3)}{(x+3)(x-3)} = \dfrac{x}{x+3}$ if $x \neq -3, 3$

    d. Simplifying: $\dfrac{x^3-27}{x^2-9} = \dfrac{(x-3)(x^2+3x+9)}{(x+3)(x-3)} = \dfrac{x^2+3x+9}{x+3}$ if $x \neq -3, 3$

3. Finding each function value:

    $h(0) = \dfrac{0-3}{0+1} = \dfrac{-3}{1} = -3$ $\qquad$ $h(-3) = \dfrac{-3-3}{-3+1} = \dfrac{-6}{-2} = 3$

    $h(3) = \dfrac{3-3}{3+1} = \dfrac{0}{4} = 0$ $\qquad$ $h(-1) = \dfrac{-1-3}{-1+1} = \dfrac{-4}{0}$, which is undefined

    $h(1) = \dfrac{1-3}{1+1} = \dfrac{-2}{2} = -1$

5. The domain is $\{x \mid x \neq 1\}$.

7. Setting the denominator equal to 0: $\qquad\qquad$ 9. Setting the denominator equal to 0:

    $$t^2 - 16 = 0$$
    $$(t+4)(t-4) = 0$$
    $$t = -4, 4$$

    The domain is $\{t \mid t \neq -4, t \neq 4\}$.

    $$3x - 15 = 0$$
    $$3x = 15$$
    $$x = 5$$

    The domain is $\{x \mid x \neq 5\}$.

11. The domain is all real numbers. $\qquad\qquad$ 13. The domain is $\{x \mid x \neq 0\}$.

15. Setting the denominator equal to 0:

    $$x^2 - x - 20 = 0$$
    $$(x+4)(x-5) = 0$$
    $$x = -4, 5$$

    The domain is $\{x \mid x \neq -4, x \neq 5\}$.

17. Reducing to lowest terms: $\dfrac{x^2-16}{6x+24} = \dfrac{(x+4)(x-4)}{6(x+4)} = \dfrac{x-4}{6}$

19. Reducing to lowest terms: $\dfrac{12x-9y}{3x^2+3xy} = \dfrac{3(4x-3y)}{3x(x+y)} = \dfrac{4x-3y}{x(x+y)}$

21. Reducing to lowest terms: $\dfrac{a^4-81}{a-3} = \dfrac{(a^2+9)(a^2-9)}{a-3} = \dfrac{(a^2+9)(a+3)(a-3)}{a-3} = (a^2+9)(a+3)$

23. Reducing to lowest terms: $\dfrac{a^2-4a-12}{a^2+8a+12} = \dfrac{(a-6)(a+2)}{(a+6)(a+2)} = \dfrac{a-6}{a+6}$

25. Reducing to lowest terms: $\dfrac{20y^2-45}{10y^2-5y-15} = \dfrac{5(4y^2-9)}{5(2y^2-y-3)} = \dfrac{5(2y+3)(2y-3)}{5(2y-3)(y+1)} = \dfrac{2y+3}{y+1}$

27. Reducing to lowest terms: $\dfrac{a^3+b^3}{a^2-b^2} = \dfrac{(a+b)(a^2-ab+b^2)}{(a+b)(a-b)} = \dfrac{a^2-ab+b^2}{a-b}$

29. Reducing to lowest terms: $\dfrac{8x^4 - 8x}{4x^4 + 4x^3 + 4x^2} = \dfrac{8x(x^3 - 1)}{4x^2(x^2 + x + 1)} = \dfrac{8x(x-1)(x^2 + x + 1)}{4x^2(x^2 + x + 1)} = \dfrac{2(x-1)}{x}$

31. Reducing to lowest terms: $\dfrac{6x^2 + 7xy - 3y^2}{6x^2 + xy - y^2} = \dfrac{(2x+3y)(3x-y)}{(2x+y)(3x-y)} = \dfrac{2x+3y}{2x+y}$

33. Reducing to lowest terms: $\dfrac{ax + 2x + 3a + 6}{ay + 2y - 4a - 8} = \dfrac{x(a+2) + 3(a+2)}{y(a+2) - 4(a+2)} = \dfrac{(a+2)(x+3)}{(a+2)(y-4)} = \dfrac{x+3}{y-4}$

35. Reducing to lowest terms: $\dfrac{x^2 + bx - 3x - 3b}{x^2 - 2bx - 3x + 6b} = \dfrac{x(x+b) - 3(x+b)}{x(x-2b) - 3(x-2b)} = \dfrac{(x+b)(x-3)}{(x-2b)(x-3)} = \dfrac{x+b}{x-2b}$

37. Reducing to lowest terms:

$$\dfrac{x^3 + 3x^2 - 4x - 12}{x^2 + x - 6} = \dfrac{x^2(x+3) - 4(x+3)}{(x+3)(x-2)} = \dfrac{(x+3)(x^2 - 4)}{(x+3)(x-2)} = \dfrac{(x+3)(x+2)(x-2)}{(x+3)(x-2)} = x+2$$

39. Reducing to lowest terms: $\dfrac{4x^4 - 25}{6x^3 - 4x^2 + 15x - 10} = \dfrac{(2x^2 + 5)(2x^2 - 5)}{2x^2(3x-2) + 5(3x-2)} = \dfrac{(2x^2 + 5)(2x^2 - 5)}{(3x-2)(2x^2 + 5)} = \dfrac{2x^2 - 5}{3x-2}$

41. Reducing to lowest terms: $\dfrac{x^3 - 8}{x^2 - 4} = \dfrac{(x-2)(x^2 + 2x + 4)}{(x-2)(x+2)} = \dfrac{x^2 + 2x + 4}{x+2}$

43. Reducing to lowest terms: $\dfrac{64 + t^3}{16 - 4t + t^2} = \dfrac{(4+t)(16 - 4t + t^2)}{16 - 4t + t^2} = 4+t$

45. Reducing to lowest terms: $\dfrac{8x^3 - 27}{4x^2 - 9} = \dfrac{(2x-3)(4x^2 + 6x + 9)}{(2x-3)(2x+3)} = \dfrac{4x^2 + 6x + 9}{2x+3}$

47. Reducing to lowest terms: $\dfrac{x-4}{4-x} = \dfrac{x-4}{-1(x-4)} = -1$

49. Reducing to lowest terms: $\dfrac{y^2 - 36}{6 - y} = \dfrac{(y+6)(y-6)}{-1(y-6)} = -(y+6)$

51. Reducing to lowest terms: $\dfrac{1 - 9a^2}{9a^2 - 6a + 1} = \dfrac{-1(9a^2 - 1)}{(3a-1)^2} = \dfrac{-1(3a+1)(3a-1)}{(3a-1)^2} = -\dfrac{3a+1}{3a-1}$

53. Reducing to lowest terms: $\dfrac{(3x-5) - (3a-5)}{x-a} = \dfrac{3x - 3a}{x-a} = \dfrac{3(x-a)}{x-a} = 3$

55. Reducing to lowest terms: $\dfrac{(x^2 - 4) - (a^2 - 4)}{x-a} = \dfrac{x^2 - a^2}{x-a} = \dfrac{(x+a)(x-a)}{x-a} = x+a$

57. Evaluating the functions:

$$f(0) = \dfrac{0^2 - 4}{0 - 2} = \dfrac{-4}{-2} = 2$$
$$g(0) = 0 + 2 = 2$$

59. Evaluating the functions:

$$f(2) = \dfrac{2^2 - 4}{2 - 2} = \dfrac{0}{0}, \text{ which is undefined}$$
$$g(2) = 2 + 2 = 4$$

61. Evaluating the functions:

$$f(0) = \dfrac{0^2 - 1}{0 - 1} = \dfrac{-1}{-1} = 1$$
$$g(0) = 0 + 1 = 1$$

63. Evaluating the functions:

$$f(2) = \dfrac{2^2 - 1}{2 - 1} = \dfrac{3}{1} = 3$$
$$g(2) = 2 + 1 = 3$$

65. The graph of $y = x + 2$ contains the point (2,4), while the other graph does not.

67. Completing the table:

Weeks x	Weight (lb) $W(x)$
0	200
1	194
4	184
12	173
24	168

69. Dividing: $\dfrac{\pi \cdot 65 \text{ feet}}{30 \text{ seconds}} \approx 6.8$ feet per second

71. **a.** The domain is $\{t \mid 20 \le t \le 50\}$.
    b. Graphing the function:

$r(t) = 204/t$

73. **a.** The domain is $\{d \mid 1 \le d \le 6\}$.
    b. Graphing the function:

$I(d) = 120/d^2$

75. Using $r = 2$ inches: $\dfrac{2(3.14) \cdot 2 \text{ inches}}{\frac{1}{300} \text{ minutes}} \approx 3,768$ inches per minute

Using $r = 1.5$ inches: $\dfrac{2(3.14) \cdot 1.5 \text{ inches}}{\frac{1}{300} \text{ minutes}} \approx 2,826$ inches per minute

77. Subtracting: $\left(4x^2 - 5x + 5\right) - \left(x^2 + 2x + 1\right) = 4x^2 - 5x + 5 - x^2 - 2x - 1 = 3x^2 - 7x + 4$

79. Subtracting: $(10x - 11) - (10x - 20) = 10x - 11 - 10x + 20 = 9$

81. Subtracting: $\left(4x^3\right) - \left(4x^3 - 8x^2\right) = 4x^3 - 4x^3 + 8x^2 = 8x^2$

83.   Multiply the first equation by 3 and the second equation by 2:

$$15x + 6y = -9$$
$$2x - 6y = -42$$

Adding:

$$17x = -51$$
$$x = -3$$

Substituting into the first equation:

$$5(-3) + 2y = -3$$
$$-15 + 2y = -3$$
$$2y = 12$$
$$y = 6$$

The solution is $(-3, 6)$.

85.   To clear each equation of fractions, multiply the first equation by 4 and the second equation by 15:

$$x + 2y = 3$$
$$5x + 12y = 2$$

Multiply the first equation by -5:

$$-5x - 10y = -15$$
$$5x + 12y = 2$$

Adding yields:

$$2y = -13$$
$$y = -\frac{13}{2}$$

Substituting into the first equation:

$$x + 2\left(-\frac{13}{2}\right) = 3$$
$$x - 13 = 3$$
$$x = 16$$

The solution is $\left(16, -\frac{13}{2}\right)$.

87.   Substituting into the first equation:

$$2x - 5(x + 2) = 2$$
$$2x - 5x - 10 = 2$$
$$-3x = 12$$
$$x = -4$$
$$y = -4 + 2 = -2$$

The solution is $(-4, -2)$.

89. Form the augmented matrix: $\begin{bmatrix} 1 & -2 & -2 & \vert & 4 \\ 3 & 3 & 1 & \vert & 2 \\ 1 & 1 & -1 & \vert & 6 \end{bmatrix}$

Add -3 times row 1 to row 2 and -1 times row 1 to row 3: $\begin{bmatrix} 1 & -2 & -2 & \vert & 4 \\ 0 & 9 & 7 & \vert & -10 \\ 0 & 3 & 1 & \vert & 2 \end{bmatrix}$

Dividing row 2 by 9: $\begin{bmatrix} 1 & -2 & -2 & \vert & 4 \\ 0 & 1 & \dfrac{7}{9} & \vert & -\dfrac{10}{9} \\ 0 & 3 & 1 & \vert & 2 \end{bmatrix}$

Adding -3 times row 2 to row 3: $\begin{bmatrix} 1 & -2 & -2 & \vert & 4 \\ 0 & 1 & \dfrac{7}{9} & \vert & -\dfrac{10}{9} \\ 0 & 0 & -\dfrac{4}{3} & \vert & \dfrac{16}{3} \end{bmatrix}$

Multiplying row 3 by $-\dfrac{3}{4}$: $\begin{bmatrix} 1 & -2 & -2 & \vert & 4 \\ 0 & 1 & \dfrac{7}{9} & \vert & -\dfrac{10}{9} \\ 0 & 0 & 1 & \vert & -4 \end{bmatrix}$

So $z = -4$. Substituting to find y:

$$y + \frac{7}{9}(-4) = -\frac{10}{9}$$

$$y - \frac{28}{9} = -\frac{10}{9}$$

$$y = 2$$

Substituting to find x:

$$x - 2(2) - 2(-4) = 4$$

$$x + 4 = 4$$

$$x = 0$$

The solution is $(0, 2, -4)$.

91. Dividing: $\dfrac{10x^5}{5x^2} = 2x^{5-2} = 2x^3$

93. Dividing: $\dfrac{4x^4 y^3}{-2x^2 y} = -2x^{4-2} y^{3-1} = -2x^2 y^2$

95. Dividing: $4{,}628 \div 25 = 185.12$

97. Multiplying: $2x^2(2x - 4) = 4x^3 - 8x^2$

99. Multiplying:

$$(2x - 4)(2x^2 + 4x + 5) = 2x(2x^2 + 4x + 5) - 4(2x^2 + 4x + 5)$$
$$= 4x^3 + 8x^2 + 10x - 8x^2 - 16x - 20$$
$$= 4x^3 - 6x - 20$$

101. Subtracting: $(2x^2 - 7x + 9) - (2x^2 - 4x) = 2x^2 - 7x + 9 - 2x^2 + 4x = -3x + 9$

103. Factoring: $x^2 - a^2 = (x + a)(x - a)$

105. Factoring: $x^2 - 6xy - 7y^2 = (x - 7y)(x + y)$

107. **a.**   From the graph: $f(2) = 2$ **b.**   From the graph: $f(-1) = -4$

c.   From the graph: $f(0)$ is undefined   **d.**   From the graph: $g(3) = 2$

e.   From the graph: $g(6) = 1$ **f.**   From the graph: $g(-1) = -6$

g.   From the graph: $f(g(6)) = f(1) = 4$ **h.**   From the graph: $g(f(-2)) = g(-2) = -3$

6.2 Division of Polynomials and Difference Quotients

1. Dividing: $\dfrac{4x^3 - 8x^2 + 6x}{2x} = \dfrac{4x^3}{2x} - \dfrac{8x^2}{2x} + \dfrac{6x}{2x} = 2x^2 - 4x + 3$

3. Dividing: $\dfrac{10x^4 + 15x^3 - 20x^2}{-5x^2} = \dfrac{10x^4}{-5x^2} + \dfrac{15x^3}{-5x^2} - \dfrac{20x^2}{-5x^2} = -2x^2 - 3x + 4$

5. Dividing: $\dfrac{8y^5 + 10y^3 - 6y}{4y^3} = \dfrac{8y^5}{4y^3} + \dfrac{10y^3}{4y^3} - \dfrac{6y}{4y^3} = 2y^2 + \dfrac{5}{2} - \dfrac{3}{2y^2}$

7. Dividing: $\dfrac{5x^3 - 8x^2 - 6x}{-2x^2} = \dfrac{5x^3}{-2x^2} - \dfrac{8x^2}{-2x^2} - \dfrac{6x}{-2x^2} = -\dfrac{5}{2}x + 4 + \dfrac{3}{x}$

9. Dividing: $\dfrac{28a^3b^5 + 42a^4b^3}{7a^2b^2} = \dfrac{28a^3b^5}{7a^2b^2} + \dfrac{42a^4b^3}{7a^2b^2} = 4ab^3 + 6a^2b$

11. Dividing: $\dfrac{10x^3y^2 - 20x^2y^3 - 30x^3y^3}{-10x^2y} = \dfrac{10x^3y^2}{-10x^2y} - \dfrac{20x^2y^3}{-10x^2y} - \dfrac{30x^3y^3}{-10x^2y} = -xy + 2y^2 + 3xy^2$

13. Dividing by factoring: $\dfrac{x^2 - x - 6}{x - 3} = \dfrac{(x-3)(x+2)}{x-3} = x + 2$

15. Dividing by factoring: $\dfrac{2a^2 - 3a - 9}{2a + 3} = \dfrac{(2a+3)(a-3)}{2a+3} = a - 3$

17. Dividing by factoring: $\dfrac{5x^2 - 14xy - 24y^2}{x - 4y} = \dfrac{(5x+6y)(x-4y)}{x-4y} = 5x + 6y$

19. Dividing by factoring: $\dfrac{x^3 - y^3}{x - y} = \dfrac{(x-y)(x^2 + xy + y^2)}{x-y} = x^2 + xy + y^2$

21. Dividing by factoring: $\dfrac{y^4 - 16}{y - 2} = \dfrac{(y^2+4)(y^2-4)}{y-2} = \dfrac{(y^2+4)(y+2)(y-2)}{y-2} = (y^2 + 4)(y + 2)$

23. Dividing by factoring:

$$\dfrac{x^3 + 2x^2 - 25x - 50}{x - 5} = \dfrac{x^2(x+2) - 25(x+2)}{x-5} = \dfrac{(x+2)(x^2-25)}{x-5} = \dfrac{(x+2)(x+5)(x-5)}{x-5} = (x+2)(x+5)$$

25. **a.**   Evaluating the formula: $\dfrac{f(x+h) - f(x)}{h} = \dfrac{4x + 4h - 4x}{h} = \dfrac{4h}{h} = 4$

b.   Evaluating the formula: $\dfrac{f(x) - f(a)}{x - a} = \dfrac{4x - 4a}{x - a} = \dfrac{4(x-a)}{x-a} = 4$

27. **a.**   Evaluating the formula: $\dfrac{f(x+h) - f(x)}{h} = \dfrac{5(x+h) + 3 - (5x+3)}{h} = \dfrac{5x + 5h + 3 - 5x - 3}{h} = \dfrac{5h}{h} = 5$

b.   Evaluating the formula: $\dfrac{f(x) - f(a)}{x - a} = \dfrac{(5x+3) - (5a+3)}{x - a} = \dfrac{5x - 5a}{x - a} = \dfrac{5(x-a)}{x-a} = 5$

29.   a.   Evaluating the formula:

$$\frac{f(x+h)-f(x)}{h}=\frac{(x+h)^2-x^2}{h}=\frac{x^2+2xh+h^2-x^2}{h}=\frac{2xh+h^2}{h}=\frac{h(2x+h)}{h}=2x+h$$

b.   Evaluating the formula: $\dfrac{f(x)-f(a)}{x-a}=\dfrac{x^2-a^2}{x-a}=\dfrac{(x+a)(x-a)}{x-a}=x+a$

31.   a.   Evaluating the formula:

$$\frac{f(x+h)-f(x)}{h}=\frac{(x+h)^2+1-\left(x^2+1\right)}{h}=\frac{x^2+2xh+h^2+1-x^2-1}{h}=\frac{2xh+h^2}{h}=\frac{h(2x+h)}{h}=2x+h$$

b.   Evaluating the formula: $\dfrac{f(x)-f(a)}{x-a}=\dfrac{\left(x^2+1\right)-\left(a^2+1\right)}{x-a}=\dfrac{x^2-a^2}{x-a}=\dfrac{(x+a)(x-a)}{x-a}=x+a$

33.   a.   Evaluating the formula:

$$\frac{f(x+h)-f(x)}{h}=\frac{(x+h)^2-3(x+h)+4-\left(x^2-3x+4\right)}{h}$$

$$=\frac{x^2+2xh+h^2-3x-3h+4-x^2+3x-4}{h}$$

$$=\frac{2xh+h^2-3h}{h}$$

$$=\frac{h(2x+h-3)}{h}$$

$$=2x+h-3$$

b.   Evaluating the formula:

$$\frac{f(x)-f(a)}{x-a}=\frac{\left(x^2-3x+4\right)-\left(a^2-3a+4\right)}{x-a}$$

$$=\frac{x^2-a^2-3x+3a}{x-a}$$

$$=\frac{(x+a)(x-a)-3(x-a)}{x-a}$$

$$=\frac{(x-a)(x+a-3)}{x-a}$$

$$=x+a-3$$

35.   a.   Evaluating the formula:

$$\frac{f(x+h)-f(x)}{h}=\frac{2(x+h)^2+3(x+h)-4-\left(2x^2+3x-4\right)}{h}$$

$$=\frac{2x^2+4xh+2h^2+3x+3h-4-2x^2-3x+4}{h}$$

$$=\frac{4xh+2h^2+3h}{h}$$

$$=\frac{h(4x+2h+3)}{h}$$

$$=4x+2h+3$$

b. Evaluating the formula:

$$\frac{f(x)-f(a)}{x-a}=\frac{\left(2x^2+3x-4\right)-\left(2a^2+3a-4\right)}{x-a}$$

$$=\frac{2x^2-2a^2+3x-3a}{x-a}$$

$$=\frac{2(x+a)(x-a)+3(x-a)}{x-a}$$

$$=\frac{(x-a)(2x+2a+3)}{x-a}$$

$$=2x+2a+3$$

37. a. Evaluating the formula:

$$\frac{f(x+h)-f(x)}{h}=\frac{-3(x+h)^2+2(x+h)-1-\left(-3x^2+2x-1\right)}{h}$$

$$=\frac{-3x^2-6xh-3h^2+2x+2h-1+3x^2-2x+1}{h}$$

$$=\frac{-6xh-3h^2+2h}{h}$$

$$=\frac{h(-6x-3h+2)}{h}$$

$$=-6x-3h+2$$

b. Evaluating the formula:

$$\frac{f(x)-f(a)}{x-a}=\frac{\left(-3x^2+2x-1\right)-\left(-3a^2+2a-1\right)}{x-a}$$

$$=\frac{-3x^2+3a^2+2x-2a}{x-a}$$

$$=\frac{-3(x+a)(x-a)+2(x-a)}{x-a}$$

$$=\frac{(x-a)(-3x-3a+2)}{x-a}$$

$$=-3x-3a+2$$

39. Dividing using long division:

$$\begin{array}{r} x-7 \\ x+2\overline{)x^2-5x-7} \\ \underline{x^2+2x} \\ -7x-7 \\ \underline{-7x-14} \\ 7 \end{array}$$

The quotient is $x-7+\dfrac{7}{x+2}$.

41. Dividing using long division:

$$\begin{array}{r} 2x^2-5x+1 \\ x+1\overline{)2x^3-3x^2-4x+5} \\ \underline{2x^3+2x^2} \\ -5x^2-4x \\ \underline{-5x^2-5x} \\ x+5 \\ \underline{x+1} \\ 4 \end{array}$$

The quotient is $2x^2-5x+1+\dfrac{4}{x+1}$.

43. Dividing using long division:

$$
\begin{array}{r}
y^2 - 3y - 13 \\
2y - 3 \overline{\smash{)}\,2y^3 - 9y^2 - 17y + 39} \\
\underline{2y^3 - 3y^2} \\
-6y^2 - 17y \\
\underline{-6y^2 + 9y} \\
-26y + 39 \\
\underline{-26y + 39} \\
0
\end{array}
$$

The quotient is $y^2 - 3y - 13$.

47. Dividing using long division:

$$
\begin{array}{r}
a^3 + 2a^2 + 4a + 6 \\
a - 2 \overline{\smash{)}\,a^4 + 0a^3 + 0a^2 - 2a + 5} \\
\underline{a^4 - 2a^3} \\
2a^3 + 0a^2 \\
\underline{2a^3 - 4a^2} \\
4a^2 - 2a \\
\underline{4a^2 - 8a} \\
6a + 5 \\
\underline{6a - 12} \\
17
\end{array}
$$

The quotient is $a^3 + 2a^2 + 4a + 6 + \dfrac{17}{a - 2}$.

45. Dividing using long division:

$$
\begin{array}{r}
3y^2 + 6y + 8 \\
2y - 4 \overline{\smash{)}\,6y^3 + 0y^2 - 8y + 5} \\
\underline{6y^3 - 12y^2} \\
12y^2 - 8y \\
\underline{12y^2 - 24y} \\
16y + 5 \\
\underline{16y - 32} \\
37
\end{array}
$$

The quotient is $3y^2 + 6y + 8 + \dfrac{37}{2y - 4}$.

49. Dividing using long division:

$$
\begin{array}{r}
y^3 + 2y^2 + 4y + 8 \\
y - 2 \overline{\smash{)}\,y^4 + 0y^3 + 0y^2 + 0y - 16} \\
\underline{y^4 - 2y^3} \\
2y^3 + 0y^2 \\
\underline{2y^3 - 4y^2} \\
4y^2 + 0y \\
\underline{4y^2 - 8y} \\
8y - 16 \\
\underline{8y - 16} \\
0
\end{array}
$$

The quotient is $y^3 + 2y^2 + 4y + 8$.

51. Dividing: $h(x) = \dfrac{x^2 - 36}{4x - 24} = \dfrac{(x+6)(x-6)}{4(x-6)} = \dfrac{x+6}{4}$. The domain is $\{x \mid x \neq 6\}$.

53. Dividing: $h(x) = \dfrac{x^2 - 16x + 64}{x^2 - 4x - 32} = \dfrac{(x-8)^2}{(x-8)(x+4)} = \dfrac{x-8}{x+4}$. The domain is $\{x \mid x \neq -4, x \neq 8\}$.

55. Dividing: $h(x) = \dfrac{x^3 - 27}{x - 3} = \dfrac{(x-3)(x^2 + 3x + 9)}{x - 3} = x^2 + 3x + 9$. The domain is $\{x \mid x \neq 3\}$.

57. **a.** Using long division:

$$
\begin{array}{r}
x^2 - x + 3 \\
x - 2 \overline{\smash{)}\,x^3 - 3x^2 + 5x - 6} \\
\underline{x^3 - 2x^2} \\
-x^2 + 5x \\
\underline{-x^2 + 2x} \\
3x - 6 \\
\underline{3x - 6} \\
0
\end{array}
$$

Since the remainder is 0, $x - 2$ is a factor of $x^3 - 3x^2 + 5x - 6$.

Also note that: $P(2) = (2)^3 - 3(2)^2 + 5(2) - 6 = 8 - 12 + 10 - 6 = 0$

b.   Using long division:

$$x-5 \overline{\smash{\big)}\ \begin{array}{c} x^3 - x + 1 \\ x^4 - 5x^3 - x^2 + 6x - 5 \end{array}}$$

$$\begin{array}{r} \underline{x^4 - 5x^3} \\ -x^2 + 6x \\ \underline{-x^2 + 5x} \\ x - 5 \\ \underline{x - 5} \\ 0 \end{array}$$

Since the remainder is 0, $x - 5$ is a factor of $x^4 - 5x^3 - x^2 + 6x - 5$.

Also note that: $P(5) = (5)^4 - 5(5)^3 - (5)^2 + 6(5) - 5 = 625 - 625 - 25 + 30 - 5 = 0$

59.   a.   Using long division:

$$x+4 \overline{\smash{\big)}\ \begin{array}{c} x^2 + 6x + 5 \\ x^3 + 10x^2 + 29x + 20 \end{array}}$$

$$\begin{array}{r} \underline{x^3 + 4x^2} \\ 6x^2 + 29x \\ \underline{6x^2 + 24x} \\ 5x + 20 \\ \underline{5x + 20} \\ 0 \end{array}$$

So $x^3 + 10x^2 + 29x + 20 = (x+4)(x^2 + 6x + 5) = (x+4)(x+5)(x+1)$.

b.   Reducing the fraction: $\dfrac{x^3 + 10x^2 + 29x + 20}{x+4} = \dfrac{(x+4)(x+5)(x+1)}{x+4} = (x+5)(x+1)$

61.   a.   Using long division:

$$x+2 \overline{\smash{\big)}\ \begin{array}{c} x^2 + x - 12 \\ x^3 + 3x^2 - 10x - 24 \end{array}}$$

$$\begin{array}{r} \underline{x^3 + 2x^2} \\ x^2 - 10x \\ \underline{x^2 + 2x} \\ -12x - 24 \\ \underline{-12x - 24} \\ 0 \end{array}$$

So $x^3 + 3x^2 - 10x - 24 = (x+2)(x^2 + x - 12) = (x+2)(x+4)(x-3)$.

b.   Reducing the fraction: $\dfrac{x^3 + 3x^2 - 10x - 24}{x+2} = \dfrac{(x+2)(x+4)(x-3)}{x+2} = (x+4)(x-3)$

63.   Evaluating the function: $P(3) = (3)^2 + 4(3) - 8 = 9 + 12 - 8 = 13$

The remainder is the same (13).

65. **a.** Completing the table:

x	1	5	10	15	20
$C(x)$	2.15	2.75	3.50	4.25	5.00

b. The average cost function is $\overline{C}(x) = \dfrac{2}{x} + 0.15$.

c. Completing the table:

x	1	5	10	15	20
$\overline{C}(x)$	2.15	0.55	0.35	0.28	0.25

d. The average cost function decreases.

67. Dividing: $\dfrac{3}{5} \div \dfrac{2}{7} = \dfrac{3}{5} \cdot \dfrac{7}{2} = \dfrac{21}{10}$

69. Dividing: $\dfrac{3}{4} \div \dfrac{6}{11} = \dfrac{3}{4} \cdot \dfrac{11}{6} = \dfrac{11}{8}$

71. Dividing: $\dfrac{4}{9} \div 8 = \dfrac{4}{9} \cdot \dfrac{1}{8} = \dfrac{1}{18}$

73. Dividing: $8 \div \dfrac{1}{4} = 8 \cdot 4 = 32$

75. Simplifying: $\left(\dfrac{1}{3}\right)^{-2} + \left(\dfrac{1}{2}\right)^{-3} = 3^2 + 2^3 = 9 + 8 = 17$

77. Simplifying: $\left(9x^{-4}y^9\right)^{-2}\left(3x^2y^{-1}\right)^4 = 9^{-2}x^8y^{-18} \cdot 3^4 x^8 y^{-4} = \dfrac{1}{81} \cdot 81 x^{16} y^{-22} = \dfrac{x^{16}}{y^{22}}$

79. Multiplying: $\dfrac{6}{7} \cdot \dfrac{14}{18} = \dfrac{6}{7} \cdot \dfrac{2 \cdot 7}{3 \cdot 6} = \dfrac{2}{3}$

81. Multiplying: $5y^2 \cdot 4x^2 = 20x^2y^2$

83. Multiplying: $9x^4 \cdot 8y^5 = 72x^4y^5$

85. Factoring: $x^2 - 4 = (x+2)(x-2)$

87. Factoring: $x^3 - x^2y = x^2(x-y)$

89. Factoring: $2y^2 - 2 = 2\left(y^2 - 1\right) = 2(y+1)(y-1)$

91. Using long division:

$$
\begin{array}{r}
4x^3 - x^2 + 3 \\
x^2 - 5 \overline{)4x^5 - x^4 - 20x^3 + 8x^2 - 15} \\
\underline{4x^5 - 20x^3 } \\
-x^4 + 8x^2 \\
\underline{-x^4 + 5x^2 } \\
3x^2 - 15 \\
\underline{3x^2 - 15} \\
0
\end{array}
$$

The quotient is $4x^3 - x^2 + 3$.

93. Using long division:

$$
\begin{array}{r}
0.5x^2 - 0.4x + 0.3 \\
x + 0.2 \overline{)0.5x^3 - 0.3x^2 + 0.22x + 0.06} \\
\underline{0.5x^3 + 0.1x^2 } \\
-0.4x^2 + 0.22x \\
\underline{-0.4x^2 - 0.08x } \\
0.3x + 0.06 \\
\underline{0.3x + 0.06} \\
0
\end{array}
$$

The quotient is $0.5x^2 - 0.4x + 0.3$.

95. Using long division:

$$
\begin{array}{r}
\frac{3}{2}x - \frac{5}{2} \\
2x + 4 \overline{)3x^2 + x - 9} \\
\underline{3x^2 + 6x } \\
-5x - 9 \\
\underline{-5x - 10} \\
1
\end{array}
$$

The quotient is $\dfrac{3}{2}x - \dfrac{5}{2} + \dfrac{1}{2x+4}$.

97. Using long division:

$$
\begin{array}{r}
\frac{2}{3}x + \frac{1}{3} \\
3x - 1 \overline{)2x^2 + \frac{1}{3}x + \frac{5}{3}} \\
\underline{2x^2 - \frac{2}{3}x } \\
x + \frac{5}{3} \\
\underline{x - \frac{1}{3}} \\
2
\end{array}
$$

The quotient is $\dfrac{2}{3}x + \dfrac{1}{3} + \dfrac{2}{3x-1}$.

6.3 Multiplication and Division of Rational Expressions

1. Performing the operations: $\dfrac{2}{9} \cdot \dfrac{3}{4} = \dfrac{2}{3 \cdot 3} \cdot \dfrac{3}{2 \cdot 2} = \dfrac{1}{2 \cdot 3} = \dfrac{1}{6}$

3. Performing the operations: $\dfrac{3}{4} \div \dfrac{1}{3} = \dfrac{3}{4} \cdot \dfrac{3}{1} = \dfrac{9}{4}$

5. Performing the operations: $\dfrac{3}{7} \cdot \dfrac{14}{24} \div \dfrac{1}{2} = \dfrac{1}{4} \div \dfrac{1}{2} = \dfrac{1}{4} \cdot \dfrac{2}{1} = \dfrac{2}{4} = \dfrac{1}{2}$

7. Performing the operations: $\dfrac{10x^2}{5y^2} \cdot \dfrac{15y^3}{2x^4} = \dfrac{150x^2 y^3}{10x^4 y^2} = \dfrac{15y}{x^2}$

9. Performing the operations: $\dfrac{11a^2 b}{5ab^2} \div \dfrac{22a^3 b^2}{10ab^4} = \dfrac{11a^2 b}{5ab^2} \cdot \dfrac{10ab^4}{22a^3 b^2} = \dfrac{110a^3 b^5}{110a^4 b^4} = \dfrac{b}{a}$

11. Performing the operations: $\dfrac{6x^2}{5y^3} \cdot \dfrac{11z^2}{2x^2} \div \dfrac{33z^5}{10y^8} = \dfrac{33z^2}{5y^3} \cdot \dfrac{10y^8}{33z^5} = \dfrac{2y^8 z^2}{y^3 z^5} = \dfrac{2y^5}{z^3}$

13. Performing the operations: $\dfrac{x^2 - 9}{x^2 - 4} \cdot \dfrac{x - 2}{x - 3} = \dfrac{(x+3)(x-3)}{(x+2)(x-2)} \cdot \dfrac{x-2}{x-3} = \dfrac{x+3}{x+2}$

15. Performing the operations: $\dfrac{y^2 - 1}{y + 2} \cdot \dfrac{y^2 + 5y + 6}{y^2 + 2y - 3} = \dfrac{(y+1)(y-1)}{y+2} \cdot \dfrac{(y+2)(y+3)}{(y+3)(y-1)} = y + 1$

17. Performing the operations: $\dfrac{3x - 12}{x^2 - 4} \cdot \dfrac{x^2 + 6x + 8}{x - 4} = \dfrac{3(x-4)}{(x+2)(x-2)} \cdot \dfrac{(x+4)(x+2)}{x-4} = \dfrac{3(x+4)}{x-2}$

19. Performing the operations: $\dfrac{5x + 2y}{25x^2 - 5xy - 6y^2} \cdot \dfrac{20x^2 - 7xy - 3y^2}{4x + y} = \dfrac{5x+2y}{(5x+2y)(5x-3y)} \cdot \dfrac{(5x-3y)(4x+y)}{4x+y} = 1$

21. Performing the operations:

$\dfrac{a^2 - 5a + 6}{a^2 - 2a - 3} \div \dfrac{a - 5}{a^2 + 3a + 2} = \dfrac{a^2 - 5a + 6}{a^2 - 2a - 3} \cdot \dfrac{a^2 + 3a + 2}{a - 5} = \dfrac{(a-3)(a-2)}{(a-3)(a+1)} \cdot \dfrac{(a+2)(a+1)}{a-5} = \dfrac{(a-2)(a+2)}{a-5}$

23. Performing the operations:

$\dfrac{4t^2 - 1}{6t^2 + t - 2} \div \dfrac{8t^3 + 1}{27t^3 + 8} = \dfrac{4t^2 - 1}{6t^2 + t - 2} \cdot \dfrac{27t^3 + 8}{8t^3 + 1} = \dfrac{(2t+1)(2t-1)}{(3t+2)(2t-1)} \cdot \dfrac{(3t+2)(9t^2 - 6t + 4)}{(2t+1)(4t^2 - 2t + 1)} = \dfrac{9t^2 - 6t + 4}{4t^2 - 2t + 1}$

25. Performing the operations:

$\dfrac{2x^2 - 5x - 12}{4x^2 + 8x + 3} \div \dfrac{x^2 - 16}{2x^2 + 7x + 3} = \dfrac{2x^2 - 5x - 12}{4x^2 + 8x + 3} \cdot \dfrac{2x^2 + 7x + 3}{x^2 - 16} = \dfrac{(2x+3)(x-4)}{(2x+1)(2x+3)} \cdot \dfrac{(2x+1)(x+3)}{(x+4)(x-4)} = \dfrac{x+3}{x+4}$

27. Performing the operations:

$\dfrac{6a^2 b + 2ab^2 - 20b^3}{4a^2 b - 16b^3} \cdot \dfrac{10a^2 - 22ab + 4b^2}{27a^3 - 125b^3} = \dfrac{2b(3a^2 + ab - 10b^2)}{4b(a^2 - 4b^2)} \cdot \dfrac{2(5a^2 - 11ab + 2b^2)}{(3a - 5b)(9a^2 + 15ab + 25b^2)}$

$= \dfrac{2b(3a - 5b)(a + 2b)}{4b(a + 2b)(a - 2b)} \cdot \dfrac{2(5a - b)(a - 2b)}{(3a - 5b)(9a^2 + 15ab + 25b^2)}$

$= \dfrac{5a - b}{9a^2 + 15ab + 25b^2}$

29.   Performing the operations:

$$\frac{360x^3 - 490x}{36x^2 + 84x + 49} \cdot \frac{30x^2 + 83x + 56}{150x^3 + 65x^2 - 280x} = \frac{10x\left(36x^2 - 49\right)}{\left(6x+7\right)^2} \cdot \frac{\left(6x+7\right)\left(5x+8\right)}{5x\left(30x^2 + 13x - 56\right)}$$

$$= \frac{10x\left(6x+7\right)\left(6x-7\right)}{\left(6x+7\right)^2} \cdot \frac{\left(6x+7\right)\left(5x+8\right)}{5x\left(6x-7\right)\left(5x+8\right)}$$

$$= 2$$

31.   Performing the operations:

$$\frac{x^5 - x^2}{5x^5 - 5x} \cdot \frac{10x^4 - 10x^2}{2x^4 + 2x^3 + 2x^2} = \frac{x^2\left(x^3 - 1\right)}{5x\left(x^4 - 1\right)} \cdot \frac{10x^2\left(x^2 - 1\right)}{2x^2\left(x^2 + x + 1\right)}$$

$$= \frac{x^2\left(x-1\right)\left(x^2 + x + 1\right)}{5x\left(x^2 + 1\right)\left(x-1\right)\left(x-1\right)} \cdot \frac{10x^2\left(x+1\right)\left(x-1\right)}{2x^2\left(x^2 + x + 1\right)}$$

$$= \frac{x\left(x-1\right)}{x^2 + 1}$$

33.   Performing the operations:

$$\frac{a^2 - 16b^2}{a^2 - 8ab + 16b^2} \cdot \frac{a^2 - 9ab + 20b^2}{a^2 - 7ab + 12b^2} \div \frac{a^2 - 25b^2}{a^2 - 6ab + 9b^2}$$

$$= \frac{a^2 - 16b^2}{a^2 - 8ab + 16b^2} \cdot \frac{a^2 - 9ab + 20b^2}{a^2 - 7ab + 12b^2} \cdot \frac{a^2 - 6ab + 9b^2}{a^2 - 25b^2}$$

$$= \frac{\left(a+4b\right)\left(a-4b\right)}{\left(a-4b\right)^2} \cdot \frac{\left(a-5b\right)\left(a-4b\right)}{\left(a-3b\right)\left(a-4b\right)} \cdot \frac{\left(a-3b\right)^2}{\left(a+5b\right)\left(a-5b\right)}$$

$$= \frac{\left(a+4b\right)\left(a-3b\right)}{\left(a-4b\right)\left(a+5b\right)}$$

35.   Performing the operations:

$$\frac{2y^2 - 7y - 15}{42y^2 - 29y - 5} \cdot \frac{12y^2 - 16y + 5}{7y^2 - 36y + 5} \div \frac{4y^2 - 9}{49y^2 - 1} = \frac{2y^2 - 7y - 15}{42y^2 - 29y - 5} \cdot \frac{12y^2 - 16y + 5}{7y^2 - 36y + 5} \cdot \frac{49y^2 - 1}{4y^2 - 9}$$

$$= \frac{\left(2y+3\right)\left(y-5\right)}{\left(6y-5\right)\left(7y+1\right)} \cdot \frac{\left(6y-5\right)\left(2y-1\right)}{\left(7y-1\right)\left(y-5\right)} \cdot \frac{\left(7y+1\right)\left(7y-1\right)}{\left(2y+3\right)\left(2y-3\right)}$$

$$= \frac{2y-1}{2y-3}$$

37.   Performing the operations:

$$\frac{xy - 2x + 3y - 6}{xy + 2x - 4y - 8} \cdot \frac{xy + x - 4y - 4}{xy - x + 3y - 3} = \frac{x\left(y-2\right) + 3\left(y-2\right)}{x\left(y+2\right) - 4\left(y+2\right)} \cdot \frac{x\left(y+1\right) - 4\left(y+1\right)}{x\left(y-1\right) + 3\left(y-1\right)}$$

$$= \frac{\left(y-2\right)\left(x+3\right)}{\left(y+2\right)\left(x-4\right)} \cdot \frac{\left(y+1\right)\left(x-4\right)}{\left(y-1\right)\left(x+3\right)}$$

$$= \frac{\left(y-2\right)\left(y+1\right)}{\left(y+2\right)\left(y-1\right)}$$

39.   Performing the operations:

$$\frac{xy^2 - y^2 + 4xy - 4y}{xy - 3y + 4x - 12} \div \frac{xy^3 + 2xy^2 + y^3 + 2y^2}{xy^2 - 3y^2 + 2xy - 6y} = \frac{xy^2 - y^2 + 4xy - 4y}{xy - 3y + 4x - 12} \cdot \frac{xy^2 - 3y^2 + 2xy - 6y}{xy^3 + 2xy^2 + y^3 + 2y^2}$$

$$= \frac{y^2(x-1) + 4y(x-1)}{y(x-3) + 4(x-3)} \cdot \frac{y^2(x-3) + 2y(x-3)}{xy^2(y+2) + y^2(y+2)}$$

$$= \frac{y(x-1)(y+4)}{(x-3)(y+4)} \cdot \frac{y(x-3)(y+2)}{y^2(y+2)(x+1)}$$

$$= \frac{x-1}{x+1}$$

41.   Performing the operations:

$$\frac{2x^3 + 10x^2 - 8x - 40}{x^3 + 4x^2 - 9x - 36} \cdot \frac{x^2 + x - 12}{2x^2 + 14x + 20} = \frac{2x^2(x+5) - 8(x+5)}{x^2(x+4) - 9(x+4)} \cdot \frac{(x+4)(x-3)}{2(x^2 + 7x + 10)}$$

$$= \frac{2(x+5)(x^2 - 4)}{(x+4)(x^2 - 9)} \cdot \frac{(x+4)(x-3)}{2(x+5)(x+2)}$$

$$= \frac{2(x+5)(x+2)(x-2)}{(x+4)(x+3)(x-3)} \cdot \frac{(x+4)(x-3)}{2(x+5)(x+2)}$$

$$= \frac{x-2}{x+3}$$

43.   **a.**   Simplifying: $\dfrac{16-1}{64-1} = \dfrac{15}{63} = \dfrac{5}{21}$

   b.   Reducing: $\dfrac{25x^2 - 9}{125x^3 - 27} = \dfrac{(5x-3)(5x+3)}{(5x-3)(25x^2 + 15x + 9)} = \dfrac{5x+3}{25x^2 + 15x + 9}$

   c.   Multiplying: $\dfrac{25x^2 - 9}{125x^3 - 27} \cdot \dfrac{5x-3}{5x+3} = \dfrac{(5x-3)(5x+3)}{(5x-3)(25x^2 + 15x + 9)} \cdot \dfrac{5x-3}{5x+3} = \dfrac{5x-3}{25x^2 + 15x + 9}$

   d.   Dividing: $\dfrac{25x^2 - 9}{125x^3 - 27} \div \dfrac{5x-3}{25x^2 + 15x + 9} = \dfrac{(5x-3)(5x+3)}{(5x-3)(25x^2 + 15x + 9)} \cdot \dfrac{25x^2 + 15x + 9}{5x-3} = \dfrac{5x+3}{5x-3}$

45.   **a.**   Multiplying: $f(x) \cdot g(x) = \dfrac{x^2 - x - 6}{x-1} \cdot \dfrac{x+2}{x^2 - 4x + 3} = \dfrac{(x-3)(x+2)}{x-1} \cdot \dfrac{x+2}{(x-3)(x-1)} = \dfrac{(x+2)^2}{(x-1)^2}$

   b.   Dividing: $f(x) \div g(x) = \dfrac{x^2 - x - 6}{x-1} \div \dfrac{x+2}{x^2 - 4x + 3} = \dfrac{(x-3)(x+2)}{x-1} \cdot \dfrac{(x-3)(x-1)}{x+2} = (x-3)^2$

47.   Multiplying:

$$f(x) \cdot g(x) = \frac{x^3 - 9x^2 - 3x + 27}{4x^2 - 12} \cdot \frac{x^2 - 2x - 8}{x^2 - 81}$$

$$= \frac{x^2(x-9) - 3(x-9)}{4(x^2 - 3)} \cdot \frac{(x-4)(x+2)}{(x-9)(x+9)}$$

$$= \frac{(x-9)(x^2 - 3)}{4(x^2 - 3)} \cdot \frac{(x-4)(x+2)}{(x-9)(x+9)}$$

$$= \frac{(x-4)(x+2)}{4(x+9)}$$

49. Multiplying:

$$f(x) \cdot g(x) = \frac{x^3 - 3x^2 - 4x + 12}{x+2} \cdot \frac{x^2 + 7x + 12}{x^2 - 5x + 6}$$

$$= \frac{x^2(x-3) - 4(x-3)}{x+2} \cdot \frac{(x+4)(x+3)}{(x-2)(x-3)}$$

$$= \frac{(x-3)(x+2)(x-2)}{x+2} \cdot \frac{(x+4)(x+3)}{(x-2)(x-3)}$$

$$= (x+4)(x+3)$$

51. Finding the product: $(3x-6) \cdot \dfrac{x}{x-2} = \dfrac{3(x-2)}{1} \cdot \dfrac{x}{x-2} = 3x$

53. Finding the product: $(x^2 - 25) \cdot \dfrac{2}{x-5} = \dfrac{(x+5)(x-5)}{1} \cdot \dfrac{2}{x-5} = 2(x+5)$

55. Finding the product: $(x^2 - 3x + 2) \cdot \dfrac{3}{3x-3} = \dfrac{(x-2)(x-1)}{1} \cdot \dfrac{3}{3(x-1)} = x-2$

57. Finding the product: $(y-3)(y-4)(y+3) \cdot \dfrac{-1}{y^2-9} = \dfrac{(y-3)(y-4)(y+3)}{1} \cdot \dfrac{-1}{(y+3)(y-3)} = -(y-4)$

59. Finding the product: $a(a+5)(a-5) \cdot \dfrac{a+1}{a^2+5a} = \dfrac{a(a+5)(a-5)}{1} \cdot \dfrac{a+1}{a(a+5)} = (a-5)(a+1)$

61. Dividing: $(x^2 - 2x - 8) \div \dfrac{x^2 - x - 6}{x-4} = \dfrac{(x-4)(x+2)}{1} \cdot \dfrac{x-4}{(x-3)(x+2)} = \dfrac{(x-4)^2}{x-3}$

63. Dividing: $(3-x) \div \dfrac{x^2 - 9}{x-1} = \dfrac{-(x-3)}{1} \cdot \dfrac{x-1}{(x+3)(x-3)} = -\dfrac{x-1}{x+3}$

65. Dividing:

$$(xy - 2x - 3y + 6) \div \frac{x^2 - 2x - 3}{x^2 - 6x - 7} = \frac{x(y-2) - 3(y-2)}{1} \cdot \frac{(x-7)(x+1)}{(x-3)(x+1)}$$

$$= \frac{(y-2)(x-3)}{1} \cdot \frac{(x-7)(x+1)}{(x-3)(x+1)}$$

$$= (y-2)(x-7)$$

67. Simplifying: $\dfrac{x^2(x+2) + 6x(x+2) + 9(x+2)}{x^2 - 2x - 8} = \dfrac{(x+2)(x^2 + 6x + 9)}{(x-4)(x+2)} = \dfrac{(x+2)(x+3)^2}{(x-4)(x+2)} = \dfrac{(x+3)^2}{x-4}$

69. Simplifying: $\dfrac{2x^2(x+3) - 5x(x+3) + 3(x+3)}{x^2 + x - 6} = \dfrac{(x+3)(2x^2 - 5x + 3)}{(x+3)(x-2)} = \dfrac{(x+3)(2x-3)(x-1)}{(x+3)(x-2)} = \dfrac{(2x-3)(x-1)}{x-2}$

71. Completing the table:

Number of Copies x	Price per Copy ($) $p(x)$
1	20.33
10	9.33
20	6.40
50	4.00
100	3.05

73. Finding the revenue: $R = 100 \cdot \dfrac{2(100 + 60)}{100 + 5} = 100 \cdot \dfrac{320}{105} \approx \$305.00$

75. Completing the table:

Number of Copies	Price per Copy ($)
1	20.50
10	10.00
20	7.20
50	4.91
100	4.00

77. The revenue is: $100(4.00) = \$400$

79. First use long division to find the length of the square base:

$$
\begin{array}{r}
x - 3 \\
x^2 + x + 1 \overline{\smash{\big)}\, x^3 - 2x^2 - 2x - 3} \\
\underline{x^3 + x^2 + x} \\
-3x^2 - 3x - 3 \\
\underline{-3x^2 - 3x - 3} \\
0
\end{array}
$$

The area of the base is therefore: $A = (x-3)^2 = x^2 - 6x + 9$

81. Multiplying the polynomials: $2x^2\left(5x^3 + 4x - 3\right) = 2x^2 \bullet 5x^3 + 2x^2 \bullet 4x - 2x^2 \bullet 3 = 10x^5 + 8x^3 - 6x^2$

83. Multiplying the polynomials: $(3a-1)(4a+5) = 12a^2 + 15a - 4a - 5 = 12a^2 + 11a - 5$

85. Multiplying the polynomials: $(3x+7)(4y-2) = 12xy + 28y - 6x - 14$

87. Multiplying the polynomials: $\left(3 - t^2\right)^2 = (3)^2 - 2(3)\left(t^2\right) + \left(t^2\right)^2 = 9 - 6t^2 + t^4$

89. Multiplying the polynomials:

$$3(x+1)(x+2)(x+3) = 3(x+1)\left(x^2 + 5x + 6\right) = 3\left(x^3 + 6x^2 + 11x + 6\right) = 3x^3 + 18x^2 + 33x + 18$$

91. Combining: $\dfrac{4}{9} + \dfrac{2}{9} = \dfrac{6}{9} = \dfrac{2 \bullet 3}{3 \bullet 3} = \dfrac{2}{3}$

93. Combining: $\dfrac{3}{14} + \dfrac{7}{30} = \dfrac{3}{14} \bullet \dfrac{15}{15} + \dfrac{7}{30} \bullet \dfrac{7}{7} = \dfrac{45}{210} + \dfrac{49}{210} = \dfrac{94}{210} = \dfrac{47}{105}$

95. Multiplying: $-1(7 - x) = -7 + x = x - 7$ **97.** Factoring: $x^2 - 1 = (x+1)(x-1)$

99. Factoring: $2x + 10 = 2(x+5)$ **101.** Factoring: $a^3 - b^3 = (a-b)\left(a^2 + ab + b^2\right)$

103. Factoring: $16x^3 + 54y^3 = 2\left(8x^3 + 27y^3\right) = 2(2x + 3y)\left(4x^2 - 6xy + 9y^2\right)$

105. Performing the operations:

$$\frac{x^6 + y^6}{x^4 + 4x^2y^2 + 3y^4} \div \frac{x^4 + 3x^2y^2 + 2y^4}{x^4 + 5x^2y^2 + 6y^4} = \frac{x^6 + y^6}{x^4 + 4x^2y^2 + 3y^4} \bullet \frac{x^4 + 5x^2y^2 + 6y^4}{x^4 + 3x^2y^2 + 2y^4}$$

$$= \frac{\left(x^2 + y^2\right)\left(x^4 - x^2y^2 + y^4\right)}{\left(x^2 + 3y^2\right)\left(x^2 + y^2\right)} \bullet \frac{\left(x^2 + 3y^2\right)\left(x^2 + 2y^2\right)}{\left(x^2 + 2y^2\right)\left(x^2 + y^2\right)}$$

$$= \frac{x^4 - x^2y^2 + y^4}{x^2 + y^2}$$

107. Performing the operations:

$$\frac{a^2(2a+b)+6a(2a+b)+5(2a+b)}{3a^2(2a+b)-2a(2a+b)+(2a+b)} \div \frac{a+1}{a-1} = \frac{a^2(2a+b)+6a(2a+b)+5(2a+b)}{3a^2(2a+b)-2a(2a+b)+(2a+b)} \cdot \frac{a-1}{a+1}$$

$$= \frac{(2a+b)(a^2+6a+5)}{(2a+b)(3a^2-2a+1)} \cdot \frac{a-1}{a+1}$$

$$= \frac{(2a+b)(a+5)(a+1)}{(2a+b)(3a^2-2a+1)} \cdot \frac{a-1}{a+1}$$

$$= \frac{(a+5)(a-1)}{3a^2-2a+1}$$

109. Performing the operations: $\dfrac{a^3-a^2b}{ac-a} \div \left(\dfrac{a-b}{c-1}\right)^2 = \dfrac{a^3-a^2b}{ac-a} \cdot \left(\dfrac{c-1}{a-b}\right)^2 = \dfrac{a^2(a-b)}{a(c-1)} \cdot \dfrac{(c-1)^2}{(a-b)^2} = \dfrac{a(c-1)}{a-b}$

111. Performing the operations: $\dfrac{x^3-x^2y}{(x-y)^2} \div \dfrac{x^2+xy}{x^2-y^2} = \dfrac{x^3-x^2y}{(x-y)^2} \cdot \dfrac{x^2-y^2}{x^2+xy} = \dfrac{x^2(x-y)}{(x-y)^2} \cdot \dfrac{(x+y)(x-y)}{x(x+y)} = x$

6.4 Addition and Subtraction of Rational Expressions

1. Combining the fractions: $\dfrac{3}{4}+\dfrac{1}{2} = \dfrac{3}{4}+\dfrac{1}{2}\cdot\dfrac{2}{2} = \dfrac{3}{4}+\dfrac{2}{4} = \dfrac{5}{4}$

3. Combining the fractions: $\dfrac{2}{5}-\dfrac{1}{15} = \dfrac{2}{5}\cdot\dfrac{3}{3}-\dfrac{1}{15} = \dfrac{6}{15}-\dfrac{1}{15} = \dfrac{5}{15} = \dfrac{1}{3}$

5. Combining the fractions: $\dfrac{5}{6}+\dfrac{7}{8} = \dfrac{5}{6}\cdot\dfrac{4}{4}+\dfrac{7}{8}\cdot\dfrac{3}{3} = \dfrac{20}{24}+\dfrac{21}{24} = \dfrac{41}{24}$

7. Combining the fractions: $\dfrac{9}{48}-\dfrac{3}{54} = \dfrac{9}{48}\cdot\dfrac{9}{9}-\dfrac{3}{54}\cdot\dfrac{8}{8} = \dfrac{81}{432}-\dfrac{24}{432} = \dfrac{57}{432} = \dfrac{19}{144}$

9. Combining the fractions: $\dfrac{3}{4}-\dfrac{1}{8}+\dfrac{2}{3} = \dfrac{3}{4}\cdot\dfrac{6}{6}-\dfrac{1}{8}\cdot\dfrac{3}{3}+\dfrac{2}{3}\cdot\dfrac{8}{8} = \dfrac{18}{24}-\dfrac{3}{24}+\dfrac{16}{24} = \dfrac{31}{24}$

11. Combining the rational expressions: $\dfrac{x}{x+3}+\dfrac{3}{x+3} = \dfrac{x+3}{x+3} = 1$

13. Combining the rational expressions: $\dfrac{4}{y-4}-\dfrac{y}{y-4} = \dfrac{4-y}{y-4} = \dfrac{-1(y-4)}{y-4} = -1$

15. Combining the rational expressions: $\dfrac{x}{x^2-y^2}-\dfrac{y}{x^2-y^2} = \dfrac{x-y}{x^2-y^2} = \dfrac{x-y}{(x+y)(x-y)} = \dfrac{1}{x+y}$

17. Combining the rational expressions: $\dfrac{2x-3}{x-2}-\dfrac{x-1}{x-2} = \dfrac{2x-3-x+1}{x-2} = \dfrac{x-2}{x-2} = 1$

19. Combining the rational expressions: $\dfrac{1}{a}+\dfrac{2}{a^2}-\dfrac{3}{a^3} = \dfrac{1}{a}\cdot\dfrac{a^2}{a^2}+\dfrac{2}{a^2}\cdot\dfrac{a}{a}-\dfrac{3}{a^3} = \dfrac{a^2+2a-3}{a^3}$

21. Combining the rational expressions: $\dfrac{7x-2}{2x+1}-\dfrac{5x-3}{2x+1} = \dfrac{7x-2-5x+3}{2x+1} = \dfrac{2x+1}{2x+1} = 1$

23.   **a.**   Multiplying: $\dfrac{3}{8}\cdot\dfrac{1}{6}=\dfrac{3}{8}\cdot\dfrac{1}{2\cdot3}=\dfrac{1}{16}$

     b.   Dividing: $\dfrac{3}{8}\div\dfrac{1}{6}=\dfrac{3}{8}\cdot\dfrac{6}{1}=\dfrac{3}{2\cdot4}\cdot\dfrac{2\cdot3}{1}=\dfrac{9}{4}$

     c.   Adding: $\dfrac{3}{8}+\dfrac{1}{6}=\dfrac{3}{8}\cdot\dfrac{3}{3}+\dfrac{1}{6}\cdot\dfrac{4}{4}=\dfrac{9}{24}+\dfrac{4}{24}=\dfrac{13}{24}$

     d.   Multiplying: $\dfrac{x+3}{x-3}\cdot\dfrac{5x+15}{x^2-9}=\dfrac{x+3}{x-3}\cdot\dfrac{5(x+3)}{(x+3)(x-3)}=\dfrac{5(x+3)}{(x-3)^2}$

     e.   Dividing: $\dfrac{x+3}{x-3}\div\dfrac{5x+15}{x^2-9}=\dfrac{x+3}{x-3}\cdot\dfrac{(x+3)(x-3)}{5(x+3)}=\dfrac{x+3}{5}$

     f.   Subtracting:

$$\dfrac{x+3}{x-3}-\dfrac{5x+15}{x^2-9}=\dfrac{x+3}{x-3}\cdot\dfrac{x+3}{x+3}-\dfrac{5x+15}{(x+3)(x-3)}$$
$$=\dfrac{x^2+6x+9}{(x+3)(x-3)}-\dfrac{5x+15}{(x+3)(x-3)}$$
$$=\dfrac{x^2+x-6}{(x+3)(x-3)}$$
$$=\dfrac{(x+3)(x-2)}{(x+3)(x-3)}$$
$$=\dfrac{x-2}{x-3}$$

25. Combining the rational expressions: $\dfrac{2}{t^2}-\dfrac{3}{2t}=\dfrac{2}{t^2}\cdot\dfrac{2}{2}-\dfrac{3}{2t}\cdot\dfrac{t}{t}=\dfrac{4}{2t^2}-\dfrac{3t}{2t^2}=\dfrac{4-3t}{2t^2}$

27. Combining the rational expressions:

$$\dfrac{3x+1}{2x-6}-\dfrac{x+2}{x-3}=\dfrac{3x+1}{2(x-3)}-\dfrac{x+2}{x-3}\cdot\dfrac{2}{2}=\dfrac{3x+1}{2(x-3)}-\dfrac{2x+4}{2(x-3)}=\dfrac{3x+1-2x-4}{2(x-3)}=\dfrac{x-3}{2(x-3)}=\dfrac{1}{2}$$

29. Combining the rational expressions:

$$\dfrac{x+1}{2x-2}-\dfrac{2}{x^2-1}=\dfrac{x+1}{2(x-1)}\cdot\dfrac{x+1}{x+1}-\dfrac{2}{(x+1)(x-1)}\cdot\dfrac{2}{2}$$
$$=\dfrac{x^2+2x+1}{2(x+1)(x-1)}-\dfrac{4}{2(x+1)(x-1)}$$
$$=\dfrac{x^2+2x-3}{2(x+1)(x-1)}$$
$$=\dfrac{(x+3)(x-1)}{2(x+1)(x-1)}$$
$$=\dfrac{x+3}{2(x+1)}$$

31.   Combining the rational expressions:

$$\frac{1}{a-b} - \frac{3ab}{a^3-b^3} = \frac{1}{a-b} \cdot \frac{a^2+ab+b^2}{a^2+ab+b^2} - \frac{3ab}{a^3-b^3}$$

$$= \frac{a^2+ab+b^2}{a^3-b^3} - \frac{3ab}{a^3-b^3}$$

$$= \frac{a^2-2ab+b^2}{a^3-b^3}$$

$$= \frac{(a-b)^2}{(a-b)(a^2+ab+b^2)}$$

$$= \frac{a-b}{a^2+ab+b^2}$$

33.   Combining the rational expressions:

$$\frac{1}{2y-3} - \frac{18y}{8y^3-27} = \frac{1}{2y-3} \cdot \frac{4y^2+6y+9}{4y^2+6y+9} - \frac{18y}{8y^3-27}$$

$$= \frac{4y^2+6y+9}{8y^3-27} - \frac{18y}{8y^3-27}$$

$$= \frac{4y^2-12y+9}{8y^3-27}$$

$$= \frac{(2y-3)^2}{(2y-3)(4y^2+6y+9)}$$

$$= \frac{2y-3}{4y^2+6y+9}$$

35.   Combining the rational expressions:

$$\frac{x}{x^2-5x+6} - \frac{3}{3-x} = \frac{x}{(x-2)(x-3)} + \frac{3}{x-3} \cdot \frac{x-2}{x-2}$$

$$= \frac{x}{(x-2)(x-3)} + \frac{3x-6}{(x-2)(x-3)}$$

$$= \frac{4x-6}{(x-2)(x-3)}$$

$$= \frac{2(2x-3)}{(x-3)(x-2)}$$

37.   Combining the rational expressions:

$$\frac{2}{4t-5} + \frac{9}{8t^2-38t+35} = \frac{2}{4t-5} \cdot \frac{2t-7}{2t-7} + \frac{9}{(4t-5)(2t-7)}$$

$$= \frac{4t-14}{(4t-5)(2t-7)} + \frac{9}{(4t-5)(2t-7)}$$

$$= \frac{4t-5}{(4t-5)(2t-7)}$$

$$= \frac{1}{2t-7}$$

39.   Combining the rational expressions:

$$\frac{1}{a^2 - 5a + 6} + \frac{3}{a^2 - a - 2} = \frac{1}{(a-2)(a-3)} \cdot \frac{a+1}{a+1} + \frac{3}{(a-2)(a+1)} \cdot \frac{a-3}{a-3}$$

$$= \frac{a+1}{(a-2)(a-3)(a+1)} + \frac{3a-9}{(a-2)(a-3)(a+1)}$$

$$= \frac{4a-8}{(a-2)(a-3)(a+1)}$$

$$= \frac{4(a-2)}{(a-2)(a-3)(a+1)}$$

$$= \frac{4}{(a-3)(a+1)}$$

41.   Combining the rational expressions:

$$\frac{1}{8x^3 - 1} - \frac{1}{4x^2 - 1} = \frac{1}{(2x-1)(4x^2 + 2x + 1)} \cdot \frac{2x+1}{2x+1} - \frac{1}{(2x+1)(2x-1)} \cdot \frac{4x^2 + 2x + 1}{4x^2 + 2x + 1}$$

$$= \frac{2x+1}{(2x+1)(2x-1)(4x^2 + 2x + 1)} - \frac{4x^2 + 2x + 1}{(2x+1)(2x-1)(4x^2 + 2x + 1)}$$

$$= \frac{2x + 1 - 4x^2 - 2x - 1}{(2x+1)(2x-1)(4x^2 + 2x + 1)}$$

$$= \frac{-4x^2}{(2x+1)(2x-1)(4x^2 + 2x + 1)}$$

43.   Combining the rational expressions:

$$\frac{4}{4x^2 - 9} - \frac{6}{8x^2 - 6x - 9} = \frac{4}{(2x+3)(2x-3)} \cdot \frac{4x+3}{4x+3} - \frac{6}{(2x-3)(4x+3)} \cdot \frac{2x+3}{2x+3}$$

$$= \frac{16x + 12}{(2x+3)(2x-3)(4x+3)} - \frac{12x + 18}{(2x+3)(2x-3)(4x+3)}$$

$$= \frac{16x + 12 - 12x - 18}{(2x+3)(2x-3)(4x+3)}$$

$$= \frac{4x - 6}{(2x+3)(2x-3)(4x+3)}$$

$$= \frac{2(2x-3)}{(2x+3)(2x-3)(4x+3)}$$

$$= \frac{2}{(2x+3)(4x+3)}$$

45.   Combining the rational expressions:

$$\frac{4a}{a^2+6a+5}-\frac{3a}{a^2+5a+4}=\frac{4a}{(a+5)(a+1)}\cdot\frac{a+4}{a+4}-\frac{3a}{(a+4)(a+1)}\cdot\frac{a+5}{a+5}$$

$$=\frac{4a^2+16a}{(a+4)(a+5)(a+1)}-\frac{3a^2+15a}{(a+4)(a+5)(a+1)}$$

$$=\frac{4a^2+16a-3a^2-15a}{(a+4)(a+5)(a+1)}$$

$$=\frac{a^2+a}{(a+4)(a+5)(a+1)}$$

$$=\frac{a(a+1)}{(a+4)(a+5)(a+1)}$$

$$=\frac{a}{(a+4)(a+5)}$$

47.   Combining the rational expressions:

$$\frac{2x-1}{x^2+x-6}-\frac{x+2}{x^2+5x+6}=\frac{2x-1}{(x+3)(x-2)}\cdot\frac{x+2}{x+2}-\frac{x+2}{(x+3)(x+2)}\cdot\frac{x-2}{x-2}$$

$$=\frac{2x^2+3x-2}{(x+3)(x+2)(x-2)}-\frac{x^2-4}{(x+3)(x+2)(x-2)}$$

$$=\frac{2x^2+3x-2-x^2+4}{(x+3)(x+2)(x-2)}$$

$$=\frac{x^2+3x+2}{(x+3)(x+2)(x-2)}$$

$$=\frac{(x+2)(x+1)}{(x+3)(x+2)(x-2)}$$

$$=\frac{x+1}{(x-2)(x+3)}$$

49.   Combining the rational expressions:

$$\frac{2x-8}{3x^2+8x+4}+\frac{x+3}{3x^2+5x+2}=\frac{2x-8}{(3x+2)(x+2)}+\frac{x+3}{(3x+2)(x+1)}$$

$$=\frac{2x-8}{(3x+2)(x+2)}\cdot\frac{x+1}{x+1}+\frac{x+3}{(3x+2)(x+1)}\cdot\frac{x+2}{x+2}$$

$$=\frac{2x^2-6x-8}{(3x+2)(x+2)(x+1)}+\frac{x^2+5x+6}{(3x+2)(x+2)(x+1)}$$

$$=\frac{3x^2-x-2}{(3x+2)(x+2)(x+1)}$$

$$=\frac{(3x+2)(x-1)}{(3x+2)(x+2)(x+1)}$$

$$=\frac{x-1}{(x+1)(x+2)}$$

51. Combining the rational expressions:

$$\frac{2}{x^2+5x+6}-\frac{4}{x^2+4x+3}+\frac{3}{x^2+3x+2}=\frac{2}{(x+3)(x+2)}-\frac{4}{(x+3)(x+1)}+\frac{3}{(x+2)(x+1)}$$

$$=\frac{2}{(x+3)(x+2)}\cdot\frac{x+1}{x+1}-\frac{4}{(x+3)(x+1)}\cdot\frac{x+2}{x+2}+\frac{3}{(x+2)(x+1)}\cdot\frac{x+3}{x+3}$$

$$=\frac{2x+2}{(x+3)(x+2)(x+1)}-\frac{4x+8}{(x+3)(x+2)(x+1)}+\frac{3x+9}{(x+3)(x+2)(x+1)}$$

$$=\frac{2x+2-4x-8+3x+9}{(x+3)(x+2)(x+1)}$$

$$=\frac{x+3}{(x+3)(x+2)(x+1)}$$

$$=\frac{1}{(x+2)(x+1)}$$

53. Combining the rational expressions:

$$\frac{2x+8}{x^2+5x+6}-\frac{x+5}{x^2+4x+3}-\frac{x-1}{x^2+3x+2}=\frac{2x+8}{(x+3)(x+2)}-\frac{x+5}{(x+3)(x+1)}-\frac{x-1}{(x+2)(x+1)}$$

$$=\frac{2x+8}{(x+3)(x+2)}\cdot\frac{x+1}{x+1}-\frac{x+5}{(x+3)(x+1)}\cdot\frac{x+2}{x+2}-\frac{x-1}{(x+2)(x+1)}\cdot\frac{x+3}{x+3}$$

$$=\frac{2x^2+10x+8}{(x+3)(x+2)(x+1)}-\frac{x^2+7x+10}{(x+3)(x+2)(x+1)}-\frac{x^2+2x-3}{(x+3)(x+2)(x+1)}$$

$$=\frac{2x^2+10x+8-x^2-7x-10-x^2-2x+3}{(x+3)(x+2)(x+1)}$$

$$=\frac{x+1}{(x+3)(x+2)(x+1)}$$

$$=\frac{1}{(x+2)(x+3)}$$

55. Combining the rational expressions: $2+\dfrac{3}{2x+1}=\dfrac{2}{1}\cdot\dfrac{2x+1}{2x+1}+\dfrac{3}{2x+1}=\dfrac{4x+2}{2x+1}+\dfrac{3}{2x+1}=\dfrac{4x+5}{2x+1}$

57. Combining the rational expressions: $5+\dfrac{2}{4-t}=\dfrac{5}{1}\cdot\dfrac{4-t}{4-t}+\dfrac{2}{4-t}=\dfrac{20-5t}{4-t}+\dfrac{2}{4-t}=\dfrac{22-5t}{4-t}$

59. Combining the rational expressions: $x-\dfrac{4}{2x+3}=\dfrac{x}{1}\cdot\dfrac{2x+3}{2x+3}-\dfrac{4}{2x+3}=\dfrac{2x^2+3x}{2x+3}-\dfrac{4}{2x+3}=\dfrac{2x^2+3x-4}{2x+3}$

61. Combining the rational expressions:

$$\frac{x}{x+2}+\frac{1}{2x+4}-\frac{3}{x^2+2x}=\frac{x}{x+2}\cdot\frac{2x}{2x}+\frac{1}{2(x+2)}\cdot\frac{x}{x}-\frac{3}{x(x+2)}\cdot\frac{2}{2}$$

$$=\frac{2x^2}{2x(x+2)}+\frac{x}{2x(x+2)}-\frac{6}{2x(x+2)}$$

$$=\frac{2x^2+x-6}{2x(x+2)}$$

$$=\frac{(2x-3)(x+2)}{2x(x+2)}$$

$$=\frac{2x-3}{2x}$$

63.   Combining the rational expressions:

$$\frac{1}{x}+\frac{x}{2x+4}-\frac{2}{x^2+2x}=\frac{1}{x}\cdot\frac{2(x+2)}{2(x+2)}+\frac{x}{2(x+2)}\cdot\frac{x}{x}-\frac{2}{x(x+2)}\cdot\frac{2}{2}$$

$$=\frac{2x+4}{2x(x+2)}+\frac{x^2}{2x(x+2)}-\frac{4}{2x(x+2)}$$

$$=\frac{x^2+2x}{2x(x+2)}$$

$$=\frac{x(x+2)}{2x(x+2)}$$

$$=\frac{1}{2}$$

65.   Finding the difference:

$$f(x)-g(x)=\frac{2x-1}{4x-16}-\frac{x-3}{x-4}$$

$$=\frac{2x-1}{4(x-4)}-\frac{x-3}{x-4}\cdot\frac{4}{4}$$

$$=\frac{2x-1}{4(x-4)}-\frac{4x-12}{4(x-4)}$$

$$=\frac{2x-1-4x+12}{4(x-4)}$$

$$=\frac{-2x+11}{4(x-4)}$$

$$=-\frac{2x-11}{4(x-4)}$$

67.   Finding the sum:

$$f(x)+g(x)=\frac{2}{x+4}+\frac{x-1}{x^2+3x-4}$$

$$=\frac{2}{x+4}\cdot\frac{x-1}{x-1}+\frac{x-1}{(x+4)(x-1)}$$

$$=\frac{2x-2}{(x+4)(x-1)}+\frac{x-1}{(x+4)(x-1)}$$

$$=\frac{3x-3}{(x+4)(x-1)}$$

$$=\frac{3(x-1)}{(x+4)(x-1)}$$

$$=\frac{3}{x+4}$$

69.   Finding the sum:

$$f(x)+g(x)=\frac{2x}{x^2-x-2}+\frac{5}{x^2+x-6}$$

$$=\frac{2x}{(x-2)(x+1)}\cdot\frac{x+3}{x+3}+\frac{5}{(x+3)(x-2)}\cdot\frac{x+1}{x+1}$$

$$=\frac{2x^2+6x}{(x+2)(x-1)(x+3)}+\frac{5x+5}{(x+2)(x-1)(x+3)}$$

$$=\frac{2x^2+11x+5}{(x+2)(x-1)(x+3)}$$

$$=\frac{(2x+1)(x+5)}{(x+2)(x-1)(x+3)}$$

71.   Finding the sum:

$$f(x)+g(x)=\frac{x}{9x^2-4}+\frac{1}{3x^2-4x-4}$$

$$=\frac{x}{(3x+2)(3x-2)}\cdot\frac{x-2}{x-2}+\frac{1}{(3x+2)(x-2)}\cdot\frac{3x-2}{3x-2}$$

$$=\frac{x^2-2x}{(3x+2)(3x-2)(x-2)}+\frac{3x-2}{(3x+2)(3x-2)(x-2)}$$

$$=\frac{x^2+x-2}{(3x+2)(3x-2)(x-2)}$$

$$=\frac{(x+2)(x-1)}{(3x+2)(3x-2)(x-2)}$$

73. Finding the sum:

$$f(x) + g(x) = \frac{3x}{2x^2 - x - 1} + \frac{6}{2x^2 - 5x - 3}$$

$$= \frac{3x}{(2x+1)(x-1)} \cdot \frac{x-3}{x-3} + \frac{6}{(2x+1)(x-3)} \cdot \frac{x-1}{x-1}$$

$$= \frac{3x^2 - 9x}{(2x+1)(x-1)(x-3)} + \frac{6x-6}{(2x+1)(x-1)(x-3)}$$

$$= \frac{3x^2 - 3x - 6}{(2x+1)(x-1)(x-3)}$$

$$= \frac{3(x^2 - x - 2)}{(2x+1)(x-1)(x-3)}$$

$$= \frac{3(x-2)(x+1)}{(2x+1)(x-1)(x-3)}$$

75. Finding the difference:

$$f(x) - g(x) = \frac{5x}{x^2 - 13x + 36} - \frac{3x}{x^2 - 11x + 28}$$

$$= \frac{5x}{(x-9)(x-4)} \cdot \frac{x-7}{x-7} - \frac{3x}{(x-7)(x-4)} \cdot \frac{x-9}{x-9}$$

$$= \frac{5x^2 - 35x}{(x-9)(x-4)(x-7)} - \frac{3x^2 - 27x}{(x-9)(x-4)(x-7)}$$

$$= \frac{2x^2 - 8x}{(x-9)(x-4)(x-7)}$$

$$= \frac{2x(x-4)}{(x-9)(x-4)(x-7)}$$

$$= \frac{2x}{(x-9)(x-7)}$$

77. Writing the expression and simplifying: $x + \dfrac{4}{x} = \dfrac{x^2 + 4}{x}$

79. Substituting the values: $P = \dfrac{1}{10} + \dfrac{1}{0.2} = 0.1 + 5 = 5.1$

81. **a.** Substituting the values:

$$\frac{1}{T} = \frac{1}{24} - \frac{1}{30} = \frac{5}{120} - \frac{4}{120} = \frac{1}{120}$$
$$T = 120$$

The two objects will meet in 120 months.

b. If $t_A = t_B$, then $\dfrac{1}{T} = \dfrac{1}{t_A} - \dfrac{1}{t_A} = 0$. Since this is impossible, the two objects will never meet.

83. Writing in scientific notation: $54,000 = 5.4 \times 10^4$ **85.** Writing in scientific notation: $0.00034 = 3.4 \times 10^{-4}$

87. Writing in expanded form: $6.44 \times 10^3 = 6,440$ **89.** Writing in expanded form: $6.44 \times 10^{-3} = 0.00644$

91. Simplifying: $\left(3 \times 10^8\right)\left(4 \times 10^{-5}\right) = 12 \times 10^3 = 1.2 \times 10^4$

93. Simplifying: $4(2x - 3) + 7 = 8x - 12 + 7 = 8x - 5$

95. Simplifying: $10 \div 5(x - 4) + 7 = 2(x - 4) + 7 = 2x - 8 + 7 = 2x - 1$

97. Solving the equation:
$$5 - 3(x - 4) = 4x + 3$$
$$5 - 3x + 12 = 4x + 3$$
$$-3x + 17 = 4x + 3$$
$$-7x = -14$$
$$x = 2$$

99. Dividing: $\dfrac{3}{4} \div \dfrac{5}{8} = \dfrac{3}{4} \cdot \dfrac{8}{5} = \dfrac{24}{20} = \dfrac{4 \cdot 6}{4 \cdot 5} = \dfrac{6}{5}$

101. Dividing: $\dfrac{2}{3} \div \dfrac{8}{9} = \dfrac{2}{3} \cdot \dfrac{9}{8} = \dfrac{18}{24} = \dfrac{6 \cdot 3}{6 \cdot 4} = \dfrac{3}{4}$

103. Multiplying: $x\left(1 + \dfrac{2}{x}\right) = x \cdot 1 + x \cdot \dfrac{2}{x} = x + 2$

105. Multiplying: $3x\left(\dfrac{1}{x} - \dfrac{1}{3}\right) = 3x \cdot \dfrac{1}{x} - 3x \cdot \dfrac{1}{3} = 3 - x$

107. Factoring: $x^2 - 4 = (x + 2)(x - 2)$

109. Simplifying the expression:
$$\left(1 - \frac{1}{x}\right)\left(1 - \frac{1}{x+1}\right)\left(1 - \frac{1}{x+2}\right)\left(1 - \frac{1}{x+3}\right) = \left(\frac{x}{x} - \frac{1}{x}\right)\left(\frac{x+1}{x+1} - \frac{1}{x+1}\right)\left(\frac{x+2}{x+2} - \frac{1}{x+2}\right)\left(\frac{x+3}{x+3} - \frac{1}{x+3}\right)$$
$$= \left(\frac{x-1}{x}\right)\left(\frac{x}{x+1}\right)\left(\frac{x+1}{x+2}\right)\left(\frac{x+2}{x+3}\right)$$
$$= \frac{x-1}{x+3}$$

111. Simplifying the expression:
$$\left(\frac{a^2 - b^2}{u^2 - v^2}\right)\left(\frac{av - au}{b - a}\right) + \left(\frac{a^2 - av}{u + v}\right)\left(\frac{1}{a}\right) = \frac{(a+b)(a-b)}{(u+v)(u-v)} \cdot \frac{a(v-u)}{b-a} + \frac{a(a-v)}{u+v} \cdot \frac{1}{a}$$
$$= \frac{a(a+b)}{u+v} + \frac{a-v}{u+v}$$
$$= \frac{a^2 + ab + a - v}{u+v}$$

113. Simplifying the expression:
$$\frac{18x - 19}{4x^2 + 27x - 7} - \frac{12x - 41}{3x^2 + 17x - 28} = \frac{18x - 19}{(4x-1)(x+7)} \cdot \frac{3x-4}{3x-4} - \frac{12x - 41}{(3x-4)(x+7)} \cdot \frac{4x-1}{4x-1}$$
$$= \frac{54x^2 - 129x + 76}{(4x-1)(x+7)(3x-4)} - \frac{48x^2 - 176x + 41}{(4x-1)(x+7)(3x-4)}$$
$$= \frac{54x^2 - 129x + 76 - 48x^2 + 176x - 41}{(4x-1)(x+7)(3x-4)}$$
$$= \frac{6x^2 + 47x + 35}{(4x-1)(x+7)(3x-4)}$$
$$= \frac{(6x+5)(x+7)}{(4x-1)(x+7)(3x-4)}$$
$$= \frac{6x+5}{(4x-1)(3x-4)}$$

115. Simplifying the expression:

$$\left(\frac{1}{y^2-1} \div \frac{1}{y^2+1}\right)\left(\frac{y^3+1}{y^4-1}\right) + \frac{1}{(y+1)^2(y-1)} = \frac{y^2+1}{y^2-1} \cdot \frac{y^3+1}{y^4-1} + \frac{1}{(y+1)^2(y-1)}$$

$$= \frac{y^2+1}{(y+1)(y-1)} \cdot \frac{(y+1)(y^2-y+1)}{(y^2+1)(y+1)(y-1)} + \frac{1}{(y+1)^2(y-1)}$$

$$= \frac{y^2-y+1}{(y+1)(y-1)^2} \cdot \frac{y+1}{y+1} + \frac{1}{(y+1)^2(y-1)} \cdot \frac{y-1}{y-1}$$

$$= \frac{y^3+1+y-1}{(y+1)^2(y-1)^2}$$

$$= \frac{y^3+y}{(y+1)^2(y-1)^2}$$

$$= \frac{y(y^2+1)}{(y+1)^2(y-1)^2}$$

6.5 Complex Fractions

1. Simplifying the complex fraction: $\dfrac{\frac{3}{4}}{\frac{2}{3}} = \dfrac{\frac{3}{4} \cdot 12}{\frac{2}{3} \cdot 12} = \dfrac{9}{8}$

3. Simplifying the complex fraction: $\dfrac{\frac{1}{3}-\frac{1}{4}}{\frac{1}{2}+\frac{1}{8}} = \dfrac{\left(\frac{1}{3}-\frac{1}{4}\right) \cdot 24}{\left(\frac{1}{2}+\frac{1}{8}\right) \cdot 24} = \dfrac{8-6}{12+3} = \dfrac{2}{15}$

5. Simplifying the complex fraction: $\dfrac{3+\frac{2}{5}}{1-\frac{3}{7}} = \dfrac{\left(3+\frac{2}{5}\right) \cdot 35}{\left(1-\frac{3}{7}\right) \cdot 35} = \dfrac{105+14}{35-15} = \dfrac{119}{20}$

7. Simplifying the complex fraction: $\dfrac{\frac{1}{x}}{1+\frac{1}{x}} = \dfrac{\left(\frac{1}{x}\right) \cdot x}{\left(1+\frac{1}{x}\right) \cdot x} = \dfrac{1}{x+1}$

9. Simplifying the complex fraction: $\dfrac{1+\frac{1}{a}}{1-\frac{1}{a}} = \dfrac{\left(1+\frac{1}{a}\right) \cdot a}{\left(1-\frac{1}{a}\right) \cdot a} = \dfrac{a+1}{a-1}$

11. Simplifying the complex fraction: $\dfrac{\frac{1}{x}-\frac{1}{y}}{\frac{1}{x}+\frac{1}{y}} = \dfrac{\left(\frac{1}{x}-\frac{1}{y}\right) \cdot xy}{\left(\frac{1}{x}+\frac{1}{y}\right) \cdot xy} = \dfrac{y-x}{y+x}$

13.   Simplifying the complex fraction:

$$\frac{\dfrac{x-5}{x^2-4}}{\dfrac{x^2-25}{x+2}} = \frac{\dfrac{x-5}{(x+2)(x-2)} \bullet (x+2)(x-2)}{\dfrac{(x+5)(x-5)}{x+2} \bullet (x+2)(x-2)} = \frac{x-5}{(x+5)(x-5)(x-2)} = \frac{1}{(x+5)(x-2)}$$

15.   Simplifying the complex fraction:

$$\frac{\dfrac{4a}{2a^3+2}}{\dfrac{8a}{4a+4}} = \frac{\dfrac{4a}{2(a+1)(a^2-a+1)} \bullet 2(a+1)(a^2-a+1)}{\dfrac{8a}{4(a+1)} \bullet 2(a+1)(a^2-a+1)} = \frac{4a}{4a(a^2-a+1)} = \frac{1}{a^2-a+1}$$

17.   Simplifying the complex fraction: $\dfrac{1-\dfrac{9}{x^2}}{1-\dfrac{1}{x}-\dfrac{6}{x^2}} = \dfrac{\left(1-\dfrac{9}{x^2}\right) \bullet x^2}{\left(1-\dfrac{1}{x}-\dfrac{6}{x^2}\right) \bullet x^2} = \dfrac{x^2-9}{x^2-x-6} = \dfrac{(x+3)(x-3)}{(x+2)(x-3)} = \dfrac{x+3}{x+2}$

19.   Simplifying the complex fraction: $\dfrac{2+\dfrac{5}{a}-\dfrac{3}{a^2}}{2-\dfrac{5}{a}+\dfrac{2}{a^2}} = \dfrac{\left(2+\dfrac{5}{a}-\dfrac{3}{a^2}\right) \bullet a^2}{\left(2-\dfrac{5}{a}+\dfrac{2}{a^2}\right) \bullet a^2} = \dfrac{2a^2+5a-3}{2a^2-5a+2} = \dfrac{(2a-1)(a+3)}{(2a-1)(a-2)} = \dfrac{a+3}{a-2}$

21.   Simplifying the complex fraction:

$$\frac{27-\dfrac{8}{x^3}}{3+\dfrac{1}{x}-\dfrac{2}{x^2}} = \frac{\left(27-\dfrac{8}{x^3}\right) \bullet x^3}{\left(3+\dfrac{1}{x}-\dfrac{2}{x^2}\right) \bullet x^3} = \frac{27x^3-8}{3x^3+x^2-2x} = \frac{(3x-2)(9x^2+6x+4)}{x(3x-2)(x+1)} = \frac{9x^2+6x+4}{x(x+1)}$$

23.   Simplifying the complex fraction:

$$\frac{1+\dfrac{2}{x}+\dfrac{4}{x^2}+\dfrac{8}{x^3}}{1-\dfrac{16}{x^4}} = \frac{\left(1+\dfrac{2}{x}+\dfrac{4}{x^2}+\dfrac{8}{x^3}\right) \bullet x^4}{\left(1-\dfrac{16}{x^4}\right) \bullet x^4}$$

$$= \frac{x^4+2x^3+4x^2+8x}{x^4-16}$$

$$= \frac{x\left(x^3+2x^2+4x+8\right)}{\left(x^2+4\right)\left(x^2-4\right)}$$

$$= \frac{x(x+2)\left(x^2+4\right)}{\left(x^2+4\right)(x+2)(x-2)}$$

$$= \frac{x}{x-2}$$

25.   Simplifying the complex fraction:

$$\frac{2+\dfrac{3}{x}-\dfrac{18}{x^2}-\dfrac{27}{x^3}}{2+\dfrac{9}{x}+\dfrac{9}{x^2}} = \frac{\left(2+\dfrac{3}{x}-\dfrac{18}{x^2}-\dfrac{27}{x^3}\right) \bullet x^3}{\left(2+\dfrac{9}{x}+\dfrac{9}{x^2}\right) \bullet x^3}$$

$$= \frac{2x^3+3x^2-18x-27}{2x^3+9x^2+9x}$$

$$= \frac{x^2(2x+3)-9(2x+3)}{x\left(2x^2+9x+9\right)}$$

$$= \frac{(2x+3)\left(x^2-9\right)}{x(2x+3)(x+3)}$$

$$= \frac{(2x+3)(x+3)(x-3)}{x(2x+3)(x+3)}$$

$$= \frac{x-3}{x}$$

27.   Simplifying the complex fraction: $\dfrac{1+\dfrac{1}{x+3}}{1-\dfrac{1}{x+3}} = \dfrac{\left(1+\dfrac{1}{x+3}\right) \bullet (x+3)}{\left(1-\dfrac{1}{x+3}\right) \bullet (x+3)} = \dfrac{x+3+1}{x+3-1} = \dfrac{x+4}{x+2}$

29. Simplifying the complex fraction:

$$\frac{1-\dfrac{1}{a+1}}{1+\dfrac{1}{a-1}} = \frac{\left(1-\dfrac{1}{a+1}\right)\bullet(a+1)(a-1)}{\left(1+\dfrac{1}{a-1}\right)\bullet(a+1)(a-1)} = \frac{(a+1)(a-1)-(a-1)}{(a+1)(a-1)+(a+1)} = \frac{(a-1)(a+1-1)}{(a+1)(a-1+1)} = \frac{a(a-1)}{a(a+1)} = \frac{a-1}{a+1}$$

31. Simplifying the complex fraction: $\dfrac{\dfrac{1}{x+3}+\dfrac{1}{x-3}}{\dfrac{1}{x+3}-\dfrac{1}{x-3}} = \dfrac{\left(\dfrac{1}{x+3}+\dfrac{1}{x-3}\right)\bullet(x+3)(x-3)}{\left(\dfrac{1}{x+3}-\dfrac{1}{x-3}\right)\bullet(x+3)(x-3)} = \dfrac{(x-3)+(x+3)}{(x-3)-(x+3)} = \dfrac{2x}{-6} = -\dfrac{x}{3}$

33. Simplifying the complex fraction:

$$\frac{\dfrac{y+1}{y-1}+\dfrac{y-1}{y+1}}{\dfrac{y+1}{y-1}-\dfrac{y-1}{y+1}} = \frac{\left(\dfrac{y+1}{y-1}+\dfrac{y-1}{y+1}\right)\bullet(y+1)(y-1)}{\left(\dfrac{y+1}{y-1}-\dfrac{y-1}{y+1}\right)\bullet(y+1)(y-1)}$$

$$= \frac{(y+1)^2+(y-1)^2}{(y+1)^2-(y-1)^2}$$

$$= \frac{y^2+2y+1+y^2-2y+1}{y^2+2y+1-y^2+2y-1}$$

$$= \frac{2y^2+2}{4y}$$

$$= \frac{2\left(y^2+1\right)}{4y}$$

$$= \frac{y^2+1}{2y}$$

35. Simplifying the complex fraction: $1-\dfrac{x}{1-\dfrac{1}{x}} = 1-\dfrac{x\bullet x}{\left(1-\dfrac{1}{x}\right)\bullet x} = 1-\dfrac{x^2}{x-1} = \dfrac{x-1-x^2}{x-1} = \dfrac{-x^2+x-1}{x-1}$

37. Simplifying the complex fraction: $1+\dfrac{1}{1+\dfrac{1}{1+1}} = 1+\dfrac{1}{1+\dfrac{1}{2}} = 1+\dfrac{1}{\dfrac{3}{2}} = 1+\dfrac{2}{3} = \dfrac{5}{3}$

39. Simplifying the complex fraction:

$$\frac{1-\dfrac{1}{x+\dfrac{1}{2}}}{1+\dfrac{1}{x+\dfrac{1}{2}}} = \frac{1-\dfrac{1\bullet2}{\left(x+\dfrac{1}{2}\right)\bullet2}}{1+\dfrac{1\bullet2}{\left(x+\dfrac{1}{2}\right)\bullet2}} = \frac{1-\dfrac{2}{2x+1}}{1+\dfrac{2}{2x+1}} = \frac{\left(1-\dfrac{2}{2x+1}\right)(2x+1)}{\left(1+\dfrac{2}{2x+1}\right)(2x+1)} = \frac{2x+1-2}{2x+1+2} = \frac{2x-1}{2x+3}$$

41. Simplifying the complex fraction:

$$\frac{\dfrac{1}{x+h}-\dfrac{1}{x}}{h} = \frac{\left(\dfrac{1}{x+h}-\dfrac{1}{x}\right)\bullet x(x+h)}{h\bullet x(x+h)} = \frac{x-(x+h)}{hx(x+h)} = \frac{x-x-h}{hx(x+h)} = \frac{-h}{hx(x+h)} = -\frac{1}{x(x+h)}$$

43. Simplifying the complex fraction: $\dfrac{\dfrac{3}{ab}+\dfrac{4}{bc}-\dfrac{2}{ac}}{\dfrac{5}{abc}} = \dfrac{\left(\dfrac{3}{ab}+\dfrac{4}{bc}-\dfrac{2}{ac}\right)\bullet abc}{\left(\dfrac{5}{abc}\right)\bullet abc} = \dfrac{3c+4a-2b}{5}$

45. Simplifying the complex fraction:

$$\dfrac{\dfrac{t^2-2t-8}{t^2+7t+6}}{\dfrac{t^2-t-6}{t^2+2t+1}} = \dfrac{\dfrac{(t-4)(t+2)}{(t+6)(t+1)}\bullet(t+6)(t+1)^2}{\dfrac{(t-3)(t+2)}{(t+1)^2}\bullet(t+6)(t+1)^2} = \dfrac{(t-4)(t+2)(t+1)}{(t-3)(t+2)(t+6)} = \dfrac{(t-4)(t+1)}{(t+6)(t-3)}$$

47. Simplifying the complex fraction:

$$\dfrac{5+\dfrac{4}{b-1}}{\dfrac{7}{b+5}-\dfrac{3}{b-1}} = \dfrac{\left(5+\dfrac{4}{b-1}\right)\bullet(b+5)(b-1)}{\left(\dfrac{7}{b+5}-\dfrac{3}{b-1}\right)\bullet(b+5)(b-1)}$$
$$= \dfrac{5(b+5)(b-1)+4(b+5)}{7(b-1)-3(b+5)}$$
$$= \dfrac{(b+5)(5b-5+4)}{7b-7-3b-15}$$
$$= \dfrac{(b+5)(5b-1)}{4b-22}$$
$$= \dfrac{(5b-1)(b+5)}{2(2b-11)}$$

49. Simplifying the complex fraction:

$$\dfrac{\dfrac{3}{x^2-x-6}}{\dfrac{2}{x+2}-\dfrac{4}{x-3}} = \dfrac{\dfrac{3}{(x-3)(x+2)}\bullet(x-3)(x+2)}{\left(\dfrac{2}{x+2}-\dfrac{4}{x-3}\right)\bullet(x-3)(x+2)}$$
$$= \dfrac{3}{2(x-3)-4(x+2)}$$
$$= \dfrac{3}{2x-6-4x-8}$$
$$= \dfrac{3}{-2x-14}$$
$$= -\dfrac{3}{2x+14}$$

51. Simplifying the complex fraction: $\dfrac{\dfrac{1}{m-4}+\dfrac{1}{m-5}}{\dfrac{1}{m^2-9m+20}} = \dfrac{\left(\dfrac{1}{m-4}+\dfrac{1}{m-5}\right)\bullet(m-4)(m-5)}{\dfrac{1}{(m-4)(m-5)}\bullet(m-4)(m-5)} = \dfrac{(m-5)+(m-4)}{1} = 2m-9$

53. **a.** Simplifying the difference quotient: $\dfrac{f(x)-f(a)}{x-a} = \dfrac{\dfrac{4}{x}-\dfrac{4}{a}}{x-a} = \dfrac{\left(\dfrac{4}{x}-\dfrac{4}{a}\right)ax}{(x-a)ax} = \dfrac{4a-4x}{ax(x-a)} = \dfrac{-4(x-a)}{ax(x-a)} = -\dfrac{4}{ax}$

  b. Simplifying the difference quotient:

$$\dfrac{f(x)-f(a)}{x-a} = \dfrac{\dfrac{1}{x+1}-\dfrac{1}{a+1}}{x-a}$$
$$= \dfrac{\left(\dfrac{1}{x+1}-\dfrac{1}{a+1}\right)(x+1)(a+1)}{(x-a)(x+1)(a+1)}$$
$$= \dfrac{a+1-x-1}{(x-a)(x+1)(a+1)}$$
$$= \dfrac{a-x}{(x-a)(x+1)(a+1)}$$
$$= -\dfrac{1}{(x+1)(a+1)}$$

  c. Simplifying the difference quotient:

$$\dfrac{f(x)-f(a)}{x-a} = \dfrac{\dfrac{1}{x^2}-\dfrac{1}{a^2}}{x-a} = \dfrac{\left(\dfrac{1}{x^2}-\dfrac{1}{a^2}\right)a^2x^2}{a^2x^2(x-a)} = \dfrac{a^2-x^2}{a^2x^2(x-a)} = \dfrac{(a+x)(a-x)}{a^2x^2(x-a)} = -\dfrac{a+x}{a^2x^2}$$

55. Rewriting without negative exponents: $f=\left(a^{-1}+b^{-1}\right)^{-1} = \dfrac{1}{a^{-1}+b^{-1}} = \dfrac{1}{\dfrac{1}{a}+\dfrac{1}{b}} = \dfrac{1\bullet ab}{\left(\dfrac{1}{a}+\dfrac{1}{b}\right)\bullet ab} = \dfrac{ab}{a+b}$

57. **a.** As v approaches 0, the denominator approaches 1.

   b. Solving v:

$$h = \frac{f}{1+\dfrac{v}{s}}$$

$$h = \frac{f \bullet s}{\left(1+\dfrac{v}{s}\right)s}$$

$$h = \frac{fs}{s+v}$$

$$h(s+v) = fs$$

$$s+v = \frac{fs}{h}$$

$$v = \frac{fs}{h} - s$$

59. Solving the equation:

$$3x + 60 = 15$$
$$3x = -45$$
$$x = -15$$

61. Solving the equation:

$$3(y-3) = 2(y-2)$$
$$3y - 9 = 2y - 4$$
$$y = 5$$

63. Solving the equation:

$$10 - 2(x+3) = x+1$$
$$10 - 2x - 6 = x+1$$
$$-2x + 4 = x+1$$
$$-3x = -3$$
$$x = 1$$

65. Solving the equation:

$$x^2 - x - 12 = 0$$
$$(x-4)(x+3) = 0$$
$$x = -3, 4$$

67. Solving the equation:

$$(x+1)(x-6) = -12$$
$$x^2 - 5x - 6 = -12$$
$$x^2 - 5x + 6 = 0$$
$$(x-2)(x-3) = 0$$
$$x = 2, 3$$

69. Multiplying: $x(y-2) = xy - 2x$

71. Multiplying: $6\left(\dfrac{x}{2} - 3\right) = 6 \bullet \dfrac{x}{2} - 6 \bullet 3 = 3x - 18$

73. Multiplying: $xab \bullet \dfrac{1}{x} = ab$

75. Factoring: $y^2 - 25 = (y+5)(y-5)$

77. Factoring: $xa + xb = x(a+b)$

79. Solving the equation:

$$5x - 4 = 6$$
$$5x = 10$$
$$x = 2$$

81. Simplifying the expression: $\dfrac{\left(\dfrac{1}{3}\right) - \left(\dfrac{1}{3}\right)^2}{1 - \dfrac{1}{3}} = \dfrac{\dfrac{1}{3} - \dfrac{1}{9}}{1 - \dfrac{1}{3}} = \dfrac{\left(\dfrac{1}{3} - \dfrac{1}{9}\right) \bullet 9}{\left(1 - \dfrac{1}{3}\right) \bullet 9} = \dfrac{3-1}{9-3} = \dfrac{2}{6} = \dfrac{1}{3}$

83. Simplifying the expression: $\dfrac{\left(\dfrac{1}{9}\right) - \dfrac{1}{9}\left(\dfrac{1}{3}\right)^4}{1 - \dfrac{1}{3}} = \dfrac{\dfrac{1}{9} - \dfrac{1}{729}}{1 - \dfrac{1}{3}} = \dfrac{\left(\dfrac{1}{9} - \dfrac{1}{729}\right) \bullet 729}{\left(1 - \dfrac{1}{3}\right) \bullet 729} = \dfrac{81-1}{729-243} = \dfrac{80}{486} = \dfrac{40}{243}$

85.   Simplifying the expression:

$$\frac{1+\dfrac{1}{1-\dfrac{a}{b}}}{1-\dfrac{3}{1-\dfrac{a}{b}}}=\frac{1+\dfrac{1\bullet b}{\left(1-\dfrac{a}{b}\right)\bullet b}}{1-\dfrac{3\bullet b}{\left(1-\dfrac{a}{b}\right)\bullet b}}=\frac{1+\dfrac{b}{b-a}}{1-\dfrac{3b}{b-a}}=\frac{\left(1+\dfrac{b}{b-a}\right)(b-a)}{\left(1-\dfrac{3b}{b-a}\right)(b-a)}=\frac{b-a+b}{b-a-3b}=\frac{-a+2b}{-a-2b}=\frac{a-2b}{a+2b}$$

87.   Simplifying the expression: $\dfrac{a^{-1}+b^{-1}}{2(ab)^{-1}}=\dfrac{\dfrac{1}{a}+\dfrac{1}{b}}{\dfrac{2}{ab}}=\dfrac{\left(\dfrac{1}{a}+\dfrac{1}{b}\right)ab}{\left(\dfrac{2}{ab}\right)ab}=\dfrac{a+b}{2}$

89.   Simplifying the expression:

$$\frac{\left(q^{-2}-t^{-2}\right)^{-1}}{\left(t^{-1}-q^{-1}\right)^{-1}}=\frac{t^{-1}-q^{-1}}{q^{-2}-t^{-2}}=\frac{\dfrac{1}{t}-\dfrac{1}{q}}{\dfrac{1}{q^2}-\dfrac{1}{t^2}}=\frac{\left(\dfrac{1}{t}-\dfrac{1}{q}\right)q^2t^2}{\left(\dfrac{1}{q^2}-\dfrac{1}{t^2}\right)q^2t^2}=\frac{q^2t-qt^2}{t^2-q^2}=\frac{-qt(t-q)}{(t+q)(t-q)}=-\frac{qt}{q+t}$$

6.6 Equations and Graphs with Rational Expressions

1.   Solving the equation:

$$\frac{x}{5}+4=\frac{5}{3}$$
$$15\left(\frac{x}{5}+4\right)=15\left(\frac{5}{3}\right)$$
$$3x+60=25$$
$$3x=-35$$
$$x=-\frac{35}{3}$$

3.   Solving the equation:

$$\frac{a}{3}+2=\frac{4}{5}$$
$$15\left(\frac{a}{3}+2\right)=15\left(\frac{4}{5}\right)$$
$$5a+30=12$$
$$5a=-18$$
$$a=-\frac{18}{5}$$

5.   Solving the equation:

$$\frac{y}{2}+\frac{y}{4}+\frac{y}{6}=3$$
$$12\left(\frac{y}{2}+\frac{y}{4}+\frac{y}{6}\right)=12(3)$$
$$6y+3y+2y=36$$
$$11y=36$$
$$y=\frac{36}{11}$$

7.   Solving the equation:

$$\frac{5}{2x}=\frac{1}{x}+\frac{3}{4}$$
$$4x\left(\frac{5}{2x}\right)=4x\left(\frac{1}{x}+\frac{3}{4}\right)$$
$$10=4+3x$$
$$3x=6$$
$$x=2$$

9.   Solving the equation:

$$\frac{1}{x}=\frac{1}{3}-\frac{2}{3x}$$
$$3x\left(\frac{1}{x}\right)=3x\left(\frac{1}{3}-\frac{2}{3x}\right)$$
$$3=x-2$$
$$x=5$$

11.   Solving the equation:

$$\frac{2x}{x-3}+2=\frac{2}{x-3}$$
$$(x-3)\left(\frac{2x}{x-3}+2\right)=(x-3)\left(\frac{2}{x-3}\right)$$
$$2x+2(x-3)=2$$
$$2x+2x-6=2$$
$$4x=8$$
$$x=2$$

13. Solving the equation:

$$1 - \frac{1}{x} = \frac{12}{x^2}$$

$$x^2\left(1 - \frac{1}{x}\right) = x^2\left(\frac{12}{x^2}\right)$$

$$x^2 - x = 12$$

$$x^2 - x - 12 = 0$$

$$(x+3)(x-4) = 0$$

$$x = -3, 4$$

15. Solving the equation:

$$y - \frac{4}{3y} = -\frac{1}{3}$$

$$3y\left(y - \frac{4}{3y}\right) = 3y\left(-\frac{1}{3}\right)$$

$$3y^2 - 4 = -y$$

$$3y^2 + y - 4 = 0$$

$$(3y+4)(y-1) = 0$$

$$x = -\frac{4}{3}, 1$$

17. Solving the equation:

$$f(x) + g(x) = \frac{5}{8}$$

$$\frac{1}{x-3} + \frac{1}{x+3} = \frac{5}{8}$$

$$8(x-3)(x+3)\left(\frac{1}{x-3} + \frac{1}{x+3}\right) = 8(x-3)(x+3)\left(\frac{5}{8}\right)$$

$$8(x+3) + 8(x-3) = 5(x-3)(x+3)$$

$$8x + 24 + 8x - 24 = 5x^2 - 45$$

$$16x = 5x^2 - 45$$

$$0 = 5x^2 - 16x - 45$$

$$0 = (5x+9)(x-5)$$

$$x = -\frac{9}{5}, 5$$

19. Solving the equation:

$$\frac{f(x)}{g(x)} = 5$$

$$\frac{\frac{1}{x-3}}{\frac{1}{x+3}} = 5$$

$$\frac{x+3}{x-3} = 5$$

$$x+3 = 5(x-3)$$

$$x + 3 = 5x - 15$$

$$18 = 4x$$

$$x = \frac{9}{2}$$

21. Solving the equation:

$$f(x) = g(x)$$

$$\frac{1}{x-3} = \frac{1}{x+3}$$

$$x + 3 = x - 3$$

$$3 = -3 \quad \text{(false)}$$

There is no solution $(\varnothing)$.

23. Solving the equation:

$$f(x) + g(x) = \frac{24}{5}$$

$$\frac{4}{x+2} + \frac{4}{x-2} = \frac{24}{5}$$

$$5(x-2)(x+2)\left(\frac{4}{x+2} + \frac{4}{x-2}\right) = 5(x-2)(x+2)\left(\frac{24}{5}\right)$$

$$20(x-2) + 20(x+2) = 24(x-2)(x+2)$$

$$20x - 40 + 20x + 40 = 24x^2 - 96$$

$$40x = 24x^2 - 96$$

$$0 = 24x^2 - 40x - 96$$

$$0 = 3x^2 - 5x - 12$$

$$0 = (3x+4)(x-3)$$

$$x = -\frac{4}{3}, 3$$

25. Solving the equation:

$$\frac{f(x)}{g(x)} = -5$$

$$\frac{\dfrac{4}{x+2}}{\dfrac{4}{x-2}} = -5$$

$$\frac{x-2}{x+2} = -5$$

$$x - 2 = -5(x+2)$$

$$x - 2 = -5x - 10$$

$$6x = -8$$

$$x = -\frac{4}{3}$$

27. Solving the equation:

$$f(x) = g(x)$$

$$\frac{4}{x+2} = \frac{4}{x-2}$$

$$x - 2 = x + 2$$

$$-2 = 2 \quad \text{(false)}$$

There is no solution $(\varnothing)$.

29. **a.** Solving the equation:

$$6x - 2 = 0$$

$$6x = 2$$

$$x = \frac{1}{3}$$

b. Solving the equation:

$$\frac{6}{x} - 2 = 0$$

$$x\left(\frac{6}{x} - 2\right) = x(0)$$

$$6 - 2x = 0$$

$$6 = 2x$$

$$x = 3$$

c. Solving the equation:

$$\frac{x}{6} - 2 = -\frac{1}{2}$$

$$6\left(\frac{x}{6} - 2\right) = 6\left(-\frac{1}{2}\right)$$

$$x - 12 = -3$$

$$x = 9$$

d. Solving the equation:

$$\frac{6}{x} - 2 = -\frac{1}{2}$$

$$2x\left(\frac{6}{x} - 2\right) = 2x\left(-\frac{1}{2}\right)$$

$$12 - 4x = -x$$

$$12 = 3x$$

$$x = 4$$

e. Solving the equation:

$$\frac{6}{x^2} + 6 = \frac{20}{x}$$

$$x^2\left(\frac{6}{x^2} + 6\right) = x^2\left(\frac{20}{x}\right)$$

$$6 + 6x^2 = 20x$$

$$6x^2 - 20x + 6 = 0$$

$$3x^2 - 10x + 3 = 0$$

$$(3x - 1)(x - 3) = 0$$

$$x = \frac{1}{3}, 3$$

31. **a.** Dividing: $\dfrac{6}{x^2 - 2x - 8} \div \dfrac{x+3}{x+2} = \dfrac{6}{(x-4)(x+2)} \cdot \dfrac{x+2}{x+3} = \dfrac{6}{(x-4)(x+3)}$

b. Adding:

$$\frac{6}{x^2 - 2x - 8} + \frac{x+3}{x+2} = \frac{6}{(x-4)(x+2)} + \frac{x+3}{x+2} \cdot \frac{x-4}{x-4}$$

$$= \frac{6}{(x-4)(x+2)} + \frac{x^2 - x - 12}{(x-4)(x+2)}$$

$$= \frac{x^2 - x - 6}{(x-4)(x+2)}$$

$$= \frac{(x-3)(x+2)}{(x-4)(x+2)}$$

$$= \frac{x-3}{x-4}$$

c. Solving the equation:

$$\frac{6}{x^2 - 2x - 8} + \frac{x+3}{x+2} = 2$$

$$(x-4)(x+2)\left(\frac{6}{(x-4)(x+2)} + \frac{x+3}{x+2}\right) = (x-4)(x+2)(2)$$

$$6 + (x-4)(x+3) = 2(x-4)(x+2)$$

$$6 + x^2 - x - 12 = 2x^2 - 4x - 16$$

$$0 = x^2 - 3x - 10$$

$$0 = (x-5)(x+2)$$

$$x = -2, 5$$

Note that $x = -2$ does not check in the original equation, so the solution is $x = 5$.

33. Solving the equation:

$$\frac{x+2}{x+1} = \frac{1}{x+1} + 2$$

$$(x+1)\left(\frac{x+2}{x+1}\right) = (x+1)\left(\frac{1}{x+1} + 2\right)$$

$$x + 2 = 1 + 2(x+1)$$

$$x + 2 = 1 + 2x + 2$$

$$x + 2 = 2x + 3$$

$$x = -1 \quad (\text{does not check})$$

There is no solution (-1 does not check).

35. Solving the equation:

$$\frac{3}{a-2} = \frac{2}{a-3}$$

$$(a-2)(a-3)\left(\frac{3}{a-2}\right) = (a-2)(a-3)\left(\frac{2}{a-3}\right)$$

$$3(a-3) = 2(a-2)$$

$$3a - 9 = 2a - 4$$

$$a = 5$$

37. Solving the equation:

$$6 - \frac{5}{x^2} = \frac{7}{x}$$

$$x^2\left(6 - \frac{5}{x^2}\right) = x^2\left(\frac{7}{x}\right)$$

$$6x^2 - 5 = 7x$$

$$6x^2 - 7x - 5 = 0$$

$$(2x+1)(3x-5) = 0$$

$$x = -\frac{1}{2}, \frac{5}{3}$$

39. Solving the equation:

$$\frac{1}{x-1} - \frac{1}{x+1} = \frac{3x}{x^2-1}$$

$$(x+1)(x-1)\left(\frac{1}{x-1} - \frac{1}{x+1}\right) = (x+1)(x-1)\left(\frac{3x}{(x+1)(x-1)}\right)$$

$$(x+1) - (x-1) = 3x$$

$$x+1-x+1 = 3x$$

$$3x = 2$$

$$x = \frac{2}{3}$$

41. Solving the equation:

$$\frac{2}{x-3} + \frac{x}{x^2-9} = \frac{4}{x+3}$$

$$(x+3)(x-3)\left(\frac{2}{x-3} + \frac{x}{(x+3)(x-3)}\right) = (x+3)(x-3)\left(\frac{4}{x+3}\right)$$

$$2(x+3) + x = 4(x-3)$$

$$2x+6+x = 4x-12$$

$$3x+6 = 4x-12$$

$$-x = -18$$

$$x = 18$$

43. Solving the equation:

$$\frac{3}{2} - \frac{1}{x-4} = \frac{-2}{2x-8}$$

$$2(x-4)\left(\frac{3}{2} - \frac{1}{x-4}\right) = 2(x-4)\left(\frac{-2}{2(x-4)}\right)$$

$$3(x-4) - 2 = -2$$

$$3x-12-2 = -2$$

$$3x-14 = -2$$

$$3x = 12$$

$$x = 4 \quad \text{(does not check)}$$

There is no solution (4 does not check).

45.   Solving the equation:

$$\frac{t-4}{t^2-3t}=\frac{-2}{t^2-9}$$

$$t(t+3)(t-3)\cdot\frac{t-4}{t(t-3)}=t(t+3)(t-3)\cdot\frac{-2}{(t+3)(t-3)}$$

$$(t+3)(t-4)=-2t$$

$$t^2-t-12=-2t$$

$$t^2+t-12=0$$

$$(t+4)(t-3)=0$$

$$t=-4 \quad (t=3 \text{ does not check})$$

The solution is –4 (3 does not check).

47.   Solving the equation:

$$\frac{3}{y-4}-\frac{2}{y+1}=\frac{5}{y^2-3y-4}$$

$$(y-4)(y+1)\left(\frac{3}{y-4}-\frac{2}{y+1}\right)=(y-4)(y+1)\left(\frac{5}{(y-4)(y+1)}\right)$$

$$3(y+1)-2(y-4)=5$$

$$3y+3-2y+8=5$$

$$y+11=5$$

$$y=-6$$

49.   Solving the equation:

$$\frac{2}{1+a}=\frac{3}{1-a}+\frac{5}{a}$$

$$a(1+a)(1-a)\left(\frac{2}{1+a}\right)=a(1+a)(1-a)\left(\frac{3}{1-a}+\frac{5}{a}\right)$$

$$2a(1-a)=3a(1+a)+5(1+a)(1-a)$$

$$2a-2a^2=3a+3a^2+5-5a^2$$

$$-2a^2+2a=-2a^2+3a+5$$

$$2a=3a+5$$

$$-a=5$$

$$a=-5$$

51.   Solving the equation:

$$\frac{3}{2x-6}-\frac{x+1}{4x-12}=4$$

$$4(x-3)\left(\frac{3}{2(x-3)}-\frac{x+1}{4(x-3)}\right)=4(x-3)(4)$$

$$6-(x+1)=16x-48$$

$$5-x=16x-48$$

$$-17x=-53$$

$$x=\frac{53}{17}$$

53.   Solving the equation:

$$\frac{y+2}{y^2-y}-\frac{6}{y^2-1}=0$$

$$y(y+1)(y-1)\left(\frac{y+2}{y(y-1)}-\frac{6}{(y+1)(y-1)}\right)=y(y+1)(y-1)(0)$$

$$(y+1)(y+2)-6y=0$$

$$y^2+3y+2-6y=0$$

$$y^2-3y+2=0$$

$$(y-1)(y-2)=0$$

$$y=2 \quad (y=1 \text{ does not check})$$

The solution is 2 (1 does not check).

55. Solving the equation:

$$\frac{4}{2x-6} - \frac{12}{4x+12} = \frac{12}{x^2-9}$$

$$4(x+3)(x-3)\left(\frac{4}{2(x-3)} - \frac{12}{4(x+3)}\right) = 4(x+3)(x-3)\left(\frac{12}{(x+3)(x-3)}\right)$$

$$8(x+3) - 12(x-3) = 48$$
$$8x + 24 - 12x + 36 = 48$$
$$-4x + 60 = 48$$
$$-4x = -12$$
$$x = 3 \quad (x = 3 \text{ does not check})$$

There is no solution (3 does not check).

57. Solving the equation:

$$\frac{2}{y^2-7y+12} - \frac{1}{y^2-9} = \frac{4}{y^2-y-12}$$

$$(y+3)(y-3)(y-4)\left(\frac{2}{(y-3)(y-4)} - \frac{1}{(y+3)(y-3)}\right) = (y+3)(y-3)(y-4)\left(\frac{4}{(y-4)(y+3)}\right)$$

$$2(y+3) - (y-4) = 4(y-3)$$
$$2y + 6 - y + 4 = 4y - 12$$
$$y + 10 = 4y - 12$$
$$-3y = -22$$
$$y = \frac{22}{3}$$

59. Solving the equation:

$$6x^{-1} + 4 = 7$$
$$x\left(6x^{-1} + 4\right) = x(7)$$
$$6 + 4x = 7x$$
$$6 = 3x$$
$$x = 2$$

61. Solving the equation:

$$1 + 5x^{-2} = 6x^{-1}$$
$$x^2\left(1 + 5x^{-2}\right) = x^2\left(6x^{-1}\right)$$
$$x^2 + 5 = 6x$$
$$x^2 - 6x + 5 = 0$$
$$(x-1)(x-5) = 0$$
$$x = 1, 5$$

63. Solving for x:

$$\frac{1}{x} = \frac{1}{b} - \frac{1}{a}$$
$$abx\left(\frac{1}{x}\right) = abx\left(\frac{1}{b} - \frac{1}{a}\right)$$
$$ab = ax - bx$$
$$ab = x(a-b)$$
$$x = \frac{ab}{a-b}$$

65. Solving for R:

$$\frac{1}{R} = \frac{1}{R_1} + \frac{1}{R_2}$$
$$RR_1R_2\left(\frac{1}{R}\right) = RR_1R_2\left(\frac{1}{R_1} + \frac{1}{R_2}\right)$$
$$R_1R_2 = RR_2 + RR_1$$
$$R_1R_2 = R(R_1 + R_2)$$
$$R = \frac{R_1R_2}{R_1 + R_2}$$

67. Solving for y:

$$x = \frac{y-3}{y-1}$$
$$x(y-1) = y-3$$
$$xy - x = y - 3$$
$$xy - y - x - 3$$
$$y(x-1) = x - 3$$
$$y = \frac{x-3}{x-1}$$

69. Solving for y:

$$x = \frac{2y+1}{3y+1}$$
$$x(3y+1) = 2y+1$$
$$3xy + x = 2y + 1$$
$$3xy - 2y = -x + 1$$
$$y(3x-2) = -x + 1$$
$$y = \frac{1-x}{3x-2}$$

71.   Graphing the function:

$$f(x) = \frac{2}{x-1}$$

73.   Graphing the function:

$$f(x) = \frac{3}{x+2}$$

75.   Graphing the function:

$$f(x) = \frac{-2}{x-3}$$

77.   Graphing the function:

$$f(x) = \frac{4}{3x-2}$$

79.   Graphing the function:

$$f(x) = \frac{-5}{2x+1}$$

81.   Graphing the function:

$$f(x) = \frac{x-1}{x+1}$$

83.   Graphing the function:

$$f(x) = \frac{2x}{x-1}$$

85.   Graphing the function:

$$f(x) = \frac{-2x}{x+1}$$

87. Graphing the function:

89. Graphing the function:

91. Simplifying the left-hand side: $\dfrac{2}{x-y} - \dfrac{1}{y-x} = \dfrac{2}{x-y} - \dfrac{-1}{x-y} = \dfrac{2}{x-y} + \dfrac{1}{x-y} = \dfrac{3}{x-y}$

93. Completing the table:

Time t (sec)	Speed of Kayak in Still Water v (m/sec)	Current of the River c (m/sec)
240	4	1
300	4	2
514	4	3
338	3	1
540	3	2
impossible	3	3

95. Let x represent the number. The equation is:

$$2(x+3) = 16$$
$$2x + 6 = 16$$
$$2x = 10$$
$$x = 5$$

The number is 5.

97. Let w represent the width and $2w - 3$ represent the length. The equation is:

$$2w + 2(2w - 3) = 42$$
$$2w + 4w - 6 = 42$$
$$6w = 48$$
$$w = 8$$
$$2w - 3 = 13$$

The width is 8 meters and the length is 13 meters.

99. Let x and $x + 1$ represent the two integers. The equation is:

$$x^2 + (x+1)^2 = 61$$
$$x^2 + x^2 + 2x + 1 = 61$$
$$2x^2 + 2x - 60 = 0$$
$$x^2 + x - 30 = 0$$
$$(x+6)(x-5) = 0$$
$$x = -6, 5$$
$$x + 1 = -5, 6$$

The integers are either –6 and –5, or 5 and 6.

101. Let x, $x + 1$, and $x + 2$ represent the three sides. The equation is:

$$x^2 + (x+1)^2 = (x+2)^2$$
$$x^2 + x^2 + 2x + 1 = x^2 + 4x + 4$$
$$x^2 - 2x - 3 = 0$$
$$(x+1)(x-3) = 0$$
$$x = -1, 3$$
$$x + 1 = 0, 4$$
$$x + 2 = 1, 5$$

Since the sides of the triangle must be positive, the sides are 3, 4, and 5.

103. Multiplying: $39.3 \cdot 60 = 2{,}358$

105. Dividing: $65{,}000 \div 5{,}280 \approx 12.3$

107. Multiplying: $2x\left(\dfrac{1}{x} + \dfrac{1}{2x}\right) = 2x \cdot \dfrac{1}{x} + 2x \cdot \dfrac{1}{2x} = 2 + 1 = 3$

109. Solving the equation:

$$12(x+3) + 12(x-3) = 3(x^2 - 9)$$
$$12x + 36 + 12x - 36 = 3x^2 - 27$$
$$24x = 3x^2 - 27$$
$$3x^2 - 24x - 27 = 0$$
$$3(x^2 - 8x - 9) = 0$$
$$3(x-9)(x+1) = 0$$
$$x = -1, 9$$

111. Solving the equation:

$$\frac{1}{10} - \frac{1}{12} = \frac{1}{x}$$
$$60x\left(\frac{1}{10} - \frac{1}{12}\right) = 60x\left(\frac{1}{x}\right)$$
$$6x - 5x = 60$$
$$x = 60$$

113. Solving the equation:

$$\frac{12}{x} + \frac{8}{x^2} - \frac{75}{x^3} - \frac{50}{x^4} = 0$$
$$x^4\left(\frac{12}{x} + \frac{8}{x^2} - \frac{75}{x^3} - \frac{50}{x^4}\right) = x^4(0)$$
$$12x^3 + 8x^2 - 75x - 50 = 0$$
$$4x^2(3x+2) - 25(3x+2) = 0$$
$$(3x+2)(4x^2 - 25) = 0$$
$$(3x+2)(2x+5)(2x-5) = 0$$
$$x = -\frac{5}{2}, -\frac{2}{3}, \frac{5}{2}$$

115. Solving the equation:

$$\frac{1}{x^3} - \frac{1}{3x^2} - \frac{1}{4x} + \frac{1}{12} = 0$$
$$12x^3\left(\frac{1}{x^3} - \frac{1}{3x^2} - \frac{1}{4x} + \frac{1}{12}\right) = 12x^3(0)$$
$$12 - 4x - 3x^2 + x^3 = 0$$
$$x^2(x-3) - 4(x-3) = 0$$
$$(x-3)(x^2 - 4) = 0$$
$$(x-3)(x+2)(x-2) = 0$$
$$x = -2, 2, 3$$

117. Solving for x:

$$\frac{2}{x} + \frac{4}{x+a} = \frac{-6}{a-x}$$
$$x(x+a)(a-x)\left(\frac{2}{x} + \frac{4}{x+a}\right) = x(x+a)(a-x)\left(\frac{-6}{a-x}\right)$$
$$2(x+a)(a-x) + 4x(a-x) = -6x(x+a)$$
$$2a^2 - 2x^2 + 4ax - 4x^2 = -6x^2 - 6ax$$
$$2a^2 = -10ax$$
$$x = -\frac{a}{5}$$

119. Solving for v:

$$\frac{s - vt}{t^2} = -16$$
$$s - vt = -16t^2$$
$$s + 16t^2 = vt$$
$$v = \frac{16t^2 + s}{t}$$

121. Solving for f:

$$\frac{1}{p} = \frac{1}{f} + \frac{1}{g}$$
$$pfg\left(\frac{1}{p}\right) = pfg\left(\frac{1}{f} + \frac{1}{g}\right)$$
$$fg = pg + pf$$
$$fg - pf = pg$$
$$f(g - p) = pg$$
$$f = \frac{pg}{g - p}$$

6.7 Applications

1. Let x and $3x$ represent the two numbers. The equation is:

$$\frac{1}{x} + \frac{1}{3x} = \frac{20}{3}$$
$$3x\left(\frac{1}{x} + \frac{1}{3x}\right) = 3x\left(\frac{20}{3}\right)$$
$$3 + 1 = 20x$$
$$20x = 4$$
$$x = \frac{1}{5}$$

The numbers are $\frac{1}{5}$ and $\frac{3}{5}$.

3. Let x represent the number. The equation is:

$$x + \frac{1}{x} = \frac{10}{3}$$
$$3x\left(x + \frac{1}{x}\right) = 3x\left(\frac{10}{3}\right)$$
$$3x^2 + 3 = 10x$$
$$3x^2 - 10x + 3 = 0$$
$$(3x - 1)(x - 3) = 0$$
$$x = \frac{1}{3}, 3$$

The number is either 3 or $\frac{1}{3}$.

5. Let x and $x + 1$ represent the two integers. The equation is:

$$\frac{1}{x} + \frac{1}{x+1} = \frac{7}{12}$$
$$12x(x+1)\left(\frac{1}{x} + \frac{1}{x+1}\right) = 12x(x+1)\left(\frac{7}{12}\right)$$
$$12(x+1) + 12x = 7x(x+1)$$
$$12x + 12 + 12x = 7x^2 + 7x$$
$$0 = 7x^2 - 17x - 12$$
$$0 = (7x + 4)(x - 3)$$
$$x = 3 \quad \left(x = -\frac{4}{7} \text{ is not an integer}\right)$$

The two integers are 3 and 4.

7. Let x represent the number. The equation is:

$$\frac{7+x}{9+x}=\frac{5}{6}$$

$$6(9+x)\left(\frac{7+x}{9+x}\right)=6(9+x)\left(\frac{5}{6}\right)$$

$$6(7+x)=5(9+x)$$

$$42+6x=45+5x$$

$$x=3$$

The number is 3.

9. Let x represent the number. The equation is:

$$\frac{9+x}{11+x}=\frac{6}{7}$$

$$7(11+x)\left(\frac{9+x}{11+x}\right)=7(11+x)\left(\frac{6}{7}\right)$$

$$7(9+x)=6(11+x)$$

$$63+7x=66+6x$$

$$x=3$$

The number is 3.

11. Let x represent the speed of the current. Setting the times equal:

$$\frac{3}{5+x}=\frac{1.5}{5-x}$$

$$3(5-x)=1.5(5+x)$$

$$15-3x=7.5+1.5x$$

$$7.5=4.5x$$

$$x=\frac{75}{45}=\frac{5}{3}$$

The speed of the current is $\frac{5}{3}$ mph.

13. Let x represent the speed of the boat. Since the total time is 3 hours:

$$\frac{8}{x-2}+\frac{8}{x+2}=3$$

$$(x+2)(x-2)\left(\frac{8}{x-2}+\frac{8}{x+2}\right)=3(x+2)(x-2)$$

$$8(x+2)+8(x-2)=3x^2-12$$

$$16x=3x^2-12$$

$$0=3x^2-16x-12$$

$$0=(3x+2)(x-6)$$

$$x=6 \quad \left(x=-\frac{2}{3}\text{ is impossible}\right)$$

The speed of the boat is 6 mph.

15. Let r represent the speed of train B and $r+15$ represent the speed of train A. Since the times are equal:

$$\frac{150}{r+15}=\frac{120}{r}$$

$$150r=120(r+15)$$

$$150r=120r+1800$$

$$30r=1800$$

$$r=60$$

The speed of train A is 75 mph and the speed of train B is 60 mph.

17. The smaller plane makes the trip in 3 hours, so the 747 must take $1\frac{1}{2}$ hours to complete the trip. Thus the average

speed is given by: $\dfrac{810 \text{ miles}}{1\frac{1}{2} \text{ hours}} = 540$ miles per hour

19. Let r represent the bus's usual speed. The difference of the two times is $\frac{1}{2}$ hour, therefore:

$$\frac{270}{r} - \frac{270}{r+6} = \frac{1}{2}$$
$$2r(r+6)\left(\frac{270}{r} - \frac{270}{r+6}\right) = 2r(r+6)\left(\frac{1}{2}\right)$$
$$540(r+6) - 540(r) = r(r+6)$$
$$540r + 3240 - 540r = r^2 + 6r$$
$$0 = r^2 + 6r - 3240$$
$$0 = (r-54)(r+60)$$
$$r = 54 \quad (r = -60 \text{ is impossible})$$

The usual speed is 54 mph.

21. The time for a person traveling 70 mph is: $\dfrac{205 \text{ miles}}{70 \text{ mph}} \approx 2.93$ hours ≈ 175.7 minutes

The time for a person traveling 75 mph is: $\dfrac{205 \text{ miles}}{75 \text{ mph}} \approx 2.73$ hours $= 164$ minutes

The time for a person traveling 80 mph is: $\dfrac{205 \text{ miles}}{80 \text{ mph}} \approx 2.56$ hours ≈ 153.8 minutes

The person traveling 75 mph arrives 12 minutes faster, and the person traveling 80 mph arrives 22 minutes faster.

23. Let x represent the time to fill the tank if both pipes are open. The rate equation is:

$$\frac{1}{8} - \frac{1}{16} = \frac{1}{x}$$
$$16x\left(\frac{1}{8} - \frac{1}{16}\right) = 16x\left(\frac{1}{x}\right)$$
$$2x - x = 16$$
$$x = 16$$

It will take 16 hours to fill the tank if both pipes are open.

25. Let x represent the time to empty the sink with the drain open. The rate equation is:

$$\frac{1}{3} - \frac{1}{5} = \frac{1}{x}$$
$$15x\left(\frac{1}{3} - \frac{1}{5}\right) = 15x\left(\frac{1}{x}\right)$$
$$5x - 3x = 15$$
$$2x = 15$$
$$x = 7\frac{1}{2}$$

It will take $7\frac{1}{2}$ minutes to empty the sink if both the faucet and drain are open.

27. Let x represent the time to fill the sink with the hot water faucet. The rate equation is:

$$\frac{1}{3.5}+\frac{1}{x}=\frac{1}{2.1}$$

$$7.35x\left(\frac{1}{3.5}+\frac{1}{x}\right)=7.35x\left(\frac{1}{2.1}\right)$$

$$2.1x+7.35=3.5x$$

$$7.35=1.4x$$

$$x=5.25$$

It will take $5\frac{1}{4}$ minutes to fill the sink with the hot water faucet.

29. Let x represent the time to fill the tank with pipe A and $3x$ represent the time to fill the tank with pipe B. The rate equation is:

$$\frac{1}{x}+\frac{1}{3x}=\frac{1}{4}$$

$$12x\left(\frac{1}{x}+\frac{1}{3x}\right)=12x\left(\frac{1}{4}\right)$$

$$12+4=3x$$

$$3x=16$$

$$x=\frac{16}{3}$$

$$3x=16$$

Pipe B will take 16 hours to fill the tank working alone.

31. Converting to acres: $\dfrac{2,700,000 \text{ sq. ft.}}{43,560 \text{ sq. ft./acre}}\approx 62.0$ acres

33. Converting the speed: $\dfrac{5750 \text{ feet}}{11 \text{ minutes}}\cdot\dfrac{1 \text{ mile}}{5280 \text{ feet}}\cdot\dfrac{60 \text{ minutes}}{1 \text{ hour}}\approx 5.9$ mph

35. Converting the speed: $\dfrac{100 \text{ meters}}{10.8 \text{ seconds}}\cdot\dfrac{3.28 \text{ feet}}{1 \text{ meter}}\cdot\dfrac{1 \text{ mile}}{5280 \text{ feet}}\cdot\dfrac{60 \text{ seconds}}{1 \text{ minute}}\cdot\dfrac{60 \text{ minutes}}{1 \text{ hour}}\approx 20.7$ mph

37. Converting the speed: $\dfrac{\pi\cdot 65 \text{ feet}}{30 \text{ seconds}}\cdot\dfrac{1 \text{ mile}}{5280 \text{ feet}}\cdot\dfrac{60 \text{ seconds}}{1 \text{ minute}}\cdot\dfrac{60 \text{ minutes}}{1 \text{ hour}}\approx 4.6$ mph

39. Converting the speed: $\dfrac{2\pi\cdot 2 \text{ inches}}{\frac{1}{300} \text{ minutes}}\cdot\dfrac{1 \text{ foot}}{12 \text{ inches}}\cdot\dfrac{1 \text{ mile}}{5280 \text{ feet}}\cdot\dfrac{60 \text{ minutes}}{1 \text{ hour}}\approx 3.6$ mph

41. Converting the speed: $\dfrac{2\pi\cdot 116 \text{ feet}}{\frac{1}{1000} \text{ minutes}}\cdot\dfrac{1 \text{ mile}}{5280 \text{ feet}}\cdot\dfrac{60 \text{ minutes}}{1 \text{ hour}}\approx 8,278$ mph

43. Solving the equation:

$$\frac{1}{3}\left[\left(x+\frac{2}{3}x\right)+\frac{1}{3}\left(x+\frac{2}{3}x\right)\right]=10$$

$$\left(x+\frac{2}{3}x\right)+\frac{1}{3}\left(x+\frac{2}{3}x\right)=30$$

$$x+\frac{2}{3}x+\frac{1}{3}x+\frac{2}{9}x=30$$

$$\frac{20}{9}x=30$$

$$20x=270$$

$$x=\frac{27}{2}$$

45. Performing the operations: $\dfrac{2a+10}{a^3}\cdot\dfrac{a^2}{3a+15}=\dfrac{2(a+5)}{a^3}\cdot\dfrac{a^2}{3(a+5)}=\dfrac{2}{3a}$

47. Performing the operations: $\left(x^2-9\right)\left(\dfrac{x+2}{x+3}\right)=(x+3)(x-3)\left(\dfrac{x+2}{x+3}\right)=(x-3)(x+2)$

49. Performing the operations: $\dfrac{2x-7}{x-2}-\dfrac{x-5}{x-2}=\dfrac{2x-7-x+5}{x-2}=\dfrac{x-2}{x-2}=1$

51. Simplifying the expression: $\dfrac{\dfrac{1}{x}-\dfrac{1}{3}}{\dfrac{1}{x}+\dfrac{1}{3}}=\dfrac{\left(\dfrac{1}{x}-\dfrac{1}{3}\right)\cdot 3x}{\left(\dfrac{1}{x}+\dfrac{1}{3}\right)\cdot 3x}=\dfrac{3-x}{3+x}$

53. Solving the equation:

$$\dfrac{x}{x-3}+\dfrac{3}{2}=\dfrac{3}{x-3}$$
$$2(x-3)\left(\dfrac{x}{x-3}+\dfrac{3}{2}\right)=2(x-3)\left(\dfrac{3}{x-3}\right)$$
$$2x+3(x-3)=6$$
$$2x+3x-9=6$$
$$5x=15$$
$$x=3\quad\text{(does not check)}$$

There is no solution (3 does not check).

Chapter 6 Review

1. Reducing the fraction: $\dfrac{125x^4yz^3}{35x^2y^4z^3}=\dfrac{25x^2}{7y^3}$

3. Reducing the fraction: $\dfrac{x^2-25}{x^2+10x+25}=\dfrac{(x+5)(x-5)}{(x+5)^2}=\dfrac{x-5}{x+5}$

5. Dividing: $\dfrac{12x^3+8x^2+16x}{4x^2}=\dfrac{12x^3}{4x^2}+\dfrac{8x^2}{4x^2}+\dfrac{16x}{4x^2}=3x+2+\dfrac{4}{x}$

7. Dividing: $\dfrac{x^{6n}-x^{5n}}{x^{3n}}=\dfrac{x^{6n}}{x^{3n}}-\dfrac{x^{5n}}{x^{3n}}=x^{3n}-x^{2n}$

9. Dividing by factoring: $\dfrac{5x^2-14xy-24y^2}{x-4y}=\dfrac{(5x+6y)(x-4y)}{x-4y}=5x+6y$

11. Dividing using long division:

$$\begin{array}{r}4x+1\\2x-7\overline{)8x^2-26x-9}\\\underline{8x^2-28x}\\2x-9\\\underline{2x-7}\\-2\end{array}$$

The quotient is $4x+1-\dfrac{2}{2x-7}$.

13. Performing the operations: $\dfrac{3}{4}\cdot\dfrac{12}{15}\div\dfrac{1}{3}=\dfrac{3}{4}\cdot\dfrac{12}{15}\cdot\dfrac{3}{1}=\dfrac{9}{5}$

15. Performing the operations: $\dfrac{x^3-1}{x^4-1}\cdot\dfrac{x^2-1}{x^2+x+1}=\dfrac{(x-1)\left(x^2+x+1\right)}{\left(x^2+1\right)\left(x^2-1\right)}\cdot\dfrac{x^2-1}{x^2+x+1}=\dfrac{x-1}{x^2+1}$

17. Performing the operations:

$$\frac{ax+bx+2a+2b}{ax-3a+bx-3b} \div \frac{ax-bx-2a+2b}{ax-bx-3a+3b} = \frac{ax+bx+2a+2b}{ax-3a+bx-3b} \cdot \frac{ax-bx-3a+3b}{ax-bx-2a+2b}$$

$$= \frac{x(a+b)+2(a+b)}{a(x-3)+b(x-3)} \cdot \frac{x(a-b)-3(a-b)}{x(a-b)-2(a-b)}$$

$$= \frac{(a+b)(x+2)}{(x-3)(a+b)} \cdot \frac{(a-b)(x-3)}{(a-b)(x-2)}$$

$$= \frac{x+2}{x-2}$$

19. Performing the operations: $\dfrac{3}{5}-\dfrac{1}{10}+\dfrac{8}{15}=\dfrac{3}{5}\cdot\dfrac{6}{6}-\dfrac{1}{10}\cdot\dfrac{3}{3}+\dfrac{8}{15}\cdot\dfrac{2}{2}=\dfrac{18}{30}-\dfrac{3}{30}+\dfrac{16}{30}=\dfrac{31}{30}$

21. Performing the operations: $\dfrac{1}{x}+\dfrac{1}{x^2}+\dfrac{1}{x^3}=\dfrac{1}{x}\cdot\dfrac{x^2}{x^2}+\dfrac{1}{x^2}\cdot\dfrac{x}{x}+\dfrac{1}{x^3}=\dfrac{x^2+x+1}{x^3}$

23. Performing the operations:

$$\frac{x-2}{x^2+5x+4} - \frac{x-4}{2x^2+12x+16} = \frac{x-2}{(x+4)(x+1)} \cdot \frac{2(x+2)}{2(x+2)} - \frac{x-4}{2(x+4)(x+2)} \cdot \frac{x+1}{x+1}$$

$$= \frac{2x^2-8}{2(x+4)(x+1)(x+2)} - \frac{x^2-3x-4}{2(x+4)(x+1)(x+2)}$$

$$= \frac{2x^2-8-x^2+3x+4}{2(x+4)(x+1)(x+2)}$$

$$= \frac{x^2+3x-4}{2(x+4)(x+1)(x+2)}$$

$$= \frac{(x+4)(x-1)}{2(x+4)(x+1)(x+2)}$$

$$= \frac{x-1}{2(x+1)(x+2)}$$

25. Simplifying the complex fraction: $\dfrac{1+\dfrac{2}{3}}{1-\dfrac{2}{3}} = \dfrac{\left(1+\dfrac{2}{3}\right)\cdot 3}{\left(1-\dfrac{2}{3}\right)\cdot 3} = \dfrac{3+2}{3-2} = 5$

27. Simplifying the complex fraction: $1+\dfrac{1}{x+\dfrac{1}{x}} = 1+\dfrac{1\cdot x}{\left(x+\dfrac{1}{x}\right)\cdot x} = 1+\dfrac{x}{x^2+1} = \dfrac{x^2+x+1}{x^2+1}$

29. Solving the equation:

$$\frac{3}{x-1}=\frac{3}{5}$$
$$5(x-1)\left(\frac{3}{x-1}\right)=5(x-1)\left(\frac{3}{5}\right)$$
$$15=3x-3$$
$$18=3x$$
$$x=6$$

31. Solving the equation:

$$\frac{5}{y+1}=\frac{4}{y+2}$$
$$5(y+2)=4(y+1)$$
$$5y+10=4y+4$$
$$y=-6$$

33. Solving the equation:

$$\frac{4}{x^2 - x - 12} + \frac{1}{x^2 - 9} = \frac{2}{x^2 - 7x + 12}$$

$$(x-4)(x+3)(x-3)\left(\frac{4}{(x-4)(x+3)} + \frac{1}{(x+3)(x-3)}\right) = (x-4)(x+3)(x-3)\left(\frac{2}{(x-4)(x-3)}\right)$$

$$4(x-3) + x - 4 = 2(x+3)$$
$$4x - 12 + x - 4 = 2x + 6$$
$$5x - 16 = 2x + 6$$
$$3x = 22$$
$$x = \frac{22}{3}$$

35. Let x represent the rate of the truck and $x + 10$ represent the rate of the car. The equation is:

$$\frac{120}{x} - \frac{120}{x+10} = 2$$

$$x(x+10)\left(\frac{120}{x} - \frac{120}{x+10}\right) = 2x(x+10)$$

$$120(x+10) - 120x = 2x^2 + 20x$$
$$1200 = 2x^2 + 20x$$
$$0 = 2(x^2 + 10x - 600)$$
$$0 = 2(x+30)(x-20)$$
$$x = 20 \quad (x = -30 \text{ is impossible})$$

The car's rate is 30 mph and the truck's rate is 20 mph.

37. Converting the speed: $\dfrac{1088 \text{ feet}}{1 \text{ second}} \cdot \dfrac{1 \text{ mile}}{5280 \text{ feet}} \cdot \dfrac{60 \text{ seconds}}{1 \text{ minute}} \cdot \dfrac{60 \text{ minutes}}{1 \text{ hour}} \approx 742 \text{ mph}$

Chapter 6 Cumulative Review

1. Simplifying: $\left(-\dfrac{3}{2}\right)^3 = \left(-\dfrac{3}{2}\right)\left(-\dfrac{3}{2}\right)\left(-\dfrac{3}{2}\right) = -\dfrac{27}{8}$

3. Simplifying: $|-28 - 36| - 17 = |-64| - 17 = 64 - 17 = 47$

5. Simplifying: $64 \div 16 \cdot 4 = 4 \cdot 4 = 16$

7. Simplifying: $\dfrac{3^4 - 6}{25(3^2 - 2^2)} = \dfrac{81 - 6}{25(9 - 4)} = \dfrac{75}{25(5)} = \dfrac{75}{125} = \dfrac{3}{5}$

9. Computing the value: $-3 \cdot \dfrac{5}{12} - \dfrac{3}{4} = -\dfrac{5}{4} - \dfrac{3}{4} = -\dfrac{8}{4} = -2$

11. Evaluating: $P(10) = 16(10) - 0.01(10)^2 = 160 - 0.01(100) = 160 - 1 = 159$

13. Evaluating: $Q(10) = \dfrac{3}{5}(10)^2 + \dfrac{1}{2}(10) + 25 = \dfrac{3}{5}(100) + \dfrac{1}{2}(10) + 25 = 60 + 5 + 25 = 90$

15. Simplifying: $-3(5x + 4) + 12x = -15x - 12 + 12x = -3x - 12$

17. Subtracting:

$$\frac{6}{y^2-9}-\frac{5}{y^2-y-6}=\frac{6}{(y+3)(y-3)}\cdot\frac{y+2}{y+2}-\frac{5}{(y+2)(y-3)}\cdot\frac{y+3}{y+3}$$

$$=\frac{6y+12}{(y+3)(y-3)(y+2)}-\frac{5y+15}{(y+3)(y-3)(y+2)}$$

$$=\frac{6y+12-5y-15}{(y+3)(y-3)(y+2)}$$

$$=\frac{y-3}{(y+3)(y-3)(y+2)}$$

$$=\frac{1}{(y+3)(y+2)}$$

19. Multiplying: $\left(4t^2+\frac{1}{3}\right)\left(3t^2-\frac{1}{4}\right)=12t^4-t^2+t^2-\frac{1}{12}=12t^4-\frac{1}{12}$

21. Dividing: $\dfrac{10x^{3n}-15x^{4n}}{5x^n}=\dfrac{10x^{3n}}{5x^n}-\dfrac{15x^{4n}}{5x^n}=2x^{2n}-3x^{3n}$

23. Simplifying: $\dfrac{x^{-5}}{x^{-8}}=x^{-5+8}=x^3$

25. Simplifying: $\dfrac{x^3+2x^2-9x-18}{x^2-x-6}=\dfrac{x^2(x+2)-9(x+2)}{(x-3)(x+2)}=\dfrac{(x+2)(x+3)(x-3)}{(x-3)(x+2)}=x+3$

27. First find the slope: $m=\dfrac{-5-(-1)}{-3-(-6)}=\dfrac{-4}{3}=-\dfrac{4}{3}$. Using the point-slope formula:

$$y+5=-\frac{4}{3}(x+3)$$

$$y+5=-\frac{4}{3}x-4$$

$$y=-\frac{4}{3}x-9$$

29. Factoring: $168=4\cdot42=(2\cdot2)\cdot(6\cdot7)=(2\cdot2)\cdot(2\cdot3\cdot7)=2^3\cdot3\cdot7$

31. Factoring: $x^2+10x+25-y^2=(x+5)^2-y^2=(x+5+y)(x+5-y)$

33. Solving the equation:

$$-\frac{3}{5}a+3=15$$

$$-\frac{3}{5}a=12$$

$$-3a=60$$

$$a=-20$$

35. Solving the equation:

$$\frac{2}{5}(15x-2)-\frac{1}{5}=5$$

$$6x-\frac{4}{5}-\frac{1}{5}=5$$

$$6x-1=5$$

$$6x=6$$

$$x=1$$

37. Solving the equation:

$$\frac{3}{y-2}=\frac{2}{y-3}$$

$$3(y-3)=2(y-2)$$

$$3y-9=2y-4$$

$$y=5$$

39. Simplifying the second equation:
$$5x - 2y = -1$$
$$2x - 2y = 2$$

Multiply the second equation by -1:
$$5x - 2y = -1$$
$$-2x + 2y = -2$$

Adding yields:
$$3x = -3$$
$$x = -1$$

Substituting into the first equation:
$$5(-1) - 2y = -1$$
$$-5 - 2y = -1$$
$$-2y = 4$$
$$y = -2$$

The solution is $(-1, -2)$.

41. Multiplying the second equation by 2:
$$-3x + 2y = -1$$
$$8x - 2y = -32$$

Adding yields:
$$5x = -33$$
$$x = -\frac{33}{5}$$

Substituting into the first equation:
$$-3\left(-\frac{33}{5}\right) + 2y = -1$$
$$\frac{99}{5} + 2y = -1$$
$$2y = -\frac{104}{5}$$
$$y = -\frac{52}{5}$$

The solution is $\left(-\frac{33}{5}, -\frac{52}{5}\right)$.

43. Solving the inequality:
$$-3(3x - 1) \le -2(3x - 3)$$
$$-9x + 3 \le -6x + 6$$
$$-3x \le 3$$
$$x \ge -1$$

Graphing the solution set:

45. Graphing the line:

47. Evaluating the function: $(f-g)(-2) = f(-2) - g(-2) = 2 - 8 = -6$

49. Evaluating the function: $(f \circ g)(0) = f[g(0)] = f(0) = 4$

Chapter 6 Test

1. Reducing the fraction: $\dfrac{x^2 - y^2}{x - y} = \dfrac{(x+y)(x-y)}{x-y} = x + y$

2. Reducing the fraction: $\dfrac{2x^2 - 5x + 3}{2x^2 - x - 3} = \dfrac{(2x-3)(x-1)}{(2x-3)(x+1)} = \dfrac{x-1}{x+1}$

3. Finding the speed: $\dfrac{(3.14)(75 \text{ feet})}{40 \text{ seconds}} = \dfrac{235.5 \text{ feet}}{40 \text{ seconds}} = 5.9 \text{ feet/second}$

4. Dividing: $\dfrac{24x^3y + 12x^2y^2 - 16xy^3}{4xy} = \dfrac{24x^3y}{4xy} + \dfrac{12x^2y^2}{4xy} - \dfrac{16xy^3}{4xy} = 6x^2 + 3xy - 4y^2$

5. Dividing using long division:

$$
\begin{array}{r}
x^2 - 4x - 2 \\
2x-1 \overline{\smash{)}\,2x^3 - 9x^2 + 0x + 10} \\
\underline{2x^3 - x^2} \\
-8x^2 + 0x \\
\underline{-8x^2 + 4x} \\
-4x + 10 \\
\underline{-4x + 2} \\
8
\end{array}
$$

The quotient is $x^2 - 4x - 2 + \dfrac{8}{2x - 1}$.

6. Performing the operations: $\dfrac{a^2 - 16}{5a - 15} \cdot \dfrac{10(a-3)^2}{a^2 - 7a + 12} = \dfrac{(a+4)(a-4)}{5(a-3)} \cdot \dfrac{10(a-3)^2}{(a-4)(a-3)} = 2(a+4)$

7. Performing the operations:

$$\dfrac{a^4 - 81}{a^2 + 9} \div \dfrac{a^2 - 8a + 15}{4a - 20} = \dfrac{a^4 - 81}{a^2 + 9} \cdot \dfrac{4a - 20}{a^2 - 8a + 15} = \dfrac{(a^2 + 9)(a+3)(a-3)}{a^2 + 9} \cdot \dfrac{4(a-5)}{(a-5)(a-3)} = 4(a+3)$$

8. Performing the operations:

$$\dfrac{x^3 - 8}{2x^2 - 9x + 10} \div \dfrac{x^2 + 2x + 4}{2x^2 + x - 15} = \dfrac{x^3 - 8}{2x^2 - 9x + 10} \cdot \dfrac{2x^2 + x - 15}{x^2 + 2x + 4} = \dfrac{(x-2)(x^2 + 2x + 4)}{(2x-5)(x-2)} \cdot \dfrac{(2x-5)(x+3)}{x^2 + 2x + 4} = x + 3$$

9. Performing the operations: $\dfrac{4}{21} + \dfrac{6}{35} = \dfrac{4}{21} \cdot \dfrac{5}{5} + \dfrac{6}{35} \cdot \dfrac{3}{3} = \dfrac{20}{105} + \dfrac{18}{105} = \dfrac{38}{105}$

10. Performing the operations: $\dfrac{3}{4} - \dfrac{1}{2} + \dfrac{5}{8} = \dfrac{3}{4} \cdot \dfrac{2}{2} - \dfrac{1}{2} \cdot \dfrac{4}{4} + \dfrac{5}{8} = \dfrac{6}{8} - \dfrac{4}{8} + \dfrac{5}{8} = \dfrac{7}{8}$

11. Performing the operations: $\dfrac{a}{a^2 - 9} + \dfrac{3}{a^2 - 9} = \dfrac{a+3}{a^2 - 9} = \dfrac{a+3}{(a+3)(a-3)} = \dfrac{1}{a-3}$

12. Performing the operations: $\dfrac{1}{x} + \dfrac{2}{x-3} = \dfrac{1}{x} \cdot \dfrac{x-3}{x-3} + \dfrac{2}{x-3} \cdot \dfrac{x}{x} = \dfrac{x-3}{x(x-3)} + \dfrac{2x}{x(x-3)} = \dfrac{3x-3}{x(x-3)} = \dfrac{3(x-1)}{x(x-3)}$

13. Performing the operations:

$$\frac{4x}{x^2+6x+5} - \frac{3x}{x^2+5x+4} = \frac{4x}{(x+5)(x+1)} - \frac{3x}{(x+4)(x+1)}$$

$$= \frac{4x}{(x+5)(x+1)} \cdot \frac{x+4}{x+4} - \frac{3x}{(x+4)(x+1)} \cdot \frac{x+5}{x+5}$$

$$= \frac{4x^2+16x}{(x+5)(x+1)(x+4)} - \frac{3x^2+15x}{(x+5)(x+1)(x+4)}$$

$$= \frac{4x^2+16x-3x^2-15x}{(x+5)(x+1)(x+4)}$$

$$= \frac{x^2+x}{(x+5)(x+1)(x+4)}$$

$$= \frac{x(x+1)}{(x+5)(x+1)(x+4)}$$

$$= \frac{x}{(x+4)(x+5)}$$

14. Performing the operations:

$$\frac{2x+8}{x^2+4x+3} - \frac{x+4}{x^2+5x+6} = \frac{2x+8}{(x+3)(x+1)} - \frac{x+4}{(x+3)(x+2)}$$

$$= \frac{2x+8}{(x+3)(x+1)} \cdot \frac{x+2}{x+2} - \frac{x+4}{(x+3)(x+2)} \cdot \frac{x+1}{x+1}$$

$$= \frac{2x^2+12x+16}{(x+1)(x+2)(x+3)} - \frac{x^2+5x+4}{(x+1)(x+2)(x+3)}$$

$$= \frac{2x^2+12x+16-x^2-5x-4}{(x+1)(x+2)(x+3)}$$

$$= \frac{x^2+7x+12}{(x+1)(x+2)(x+3)}$$

$$= \frac{(x+3)(x+4)}{(x+1)(x+2)(x+3)}$$

$$= \frac{x+4}{(x+1)(x+2)}$$

15. Simplifying the complex fraction: $\dfrac{3-\dfrac{1}{a+3}}{3+\dfrac{1}{a+3}} = \dfrac{\left(3-\dfrac{1}{a+3}\right)(a+3)}{\left(3+\dfrac{1}{a+3}\right)(a+3)} = \dfrac{3(a+3)-1}{3(a+3)+1} = \dfrac{3a+9-1}{3a+9+1} = \dfrac{3a+8}{3a+10}$

16. Simplifying the complex fraction: $\dfrac{1-\dfrac{9}{x^2}}{1+\dfrac{1}{x}-\dfrac{6}{x^2}} = \dfrac{\left(1-\dfrac{9}{x^2}\right)\cdot x^2}{\left(1+\dfrac{1}{x}-\dfrac{6}{x^2}\right)\cdot x^2} = \dfrac{x^2-9}{x^2+x-6} = \dfrac{(x+3)(x-3)}{(x+3)(x-2)} = \dfrac{x-3}{x-2}$

17. **a.** Simplifying the difference quotient:

$$\frac{f(x+h)-f(x)}{h} = \frac{[2(x+h)-4]-[2x-4]}{h}$$

$$= \frac{2x+2h-4-2x+4}{h}$$

$$= \frac{2h}{h}$$

$$= 2$$

b. Simplifying the difference quotient:

$$\frac{f(x)-f(a)}{x-a} = \frac{(2x-4)-(2a-4)}{x-a}$$

$$= \frac{2x-2a}{x-a}$$

$$= \frac{2(x-a)}{x-a}$$

$$= 2$$

18.   **a.**   Simplifying the difference quotient:

$$\frac{f(x+h)-f(x)}{h} = \frac{\left[(x+h)^2 + 5(x+h)-1\right] - \left[x^2 + 5x - 1\right]}{h}$$

$$= \frac{x^2 + 2xh + h^2 + 5x + 5h - 1 - x^2 - 5x + 1}{h}$$

$$= \frac{2xh + h^2 + 5h}{h}$$

$$= 2x + h + 5$$

    b.   Simplifying the difference quotient:

$$\frac{f(x)-f(a)}{x-a} = \frac{\left(x^2 + 5x - 1\right) - \left(a^2 + 5a - 1\right)}{x-a}$$

$$= \frac{x^2 - a^2 + 5x - 5a}{x-a}$$

$$= \frac{(x+a)(x-a) + 5(x-a)}{x-a}$$

$$= \frac{(x-a)(x+a+5)}{x-a}$$

$$= x + a + 5$$

19.   Solving the equation:

$$\frac{1}{x} + 3 = \frac{4}{3}$$

$$3x\left(\frac{1}{x} + 3\right) = 3x\left(\frac{4}{3}\right)$$

$$3 + 9x = 4x$$

$$5x = -3$$

$$x = -\frac{3}{5}$$

20.   Solving the equation:

$$\frac{x}{x-3} + 3 = \frac{3}{x-3}$$

$$(x-3)\left(\frac{x}{x-3} + 3\right) = (x-3)\left(\frac{3}{x-3}\right)$$

$$x + 3(x-3) = 3$$

$$x + 3x - 9 = 3$$

$$4x = 12$$

$$x = 3 \quad (x = 3 \text{ does not check})$$

There is no solution (3 does not check).

21.   Solving the equation:

$$\frac{y+3}{2y} + \frac{5}{y-1} = \frac{1}{2}$$

$$2y(y-1)\left(\frac{y+3}{2y} + \frac{5}{y-1}\right) = 2y(y-1)\left(\frac{1}{2}\right)$$

$$(y-1)(y+3) + 10y = y(y-1)$$

$$y^2 + 2y - 3 + 10y = y^2 - y$$

$$12y - 3 = -y$$

$$13y = 3$$

$$y = \frac{3}{13}$$

22.   Solving the equation:

$$1 - \frac{1}{x} = \frac{6}{x^2}$$

$$x^2\left(1 - \frac{1}{x}\right) = x^2\left(\frac{6}{x^2}\right)$$

$$x^2 - x = 6$$

$$x^2 - x - 6 = 0$$

$$(x+2)(x-3) = 0$$

$$x = -2, 3$$

23.   Let x represent the number. The equation is:

$$\frac{10}{23-x} = \frac{1}{3}$$

$$30 = 23 - x$$

$$x = -7$$

The number is -7.

24. Let x represent the speed of the boat. Since the total time is 3 hours:

$$\frac{8}{x-2}+\frac{8}{x+2}=3$$

$$(x-2)(x+2)\left(\frac{8}{x-2}+\frac{8}{x+2}\right)=3(x-2)(x+2)$$

$$8(x+2)+8(x-2)=3\left(x^2-4\right)$$

$$16x=3x^2-12$$

$$0=3x^2-16x-12$$

$$0=(3x+2)(x-6)$$

$$x=6 \qquad \left(x=-\frac{2}{3} \text{ is impossible}\right)$$

The speed of the boat is 6 mph.

25. Let x represent the time to fill the pool with both the pipe and drain open. The rate equation is:

$$\frac{1}{10}-\frac{1}{15}=\frac{1}{2}\bullet\frac{1}{x}$$

$$30x\left(\frac{1}{10}-\frac{1}{15}\right)=30x\left(\frac{1}{2x}\right)$$

$$3x-2x=15$$

$$x=15$$

The pool can be filled in 15 hours with both the pipe and drain open.

26. Converting the height: $14,494$ feet $\bullet \dfrac{1 \text{ mile}}{5,280 \text{ feet}}\approx 2.7$ miles

27. Converting each height:

K2: $28,238$ feet $\bullet \dfrac{1 \text{ mile}}{5,280 \text{ feet}}\approx 5.3$ miles

Mount Everest: $29,035$ feet $\bullet \dfrac{1 \text{ mile}}{5,280 \text{ feet}}\approx 5.5$ miles

Kangchenjunga: $28,208$ feet $\bullet \dfrac{1 \text{ mile}}{5,280 \text{ feet}}\approx 5.3$ miles

28. Converting the speed: $\dfrac{4,750 \text{ feet}}{3.2 \text{ seconds}}\bullet\dfrac{1 \text{ mile}}{5280 \text{ feet}}\bullet\dfrac{60 \text{ seconds}}{1 \text{ minute}}\bullet\dfrac{60 \text{ minutes}}{1 \text{ hour}}\approx 1,012$ miles per hour

Chapter 7
Rational Exponents and Roots

Getting Ready for Chapter 7

1. Simplifying: $8^2 = 8 \cdot 8 = 64$

2. Simplifying: $(-8)^2 = (-8)(-8) = 64$

3. Simplifying: $-8^2 = -8 \cdot 8 = -64$

4. Simplifying: $-(-8)^2 = -(-8)(-8) = -64$

5. Simplifying: $(-2)^3 = (-2)(-2)(-2) = -8$

6. Simplifying: $-2^3 = -2 \cdot 2 \cdot 2 = -8$

7. Simplifying: $12\left(\dfrac{1}{3}\right) = 4$

8. Simplifying: $8\left(\dfrac{1}{4}\right) = 2$

9. Simplifying: $\dfrac{1}{3} + \dfrac{5}{6} = \dfrac{1}{3} \cdot \dfrac{2}{2} + \dfrac{5}{6} = \dfrac{2}{6} + \dfrac{5}{6} = \dfrac{7}{6}$

10. Simplifying: $\dfrac{3}{5} \cdot \dfrac{5}{9} = \dfrac{3}{9} = \dfrac{1}{3}$

11. Simplifying: $-12 - 10 = -22$

12. Simplifying: $-7 - (-11) = -7 + 11 = 4$

13. Simplifying: $2 - \dfrac{3}{2} = \dfrac{2}{1} \cdot \dfrac{2}{2} - \dfrac{3}{2} = \dfrac{4}{2} - \dfrac{3}{2} = \dfrac{1}{2}$

14. Simplifying: $3 + \dfrac{5}{4} = \dfrac{3}{1} \cdot \dfrac{4}{4} + \dfrac{5}{4} = \dfrac{12}{4} + \dfrac{5}{4} = \dfrac{17}{4} = 4\dfrac{1}{4}$

15. Using the rules for exponents: $x^8 \cdot x^{-3} = x^{8-3} = x^5$

16. Using the rules for exponents: $\left(y^2\right)^3 = y^{2 \cdot 3} = y^6$

17. Using the rules for exponents: $\left(3r^8 s^{20}\right)^5 = 3^5 r^{8 \cdot 5} s^{20 \cdot 5} = 243 r^{40} s^{100}$

18. Using the rules for exponents: $\dfrac{x^{-3}}{x^{-4}} = x^{-3-(-4)} = x^{-3+4} = x$

19. Using the rules for exponents: $3^{-3} = \dfrac{1}{3^3} = \dfrac{1}{3 \cdot 3 \cdot 3} = \dfrac{1}{27}$

20. Using the rules for exponents: $\left(\dfrac{3}{2}\right)^{-4} = \left(\dfrac{2}{3}\right)^4 = \dfrac{2^4}{3^4} = \dfrac{16}{81}$

7.1 Rational Exponents

1. Finding the root: $\sqrt{144} = 12$

3. Finding the root: $\sqrt{-144}$ is not a real number

5. Finding the root: $-\sqrt{49} = -7$

7. Finding the root: $\sqrt[3]{-27} = -3$

9. Finding the root: $\sqrt[4]{16} = 2$

11. Finding the root: $\sqrt[4]{-16}$ is not a real number

13. Finding the root: $\sqrt{0.04} = 0.2$

15. Finding the root: $\sqrt[3]{0.008} = 0.2$

17. Finding the root: $\sqrt[3]{125} = 5$

19. Finding the root: $-\sqrt[3]{216} = -6$

21. Finding the root: $\sqrt{\dfrac{1}{36}} = \dfrac{1}{6}$

23. Finding the root: $\sqrt[3]{\dfrac{8}{125}} = \dfrac{2}{5}$

25. Simplifying: $\sqrt{36a^8} = 6a^4$

27. Simplifying: $\sqrt[3]{27a^{12}} = 3a^4$

29. Simplifying: $\sqrt[5]{32x^{10}y^5} = 2x^2y$

31. Simplifying: $\sqrt[4]{16a^{12}b^{20}} = 2a^3b^5$

33. Writing as a root and simplifying: $36^{1/2} = \sqrt{36} = 6$

35. Writing as a root and simplifying: $-9^{1/2} = -\sqrt{9} = -3$

37. Writing as a root and simplifying: $8^{1/3} = \sqrt[3]{8} = 2$

39. Writing as a root and simplifying: $(-8)^{1/3} = \sqrt[3]{-8} = -2$

41. Writing as a root and simplifying: $32^{1/5} = \sqrt[5]{32} = 2$

43. Writing as a root and simplifying: $\left(\dfrac{81}{25}\right)^{1/2} = \sqrt{\dfrac{81}{25}} = \dfrac{9}{5}$

45. Simplifying: $27^{2/3} = \left(27^{1/3}\right)^2 = 3^2 = 9$

47. Simplifying: $25^{3/2} = \left(25^{1/2}\right)^3 = 5^3 = 125$

49. Simplifying: $27^{-1/3} = \left(27^{1/3}\right)^{-1} = 3^{-1} = \dfrac{1}{3}$

51. Simplifying: $81^{-3/4} = \left(81^{1/4}\right)^{-3} = 3^{-3} = \dfrac{1}{3^3} = \dfrac{1}{27}$

53. Simplifying: $\left(\dfrac{25}{36}\right)^{-1/2} = \left(\dfrac{36}{25}\right)^{1/2} = \dfrac{6}{5}$

55. Simplifying: $\left(\dfrac{81}{16}\right)^{-3/4} = \left(\dfrac{16}{81}\right)^{3/4} = \left[\left(\dfrac{16}{81}\right)^{1/4}\right]^3 = \left(\dfrac{2}{3}\right)^3 = \dfrac{8}{27}$

57. Simplifying: $16^{1/2} + 27^{1/3} = 4 + 3 = 7$

59. Simplifying: $8^{-2/3} + 4^{-1/2} = \left(8^{1/3}\right)^{-2} + \left(4^{1/2}\right)^{-1} = 2^{-2} + 2^{-1} = \dfrac{1}{4} + \dfrac{1}{2} = \dfrac{3}{4}$

61. Using properties of exponents: $x^{3/5} \bullet x^{1/5} = x^{3/5+1/5} = x^{4/5}$

63. Using properties of exponents: $\left(a^{3/4}\right)^{4/3} = a^{3/4 \bullet 4/3} = a$

65. Using properties of exponents: $\dfrac{x^{1/5}}{x^{3/5}} = x^{1/5-3/5} = x^{-2/5} = \dfrac{1}{x^{2/5}}$

67. Using properties of exponents: $\dfrac{x^{5/6}}{x^{2/3}} = x^{5/6-2/3} = x^{5/6-4/6} = x^{1/6}$

69. Using properties of exponents: $\left(x^{3/5}y^{5/6}z^{1/3}\right)^{3/5} = x^{3/5 \bullet 3/5}y^{5/6 \bullet 3/5}z^{1/3 \bullet 3/5} = x^{9/25}y^{1/2}z^{1/5}$

71. Using properties of exponents: $\dfrac{a^{3/4}b^2}{a^{7/8}b^{1/4}} = a^{3/4-7/8}b^{2-1/4} = a^{6/8-7/8}b^{8/4-1/4} = a^{-1/8}b^{7/4} = \dfrac{b^{7/4}}{a^{1/8}}$

73. Using properties of exponents: $\dfrac{\left(y^{2/3}\right)^{3/4}}{\left(y^{1/3}\right)^{3/5}} = \dfrac{y^{1/2}}{y^{1/5}} = y^{1/2-1/5} = y^{5/10-2/10} = y^{3/10}$

75. Using properties of exponents: $\left(\dfrac{a^{-1/4}}{b^{1/2}}\right)^8 = \dfrac{a^{-1/4 \bullet 8}}{b^{1/2 \bullet 8}} = \dfrac{a^{-2}}{b^4} = \dfrac{1}{a^2 b^4}$

77. Using properties of exponents: $\dfrac{\left(r^{-2}s^{1/3}\right)^6}{r^8s^{3/2}} = \dfrac{r^{-12}s^2}{r^8s^{3/2}} = r^{-12-8}s^{2-3/2} = r^{-20}s^{1/2} = \dfrac{s^{1/2}}{r^{20}}$

79. Using properties of exponents: $\dfrac{\left(25a^6b^4\right)^{1/2}}{\left(8a^{-9}b^3\right)^{-1/3}} = \dfrac{25^{1/2}\,a^3b^2}{8^{-1/3}\,a^3b^{-1}} = \dfrac{5}{1/2}\,a^{3-3}b^{2+1} = 10b^3$

81. Substituting $r = 250$: $v = \left(\dfrac{5 \cdot 250}{2}\right)^{1/2} = 625^{1/2} = 25$. The maximum speed is 25 mph.

83. **a.** The length of the side is $126 + 86 + 86 + 126 = 424$ pm

   b. Let d represent the diagonal. Using the Pythagorean theorem:
$$d^2 = 424^2 + 424^2 = 359{,}552$$
$$d = \sqrt{359{,}552} \approx 600 \text{ pm}$$

   c. Converting to meters: $600 \text{ pm} \cdot \dfrac{1 \text{ m}}{10^{12} \text{ pm}} = 6 \times 10^{-10}$ m

85. **a.** This graph is B.   **b.** This graph is A.

   c. This graph is C.   **d.** The points of intersection are (0,0) and (1,1).

87. Multiplying: $x^2\left(x^4 - x\right) = x^2 \bullet x^4 - x^2 \bullet x = x^6 - x^3$

89. Multiplying: $(x-3)(x+5) = x^2 + 5x - 3x - 15 = x^2 + 2x - 15$

91. Multiplying: $\left(x^2 - 5\right)^2 = \left(x^2\right)^2 - 2\left(x^2\right)(5) + 5^2 = x^4 - 10x^2 + 25$

93. Multiplying: $(x-3)\left(x^2 + 3x + 9\right) = x^3 + 3x^2 + 9x - 3x^2 - 9x - 27 = x^3 - 27$

95. Simplifying: $x^2\left(x^4 - x^3\right) = x^2 \bullet x^4 - x^2 \bullet x^3 = x^6 - x^5$

97. Simplifying: $(3a - 2b)(4a - b) = 12a^2 - 3ab - 8ab + 2b^2 = 12a^2 - 11ab + 2b^2$

99. Simplifying: $\left(x^3 - 2\right)\left(x^3 + 2\right) = \left(x^3\right)^2 - (2)^2 = x^6 - 4$

101. Simplifying: $\dfrac{15x^2y - 20x^4y^2}{5xy} = \dfrac{15x^2y}{5xy} - \dfrac{20x^4y^2}{5xy} = 3x - 4x^3y$

103. Factoring: $x^2 - 3x - 10 = (x-5)(x+2)$ **105.** Factoring: $6x^2 + 11x - 10 = (3x-2)(2x+5)$

107. Simplifying: $x^{2/3} \bullet x^{4/3} = x^{2/3+4/3} = x^2$ **109.** Simplifying: $\left(t^{1/2}\right)^2 = t^{1/2 \bullet 2} = t^1 = t$

111. Simplifying: $\dfrac{x^{2/3}}{x^{1/3}} = x^{2/3-1/3} = x^{1/3}$

113. Simplifying each expression:
$$\left(9^{1/2} + 4^{1/2}\right)^2 = (3+2)^2 = 5^2 = 25$$
$$9 + 4 = 13$$
Note that the values are not equal.

115. Rewriting with exponents: $\sqrt{\sqrt{a}} = \sqrt{a^{1/2}} = \left(a^{1/2}\right)^{1/2} = a^{1/4} = \sqrt[4]{a}$

7.2 More Expressions Involving Rational Exponents

1.   Multiplying: $x^{2/3}\left(x^{1/3} + x^{4/3}\right) = x^{2/3} \cdot x^{1/3} + x^{2/3} \cdot x^{4/3} = x + x^2$

3.   Multiplying: $a^{1/2}\left(a^{3/2} - a^{1/2}\right) = a^{1/2} \cdot a^{3/2} - a^{1/2} \cdot a^{1/2} = a^2 - a$

5.   Multiplying: $2x^{1/3}\left(3x^{8/3} - 4x^{5/3} + 5x^{2/3}\right) = 2x^{1/3} \cdot 3x^{8/3} - 2x^{1/3} \cdot 4x^{5/3} + 2x^{1/3} \cdot 5x^{2/3} = 6x^3 - 8x^2 + 10x$

7.   Multiplying:
$$4x^{1/2}y^{3/5}\left(3x^{3/2}y^{-3/5} - 9x^{-1/2}y^{7/5}\right) = 4x^{1/2}y^{3/5} \cdot 3x^{3/2}y^{-3/5} - 4x^{1/2}y^{3/5} \cdot 9x^{-1/2}y^{7/5} = 12x^2 - 36y^2$$

9.   Multiplying: $\left(x^{2/3} - 4\right)\left(x^{2/3} + 2\right) = x^{2/3} \cdot x^{2/3} + 2x^{2/3} - 4x^{2/3} - 8 = x^{4/3} - 2x^{2/3} - 8$

11.   Multiplying: $\left(a^{1/2} - 3\right)\left(a^{1/2} - 7\right) = a^{1/2} \cdot a^{1/2} - 7a^{1/2} - 3a^{1/2} + 21 = a - 10a^{1/2} + 21$

13.   Multiplying: $\left(4y^{1/3} - 3\right)\left(5y^{1/3} + 2\right) = 20y^{2/3} + 8y^{1/3} - 15y^{1/3} - 6 = 20y^{2/3} - 7y^{1/3} - 6$

15.   Multiplying: $\left(5x^{2/3} + 3y^{1/2}\right)\left(2x^{2/3} + 3y^{1/2}\right) = 10x^{4/3} + 15x^{2/3}y^{1/2} + 6x^{2/3}y^{1/2} + 9y = 10x^{4/3} + 21x^{2/3}y^{1/2} + 9y$

17.   Multiplying: $\left(t^{1/2} + 5\right)^2 = \left(t^{1/2} + 5\right)\left(t^{1/2} + 5\right) = t + 5t^{1/2} + 5t^{1/2} + 25 = t + 10t^{1/2} + 25$

19.   Multiplying: $\left(x^{3/2} + 4\right)^2 = \left(x^{3/2} + 4\right)\left(x^{3/2} + 4\right) = x^3 + 4x^{3/2} + 4x^{3/2} + 16 = x^3 + 8x^{3/2} + 16$

21.   Multiplying: $\left(a^{1/2} - b^{1/2}\right)^2 = \left(a^{1/2} - b^{1/2}\right)\left(a^{1/2} - b^{1/2}\right) = a - a^{1/2}b^{1/2} - a^{1/2}b^{1/2} + b = a - 2a^{1/2}b^{1/2} + b$

23.   Multiplying:
$$\left(2x^{1/2} - 3y^{1/2}\right)^2 = \left(2x^{1/2} - 3y^{1/2}\right)\left(2x^{1/2} - 3y^{1/2}\right)$$
$$= 4x - 6x^{1/2}y^{1/2} - 6x^{1/2}y^{1/2} + 9y$$
$$= 4x - 12x^{1/2}y^{1/2} + 9y$$

25.   Multiplying: $\left(a^{1/2} - 3^{1/2}\right)\left(a^{1/2} + 3^{1/2}\right) = \left(a^{1/2}\right)^2 - \left(3^{1/2}\right)^2 = a - 3$

27.   Multiplying: $\left(x^{3/2} + y^{3/2}\right)\left(x^{3/2} - y^{3/2}\right) = \left(x^{3/2}\right)^2 - \left(y^{3/2}\right)^2 = x^3 - y^3$

29.   Multiplying: $\left(t^{1/2} - 2^{3/2}\right)\left(t^{1/2} + 2^{3/2}\right) = \left(t^{1/2}\right)^2 - \left(2^{3/2}\right)^2 = t - 2^3 = t - 8$

31.   Multiplying: $\left(2x^{3/2} + 3^{1/2}\right)\left(2x^{3/2} - 3^{1/2}\right) = \left(2x^{3/2}\right)^2 - \left(3^{1/2}\right)^2 = 4x^3 - 3$

33.   Multiplying: $\left(x^{1/3} + y^{1/3}\right)\left(x^{2/3} - x^{1/3}y^{1/3} + y^{2/3}\right) = \left(x^{1/3}\right)^3 + \left(y^{1/3}\right)^3 = x + y$

35.   Multiplying: $\left(a^{1/3} - 2\right)\left(a^{2/3} + 2a^{1/3} + 4\right) = \left(a^{1/3}\right)^3 - (2)^3 = a - 8$

37.   Multiplying: $\left(2x^{1/3} + 1\right)\left(4x^{2/3} - 2x^{1/3} + 1\right) = \left(2x^{1/3}\right)^3 + (1)^3 = 8x + 1$

39.   Multiplying: $\left(t^{1/4} - 1\right)\left(t^{1/4} + 1\right)\left(t^{1/2} + 1\right) = \left(t^{1/2} - 1\right)\left(t^{1/2} + 1\right) = t - 1$

41.   Dividing: $\dfrac{18x^{3/4} + 27x^{1/4}}{9x^{1/4}} = \dfrac{18x^{3/4}}{9x^{1/4}} + \dfrac{27x^{1/4}}{9x^{1/4}} = 2x^{1/2} + 3$

43.   Dividing: $\dfrac{12x^{2/3}y^{1/3} - 16x^{1/3}y^{2/3}}{4x^{1/3}y^{1/3}} = \dfrac{12x^{2/3}y^{1/3}}{4x^{1/3}y^{1/3}} - \dfrac{16x^{1/3}y^{2/3}}{4x^{1/3}y^{1/3}} = 3x^{1/3} - 4y^{1/3}$

45.   Dividing: $\dfrac{21a^{7/5}b^{3/5} - 14a^{2/5}b^{8/5}}{7a^{2/5}b^{3/5}} = \dfrac{21a^{7/5}b^{3/5}}{7a^{2/5}b^{3/5}} - \dfrac{14a^{2/5}b^{8/5}}{7a^{2/5}b^{3/5}} = 3a - 2b$

47. Factoring: $12(x-2)^{3/2} - 9(x-2)^{1/2} = 3(x-2)^{1/2}\left[4(x-2)-3\right] = 3(x-2)^{1/2}(4x-8-3) = 3(x-2)^{1/2}(4x-11)$

49. Factoring: $5(x-3)^{12/5} - 15(x-3)^{7/5} = 5(x-3)^{7/5}\left[(x-3)-3\right] = 5(x-3)^{7/5}(x-6)$

51. Factoring: $9x(x+1)^{3/2} + 6(x+1)^{1/2} = 3(x+1)^{1/2}\left[3x(x+1)+2\right] = 3(x+1)^{1/2}(3x^2+3x+2)$

53. Factoring: $x^{2/3} - 5x^{1/3} + 6 = (x^{1/3}-2)(x^{1/3}-3)$ **55.** Factoring: $a^{2/5} - 2a^{1/5} - 8 = (a^{1/5}-4)(a^{1/5}+2)$

57. Factoring: $2y^{2/3} - 5y^{1/3} - 3 = (2y^{1/3}+1)(y^{1/3}-3)$ **59.** Factoring: $9t^{2/5} - 25 = (3t^{1/5}+5)(3t^{1/5}-5)$

61. Factoring: $4x^{2/7} + 20x^{1/7} + 25 = (2x^{1/7}+5)^2$ **63.** Factoring: $f(4) = 4 - 2\sqrt{4} - 8 = 4 - 4 - 8 = -8$

65. Evaluating: $f(25) = 2(25) + 9\sqrt{25} - 5 = 50 + 45 - 5 = 90$

67. Evaluating: $g\left(\dfrac{9}{4}\right) = 2\left(\dfrac{9}{4}\right) - \sqrt{\dfrac{9}{4}} - 6 = \dfrac{9}{2} - \dfrac{3}{2} - 6 = 3 - 6 = -3$

69. Evaluating: $f(-8) = (-8)^{2/3} - 2(-8)^{1/3} - 8 = 4 + 4 - 8 = 0$

71. Writing as a single fraction: $\dfrac{3}{x^{1/2}} + x^{1/2} = \dfrac{3}{x^{1/2}} + x^{1/2} \cdot \dfrac{x^{1/2}}{x^{1/2}} = \dfrac{3+x}{x^{1/2}}$

73. Writing as a single fraction: $x^{2/3} + \dfrac{5}{x^{1/3}} = x^{2/3} \cdot \dfrac{x^{1/3}}{x^{1/3}} + \dfrac{5}{x^{1/3}} = \dfrac{x+5}{x^{1/3}}$

75. Writing as a single fraction:

$$\dfrac{3x^2}{(x^3+1)^{1/2}} + (x^3+1)^{1/2} = \dfrac{3x^2}{(x^3+1)^{1/2}} + (x^3+1)^{1/2} \cdot \dfrac{(x^3+1)^{1/2}}{(x^3+1)^{1/2}} = \dfrac{3x^2+x^3+1}{(x^3+1)^{1/2}} = \dfrac{x^3+3x^2+1}{(x^3+1)^{1/2}}$$

77. Writing as a single fraction:

$$\dfrac{x^2}{(x^2+4)^{1/2}} - (x^2+4)^{1/2} = \dfrac{x^2}{(x^2+4)^{1/2}} - (x^2+4)^{1/2} \cdot \dfrac{(x^2+4)^{1/2}}{(x^2+4)^{1/2}}$$

$$= \dfrac{x^2 - (x^2+4)}{(x^2+4)^{1/2}}$$

$$= \dfrac{x^2 - x^2 - 4}{(x^2+4)^{1/2}}$$

$$= \dfrac{-4}{(x^2+4)^{1/2}}$$

79. Using the formula $r = \left(\dfrac{A}{P}\right)^{1/t} - 1$ to find the annual rate of return: $r = \left(\dfrac{900}{500}\right)^{1/4} - 1 \approx 0.158$

The annual rate of return is approximately 15.8%.

81. Reducing to lowest terms: $\dfrac{x^2-9}{x^4-81} = \dfrac{x^2-9}{(x^2+9)(x^2-9)} = \dfrac{1}{x^2+9}$

83. Dividing: $\dfrac{15x^2y - 20x^4y^2}{5xy} = \dfrac{15x^2y}{5xy} - \dfrac{20x^4y^2}{5xy} = 3x - 4x^3y$

85. Dividing using long division:

$$2x+3 \overline{)\begin{array}{r} 5x-4 \\ 10x^2 + 7x - 12 \end{array}}$$
$$\begin{array}{r} 10x^2 + 15x \\ \hline -8x - 12 \\ -8x - 12 \\ \hline 0 \end{array}$$

87. Dividing using long division:

$$x-5 \overline{)\begin{array}{r} x^2 + 5x + 25 \\ x^3 + 0x^2 + 0x - 125 \end{array}}$$
$$\begin{array}{r} x^3 - 5x^2 \\ \hline 5x^2 + 0x \\ 5x^2 - 25x \\ \hline 25x - 125 \\ 25x - 125 \\ \hline 0 \end{array}$$

89. Simplifying: $\sqrt{25} = 5$

91. Simplifying: $\sqrt{6^2} = 6$

93. Simplifying: $\sqrt{16x^4 y^2} = 4x^2 y$

95. Simplifying: $\sqrt{(5y)^2} = 5y$

97. Simplifying: $\sqrt[3]{27} = 3$

99. Simplifying: $\sqrt[3]{2^3} = 2$

101. Simplifying: $\sqrt[3]{8a^3 b^3} = 2ab$

103. Filling in the blank: $50 = 25 \cdot 2$

105. Filling in the blank: $48x^4 y^3 = 48x^4 y^2 \cdot y$

107. Filling in the blank: $12x^7 y^6 = 4x^6 y^6 \cdot 3x$

7.3 Simplified Form for Radicals

1. Simplifying the radical: $\sqrt{8} = \sqrt{4 \cdot 2} = 2\sqrt{2}$

3. Simplifying the radical: $\sqrt{98} = \sqrt{49 \cdot 2} = 7\sqrt{2}$

5. Simplifying the radical: $\sqrt{288} = \sqrt{144 \cdot 2} = 12\sqrt{2}$

7. Simplifying the radical: $\sqrt{80} = \sqrt{16 \cdot 5} = 4\sqrt{5}$

9. Simplifying the radical: $\sqrt{48} = \sqrt{16 \cdot 3} = 4\sqrt{3}$

11. Simplifying the radical: $\sqrt{675} = \sqrt{225 \cdot 3} = 15\sqrt{3}$

13. Simplifying the radical: $\sqrt[3]{54} = \sqrt[3]{27 \cdot 2} = 3\sqrt[3]{2}$

15. Simplifying the radical: $\sqrt[3]{128} = \sqrt[3]{64 \cdot 2} = 4\sqrt[3]{2}$

17. Simplifying the radical: $\sqrt[3]{432} = \sqrt[3]{216 \cdot 2} = 6\sqrt[3]{2}$

19. Simplifying the radical: $\sqrt[5]{64} = \sqrt[5]{32 \cdot 2} = 2\sqrt[5]{2}$

21. Simplifying the radical: $\sqrt{18x^3} = \sqrt{9x^2 \cdot 2x} = 3x\sqrt{2x}$

23. Simplifying the radical: $\sqrt[4]{32y^7} = \sqrt[4]{16y^4 \cdot 2y^3} = 2y\sqrt[4]{2y^3}$

25. Simplifying the radical: $\sqrt[3]{40x^4 y^7} = \sqrt[3]{8x^3 y^6 \cdot 5xy} = 2xy^2 \sqrt[3]{5xy}$

27. Simplifying the radical: $\sqrt{48a^2 b^3 c^4} = \sqrt{16a^2 b^2 c^4 \cdot 3b} = 4abc^2 \sqrt{3b}$

29. Simplifying the radical: $\sqrt[3]{48a^2 b^3 c^4} = \sqrt[3]{8b^3 c^3 \cdot 6a^2 c} = 2bc\sqrt[3]{6a^2 c}$

31. Simplifying the radical: $\sqrt[5]{64x^8 y^{12}} = \sqrt[5]{32x^5 y^{10} \cdot 2x^3 y^2} = 2xy^2 \sqrt[5]{2x^3 y^2}$

33. Simplifying the radical: $\sqrt[5]{243x^7 y^{10} z^5} = \sqrt[5]{243x^5 y^{10} z^5 \cdot x^2} = 3xy^2 z\sqrt[5]{x^2}$

35. Substituting into the expression: $\sqrt{b^2 - 4ac} = \sqrt{(-6)^2 - 4(2)(3)} = \sqrt{36 - 24} = \sqrt{12} = 2\sqrt{3}$

37. Substituting into the expression: $\sqrt{b^2 - 4ac} = \sqrt{(2)^2 - 4(1)(6)} = \sqrt{4 - 24} = \sqrt{-20}$, which is not a real number

39. Substituting into the expression: $\sqrt{b^2 - 4ac} = \sqrt{\left(-\frac{1}{2}\right)^2 - 4\left(\frac{1}{2}\right)\left(-\frac{5}{4}\right)} = \sqrt{\frac{1}{4} + \frac{5}{2}} = \sqrt{\frac{11}{4}} = \frac{\sqrt{11}}{2}$

41. Rationalizing the denominator: $\dfrac{2}{\sqrt{3}} = \dfrac{2}{\sqrt{3}} \cdot \dfrac{\sqrt{3}}{\sqrt{3}} = \dfrac{2\sqrt{3}}{3}$

43. Rationalizing the denominator: $\dfrac{5}{\sqrt{6}} = \dfrac{5}{\sqrt{6}} \cdot \dfrac{\sqrt{6}}{\sqrt{6}} = \dfrac{5\sqrt{6}}{6}$

45. Rationalizing the denominator: $\sqrt{\dfrac{1}{2}} = \dfrac{1}{\sqrt{2}} \cdot \dfrac{\sqrt{2}}{\sqrt{2}} = \dfrac{\sqrt{2}}{2}$

47. Rationalizing the denominator: $\sqrt{\dfrac{1}{5}} = \dfrac{1}{\sqrt{5}} \cdot \dfrac{\sqrt{5}}{\sqrt{5}} = \dfrac{\sqrt{5}}{5}$

49. Rationalizing the denominator: $\dfrac{4}{\sqrt[3]{2}} = \dfrac{4}{\sqrt[3]{2}} \cdot \dfrac{\sqrt[3]{4}}{\sqrt[3]{4}} = \dfrac{4\sqrt[3]{4}}{2} = 2\sqrt[3]{4}$

51. Rationalizing the denominator: $\dfrac{2}{\sqrt[3]{9}} = \dfrac{2}{\sqrt[3]{9}} \cdot \dfrac{\sqrt[3]{3}}{\sqrt[3]{3}} = \dfrac{2\sqrt[3]{3}}{3}$

53. Rationalizing the denominator: $\sqrt[4]{\dfrac{3}{2x^2}} = \dfrac{\sqrt[4]{3}}{\sqrt[4]{2x^2}} \cdot \dfrac{\sqrt[4]{8x^2}}{\sqrt[4]{8x^2}} = \dfrac{\sqrt[4]{24x^2}}{2x}$

55. Rationalizing the denominator: $\sqrt[4]{\dfrac{8}{y}} = \dfrac{\sqrt[4]{8}}{\sqrt[4]{y}} \cdot \dfrac{\sqrt[4]{y^3}}{\sqrt[4]{y^3}} = \dfrac{\sqrt[4]{8y^3}}{y}$

57. Rationalizing the denominator: $\sqrt[3]{\dfrac{4x}{3y}} = \dfrac{\sqrt[3]{4x}}{\sqrt[3]{3y}} \cdot \dfrac{\sqrt[3]{9y^2}}{\sqrt[3]{9y^2}} = \dfrac{\sqrt[3]{36xy^2}}{3y}$

59. Rationalizing the denominator: $\sqrt[3]{\dfrac{2x}{9y}} = \dfrac{\sqrt[3]{2x}}{\sqrt[3]{9y}} \cdot \dfrac{\sqrt[3]{3y^2}}{\sqrt[3]{3y^2}} = \dfrac{\sqrt[3]{6xy^2}}{3y}$

61. Rationalizing the denominator: $\sqrt[4]{\dfrac{1}{8x^3}} = \dfrac{1}{\sqrt[4]{8x^3}} \cdot \dfrac{\sqrt[4]{2x}}{\sqrt[4]{2x}} = \dfrac{\sqrt[4]{2x}}{2x}$

63. Simplifying: $\sqrt{\dfrac{27x^3}{5y}} = \dfrac{\sqrt{27x^3}}{\sqrt{5y}} \cdot \dfrac{\sqrt{5y}}{\sqrt{5y}} = \dfrac{\sqrt{135x^3y}}{5y} = \dfrac{3x\sqrt{15xy}}{5y}$

65. Simplifying: $\sqrt{\dfrac{75x^3y^2}{2z}} = \dfrac{\sqrt{75x^3y^2}}{\sqrt{2z}} \cdot \dfrac{\sqrt{2z}}{\sqrt{2z}} = \dfrac{\sqrt{150x^3y^2z}}{2z} = \dfrac{5xy\sqrt{6xz}}{2z}$

67. Simplifying: $\sqrt[3]{\dfrac{16a^4b^3}{9c}} = \dfrac{\sqrt[3]{16a^4b^3}}{\sqrt[3]{9c}} \cdot \dfrac{\sqrt[3]{3c^2}}{\sqrt[3]{3c^2}} = \dfrac{\sqrt[3]{48a^4b^3c^2}}{3c} = \dfrac{2ab\sqrt[3]{6ac^2}}{3c}$

69. Simplifying: $\sqrt[3]{\dfrac{8x^3y^6}{9z}} = \dfrac{\sqrt[3]{8x^3y^6}}{\sqrt[3]{9z}} \cdot \dfrac{\sqrt[3]{3z^2}}{\sqrt[3]{3z^2}} = \dfrac{\sqrt[3]{24x^3y^6z^2}}{3z} = \dfrac{2xy^2\sqrt[3]{3z^2}}{3z}$

71. Simplifying: $\sqrt{\sqrt{x^2}} = \sqrt{x}$ **73.** Simplifying: $\sqrt[3]{\sqrt{xy}} = \left((xy)^{1/2}\right)^{1/3} = (xy)^{1/6} = \sqrt[6]{xy}$

75. Simplifying: $\sqrt[3]{\sqrt[4]{a}} = \left(a^{1/4}\right)^{1/3} = a^{1/12} = \sqrt[12]{a}$

77. Simplifying: $\sqrt[3]{\sqrt[3]{6x^{10}}} = \left(\left(6x^{10}\right)^{1/3}\right)^{1/3} = \left(6x^{10}\right)^{1/9} = x(6x)^{1/9} = x\sqrt[9]{6x}$

79. Simplifying: $\sqrt[4]{\sqrt[3]{a^{12}b^{24}c^{14}}} = \left(\left(a^{12}b^{24}c^{14}\right)^{1/3}\right)^{1/4} = \left(a^{12}b^{24}c^{14}\right)^{1/12} = ab^2c\left(c^2\right)^{1/12} = ab^2c\left(c^{1/6}\right) = ab^2c\sqrt[6]{c}$

81. Simplifying: $\sqrt[3]{\sqrt[5]{3a^{17}b^{16}c^{30}}} = \left(\left(3a^{17}b^{16}c^{30}\right)^{1/3}\right)^{1/5} = \left(3a^{17}b^{16}c^{30}\right)^{1/15} = abc^2\left(3a^2b\right)^{1/15} = abc^2\sqrt[15]{3a^2b}$

83. Simplifying: $\left(\sqrt{\sqrt[3]{8ab^6}}\right)^2 = \sqrt[3]{8ab^6} = 2b^2\sqrt[3]{a}$ **85.** Simplifying: $\sqrt{25x^2} = 5|x|$

87. Simplifying: $\sqrt{27x^3y^2} = \sqrt{9x^2y^2 \cdot 3x} = 3|xy|\sqrt{3x}$ **89.** Simplifying: $\sqrt{x^2 - 10x + 25} = \sqrt{(x-5)^2} = |x - 5|$

91. Simplifying: $\sqrt{4x^2+12x+9} = \sqrt{(2x+3)^2} = |2x+3|$

93. Simplifying: $\sqrt{4a^4+16a^3+16a^2} = \sqrt{4a^2\left(a^2+4a+4\right)} = \sqrt{4a^2\left(a+2\right)^2} = 2|a(a+2)|$

95. Simplifying: $\sqrt{4x^3-8x^2} = \sqrt{4x^2\left(x-2\right)} = 2|x|\sqrt{x-2}$

97. Substituting $a = 9$ and $b = 16$:
$$\sqrt{a+b} = \sqrt{9+16} = \sqrt{25} = 5$$
$$\sqrt{a}+\sqrt{b} = \sqrt{9}+\sqrt{16} = 3+4 = 7$$
Thus $\sqrt{a+b} \neq \sqrt{a}+\sqrt{b}$.

99. Substituting $w = 10$ and $l = 15$: $d = \sqrt{l^2+w^2} = \sqrt{15^2+10^2} = \sqrt{225+100} = \sqrt{325} = \sqrt{25\cdot13} = 5\sqrt{13}$ feet

101. Substituting $V = 7$ and $h = 8$: $r = \sqrt{\dfrac{V}{h\pi}} = \sqrt{\dfrac{7}{8\left(\frac{22}{7}\right)}} = \dfrac{\sqrt{49}}{\sqrt{16\cdot11}} \cdot \dfrac{\sqrt{11}}{\sqrt{11}} = \dfrac{\sqrt{11\cdot49}}{\sqrt{16\cdot121}} = \dfrac{7\sqrt{11}}{4\cdot11} = \dfrac{7\sqrt{11}}{44}$ meters

103. **a.** Substituting $w = 3$, $l = 4$, and $h = 12$:
$$d = \sqrt{l^2+w^2+h^2} = \sqrt{4^2+3^2+12^2} = \sqrt{16+9+144} = \sqrt{169} = 13 \text{ feet}$$
b. Substituting $w = 2$, $l = 4$, and $h = 6$:
$$d = \sqrt{l^2+w^2+h^2} = \sqrt{4^2+2^2+6^2} = \sqrt{16+4+36} = \sqrt{56} = 2\sqrt{14} \approx 7.5 \text{ feet}$$

105. Answers will vary.

107. Simplifying: $5x-4x+6x = 7x$

109. Simplifying: $35xy^2-8xy^2 = 27xy^2$

111. Simplifying: $\dfrac{1}{2}x+\dfrac{1}{3}x = \dfrac{3}{6}x+\dfrac{2}{6}x = \dfrac{5}{6}x$

113. Simplifying: $\sqrt{18} = \sqrt{9\cdot2} = 3\sqrt{2}$

115. Simplifying: $\sqrt{75xy^3} = \sqrt{25y^2\cdot3xy} = 5y\sqrt{3xy}$

117. Simplifying: $\sqrt[3]{8a^4b^2} = \sqrt[3]{8a^3\cdot ab^2} = 2a\sqrt[3]{ab^2}$

119. Performing the operations: $\dfrac{8xy^3}{9x^2y} \div \dfrac{16x^2y^2}{18xy^3} = \dfrac{8xy^3}{9x^2y} \cdot \dfrac{18xy^3}{16x^2y^2} = \dfrac{144x^2y^6}{144x^4y^3} = \dfrac{y^3}{x^2}$

121. Performing the operations: $\dfrac{12a^2-4a-5}{2a+1} \cdot \dfrac{7a+3}{42a^2-17a-15} = \dfrac{(6a-5)(2a+1)}{2a+1} \cdot \dfrac{7a+3}{(7a+3)(6a-5)} = 1$

123. Performing the operations:
$$\dfrac{8x^3+27}{27x^3+1} \div \dfrac{6x^2+7x-3}{9x^2-1} = \dfrac{8x^3+27}{27x^3+1} \cdot \dfrac{9x^2-1}{6x^2+7x-3}$$
$$= \dfrac{(2x+3)\left(4x^2-6x+9\right)}{(3x+1)\left(9x^2-3x+1\right)} \cdot \dfrac{(3x+1)(3x-1)}{(2x+3)(3x-1)}$$
$$= \dfrac{4x^2-6x+9}{9x^2-3x+1}$$

125. The prime factorization of 8,640 is: $8640 = 2^6\cdot3^3\cdot5$. Therefore: $\sqrt[3]{8640} = \sqrt[3]{2^6\cdot3^3\cdot5} = 2^2\cdot3\sqrt[3]{5} = 12\sqrt[3]{5}$

127. The prime factorization of 10,584 is: $10584 = 2^3\cdot3^3\cdot7^2$. Therefore: $\sqrt[3]{10584} = \sqrt[3]{2^3\cdot3^3\cdot7^2} = 2\cdot3\sqrt[3]{7^2} = 6\sqrt[3]{49}$

129. Rationalizing the denominator: $\dfrac{1}{\sqrt[10]{a^3}} = \dfrac{1}{\sqrt[10]{a^3}} \cdot \dfrac{\sqrt[10]{a^7}}{\sqrt[10]{a^7}} = \dfrac{\sqrt[10]{a^7}}{a}$

131. Rationalizing the denominator: $\dfrac{1}{\sqrt[20]{a^{11}}} = \dfrac{1}{\sqrt[20]{a^{11}}} \cdot \dfrac{\sqrt[20]{a^9}}{\sqrt[20]{a^9}} = \dfrac{\sqrt[20]{a^9}}{a}$

133. Graphing each function:

Note that the two graphs are not the same.

135. When $x = 2$, the distance apart is: $(2+1) - \sqrt{2^2+1} = 3 - \sqrt{5} \approx 0.76$

137. The two expressions are equal when $x = 0$.

7.4 Addition and Subtraction of Radical Expressions

1. Combining radicals: $3\sqrt{5} + 4\sqrt{5} = 7\sqrt{5}$

3. Combining radicals: $3x\sqrt{7} - 4x\sqrt{7} = x\sqrt{7}$

5. Combining radicals: $5\sqrt[3]{10} - 4\sqrt[3]{10} = \sqrt[3]{10}$

7. Combining radicals: $8\sqrt[5]{6} - 2\sqrt[5]{6} + 3\sqrt[5]{6} = 9\sqrt[5]{6}$

9. Combining radicals: $3x\sqrt{2} - 4x\sqrt{2} + x\sqrt{2} = 0$

11. Combining radicals: $\sqrt{20} - \sqrt{80} + \sqrt{45} = 2\sqrt{5} - 4\sqrt{5} + 3\sqrt{5} = \sqrt{5}$

13. Combining radicals: $4\sqrt{8} - 2\sqrt{50} - 5\sqrt{72} = 8\sqrt{2} - 10\sqrt{2} - 30\sqrt{2} = -32\sqrt{2}$

15. Combining radicals: $5x\sqrt{8} + 3\sqrt{32x^2} - 5\sqrt{50x^2} = 10x\sqrt{2} + 12x\sqrt{2} - 25x\sqrt{2} = -3x\sqrt{2}$

17. Combining radicals: $5\sqrt[3]{16} - 4\sqrt[3]{54} = 10\sqrt[3]{2} - 12\sqrt[3]{2} = -2\sqrt[3]{2}$

19. Combining radicals: $\sqrt[3]{x^4y^2} + 7x\sqrt[3]{xy^2} = x\sqrt[3]{xy^2} + 7x\sqrt[3]{xy^2} = 8x\sqrt[3]{xy^2}$

21. Combining radicals: $5a^2\sqrt{27ab^3} - 6b\sqrt{12a^5b} = 15a^2b\sqrt{3ab} - 12a^2b\sqrt{3ab} = 3a^2b\sqrt{3ab}$

23. Combining radicals: $b\sqrt[3]{24a^5b} + 3a\sqrt[3]{81a^2b^4} = 2ab\sqrt[3]{3a^2b} + 9ab\sqrt[3]{3a^2b} = 11ab\sqrt[3]{3a^2b}$

25. Combining radicals: $5x\sqrt[4]{3y^5} + y\sqrt[4]{243x^4y} + \sqrt[4]{48x^4y^5} = 5xy\sqrt[4]{3y} + 3xy\sqrt[4]{3y} + 2xy\sqrt[4]{3y} = 10xy\sqrt[4]{3y}$

27. Combining radicals: $\dfrac{\sqrt{2}}{2} + \dfrac{1}{\sqrt{2}} = \dfrac{\sqrt{2}}{2} + \dfrac{1}{\sqrt{2}} \cdot \dfrac{\sqrt{2}}{\sqrt{2}} = \dfrac{\sqrt{2}}{2} + \dfrac{\sqrt{2}}{2} = \sqrt{2}$

29. Combining radicals: $\dfrac{\sqrt{5}}{3} + \dfrac{1}{\sqrt{5}} = \dfrac{\sqrt{5}}{3} + \dfrac{1}{\sqrt{5}} \cdot \dfrac{\sqrt{5}}{\sqrt{5}} = \dfrac{\sqrt{5}}{3} + \dfrac{\sqrt{5}}{5} = \dfrac{5\sqrt{5}}{15} + \dfrac{3\sqrt{5}}{15} = \dfrac{8\sqrt{5}}{15}$

31. Combining radicals: $\sqrt{x} - \dfrac{1}{\sqrt{x}} = \sqrt{x} - \dfrac{1}{\sqrt{x}} \cdot \dfrac{\sqrt{x}}{\sqrt{x}} = \sqrt{x} - \dfrac{\sqrt{x}}{x} = \dfrac{x\sqrt{x}}{x} - \dfrac{\sqrt{x}}{x} = \dfrac{(x-1)\sqrt{x}}{x}$

33. Combining radicals: $\dfrac{\sqrt{18}}{6} + \sqrt{\dfrac{1}{2}} + \dfrac{\sqrt{2}}{2} = \dfrac{3\sqrt{2}}{6} + \dfrac{1}{\sqrt{2}} \cdot \dfrac{\sqrt{2}}{\sqrt{2}} + \dfrac{\sqrt{2}}{2} = \dfrac{\sqrt{2}}{2} + \dfrac{\sqrt{2}}{2} + \dfrac{\sqrt{2}}{2} = \dfrac{3\sqrt{2}}{2}$

35. Combining radicals: $\sqrt{6} - \sqrt{\dfrac{2}{3}} + \sqrt{\dfrac{1}{6}} = \sqrt{6} - \dfrac{\sqrt{2}}{\sqrt{3}} \cdot \dfrac{\sqrt{3}}{\sqrt{3}} + \dfrac{1}{\sqrt{6}} \cdot \dfrac{\sqrt{6}}{\sqrt{6}} = \sqrt{6} - \dfrac{\sqrt{6}}{3} + \dfrac{\sqrt{6}}{6} = \dfrac{6\sqrt{6}}{6} - \dfrac{2\sqrt{6}}{6} + \dfrac{\sqrt{6}}{6} = \dfrac{5\sqrt{6}}{6}$

37. Combining radicals: $\sqrt[3]{25} + \dfrac{3}{\sqrt[3]{5}} = \sqrt[3]{25} + \dfrac{3}{\sqrt[3]{5}} \cdot \dfrac{\sqrt[3]{25}}{\sqrt[3]{25}} = \sqrt[3]{25} + \dfrac{3\sqrt[3]{25}}{5} = \dfrac{5\sqrt[3]{25}}{5} + \dfrac{3\sqrt[3]{25}}{5} = \dfrac{8\sqrt[3]{25}}{5}$

39.   **a.**   Finding the function: $f(x) + g(x) = \sqrt{8x} + \sqrt{72x} = 2\sqrt{2x} + 6\sqrt{2x} = 8\sqrt{2x}$

      b.   Finding the function: $f(x) - g(x) = \sqrt{8x} - \sqrt{72x} = 2\sqrt{2x} - 6\sqrt{2x} = -4\sqrt{2x}$

41.   **a.**   Finding the function: $f(x) + g(x) = 3\sqrt{2x} + \sqrt{2x} = 4\sqrt{2x}$

      b.   Finding the function: $f(x) - g(x) = 3\sqrt{2x} - \sqrt{2x} = 2\sqrt{2x}$

43.   **a.**   Finding the function: $f(x) + g(x) = x\sqrt{2} + 2x\sqrt{2} = 3x\sqrt{2}$

      b.   Finding the function: $f(x) - g(x) = x\sqrt{2} - 2x\sqrt{2} = -x\sqrt{2}$

45.   **a.**   Finding the function: $f(x) + g(x) = \left(\sqrt{2x} - 2\right) + \left(2\sqrt{2x} + 5\right) = 3\sqrt{2x} + 3$

      b.   Finding the function: $f(x) - g(x) = \left(\sqrt{2x} - 2\right) - \left(2\sqrt{2x} + 5\right) = -\sqrt{2x} - 7$

47.   Using a calculator:
$$\sqrt{12} \approx 3.464 \qquad 2\sqrt{3} \approx 3.464$$

49.   It is equal to the decimal approximation for $\sqrt{50}$:
$$\sqrt{8} + \sqrt{18} \approx 7.071 \approx \sqrt{50} \qquad \sqrt{26} \approx 5.099$$

51.   Answers will vary.

53.   Answers will vary.

55.   First use the Pythagorean theorem to find the height h:
$$x^2 + h^2 = (2x)^2$$
$$x^2 + h^2 = 4x^2$$
$$h^2 = 3x^2$$
$$h = x\sqrt{3}$$

Therefore the ratio is: $\dfrac{x\sqrt{3}}{2x} = \dfrac{\sqrt{3}}{2}$

57.   Let d represent the diagonal of the base. Using the Pythagorean theorem:
$$8^2 + 8^2 = d^2$$
$$64 + 64 = d^2$$
$$d^2 = 128$$
$$d = 8\sqrt{2}$$

Similarly, using the Pythagorean theorem on the side triangle:
$$4^2 + a^2 = 8^2$$
$$16 + a^2 = 64$$
$$a^2 = 48$$
$$a = 4\sqrt{3}$$

Now using the Pythagorean theorem on the triangle involving h and a:
$$\left(4\sqrt{2}\right)^2 + h^2 = 8^2$$
$$32 + h^2 = 64$$
$$h^2 = 32$$
$$h = 4\sqrt{2}$$

Therefore the ratio is: $\dfrac{h}{a} = \dfrac{4\sqrt{2}}{4\sqrt{3}} = \dfrac{\sqrt{2}}{\sqrt{3}} \cdot \dfrac{\sqrt{3}}{\sqrt{3}} = \dfrac{\sqrt{6}}{3}$

59.   Combining the fractions: $\dfrac{2a - 4}{a + 2} - \dfrac{a - 6}{a + 2} = \dfrac{2a - 4 - a + 6}{a + 2} = \dfrac{a + 2}{a + 2} = 1$

61.   Combining the fractions: $3 + \dfrac{4}{3 - t} = 3 \cdot \dfrac{3 - t}{3 - t} + \dfrac{4}{3 - t} = \dfrac{9 - 3t + 4}{3 - t} = \dfrac{13 - 3t}{3 - t}$

63. Combining the fractions:

$$\frac{3}{2x-5} - \frac{39}{8x^2-14x-15} = \frac{3}{2x-5} \cdot \frac{4x+3}{4x+3} - \frac{39}{(2x-5)(4x+3)}$$

$$= \frac{12x+9-39}{(2x-5)(4x+3)}$$

$$= \frac{12x-30}{(2x-5)(4x+3)}$$

$$= \frac{6(2x-5)}{(2x-5)(4x+3)}$$

$$= \frac{6}{4x+3}$$

65. Combining the fractions:

$$\frac{1}{x-y} - \frac{3xy}{x^3-y^3} = \frac{1}{x-y} \cdot \frac{x^2+xy+y^2}{x^2+xy+y^2} - \frac{3xy}{(x-y)(x^2+xy+y^2)}$$

$$= \frac{x^2+xy+y^2-3xy}{(x-y)(x^2+xy+y^2)}$$

$$= \frac{x^2-2xy+y^2}{(x-y)(x^2+xy+y^2)}$$

$$= \frac{(x-y)^2}{(x-y)(x^2+xy+y^2)}$$

$$= \frac{x-y}{x^2+xy+y^2}$$

67. Simplifying: $3 \cdot 2 = 6$

69. Simplifying: $(x+y)(4x-y) = 4x^2 - xy + 4xy - y^2 = 4x^2 + 3xy - y^2$

71. Simplifying: $(x+3)^2 = x^2 + 2(3x) + 3^2 = x^2 + 6x + 9$

73. Simplifying: $(x-2)(x+2) = x^2 - 2^2 = x^2 - 4$ **75.** Simplifying: $2\sqrt{18} = 2\sqrt{9 \cdot 2} = 2 \cdot 3\sqrt{2} = 6\sqrt{2}$

77. Simplifying: $\left(\sqrt{6}\right)^2 = 6$ **79.** Simplifying: $\left(3\sqrt{x}\right)^2 = 9x$

81. Rationalizing the denominator: $\dfrac{\sqrt{3}}{\sqrt{2}} = \dfrac{\sqrt{3}}{\sqrt{2}} \cdot \dfrac{\sqrt{2}}{\sqrt{2}} = \dfrac{\sqrt{6}}{2}$

83. Simplifying: $\sqrt[5]{32x^5y^5} - y\sqrt[3]{27x^3} = 2xy - 3xy = -xy$

85. Simplifying: $3\sqrt[9]{x^9y^{18}z^{27}} - 4\sqrt[6]{x^6y^{12}z^{18}} = 3xy^2z^3 - 4xy^2z^3 = -xy^2z^3$

87. Simplifying:

$$3c\sqrt[8]{4a^6b^{18}} + b\sqrt[4]{32a^3b^5c^4} = 3c\left(2^2a^6b^{18}\right)^{1/8} + b \cdot 2bc\sqrt[4]{2a^3b}$$

$$= 3c\left(2a^3b^9\right)^{1/4} + 2b^2c\sqrt[4]{2b}$$

$$= 3c\left(2a^3b^9\right)^{1/4} + 2b^2c\sqrt[4]{2a^3b}$$

$$= 3b^2c\sqrt[4]{2a^3b} + 2b^2c\sqrt[4]{2a^3b}$$

$$= 5b^2c\sqrt[4]{2a^3b}$$

89. Simplifying:

$$3\sqrt[9]{8a^{12}b^9} + b\sqrt[3]{16a^4} - 8\sqrt[6]{4a^8b^6} = 3\left(2^3 a^{12} b^9\right)^{1/9} + b \cdot 2a\sqrt[3]{2a} - 8\left(2^2 a^8 b^6\right)^{1/6}$$

$$= 3b\left(2a^4\right)^{1/3} + 2ab\sqrt[3]{2a} - 8b\left(2a^4\right)^{1/3}$$

$$= 3ab\sqrt[3]{2a} + 2ab\sqrt[3]{2a} - 8ab\sqrt[3]{2a}$$

$$= -3ab\sqrt[3]{2a}$$

7.5 Multiplication and Division of Radical Expressions

1. Multiplying: $\sqrt{6}\sqrt{3} = \sqrt{18} = 3\sqrt{2}$ **3.** Multiplying: $\left(2\sqrt{3}\right)\left(5\sqrt{7}\right) = 10\sqrt{21}$

5. Multiplying: $\left(4\sqrt{6}\right)\left(2\sqrt{15}\right)\left(3\sqrt{10}\right) = 24\sqrt{900} = 24 \cdot 30 = 720$

7. Multiplying: $\left(3\sqrt[3]{3}\right)\left(6\sqrt[3]{9}\right) = 18\sqrt[3]{27} = 18 \cdot 3 = 54$ **9.** Multiplying: $\sqrt{3}\left(\sqrt{2} - 3\sqrt{3}\right) = \sqrt{6} - 3\sqrt{9} = \sqrt{6} - 9$

11. Multiplying: $6\sqrt[3]{4}\left(2\sqrt[3]{2} + 1\right) = 12\sqrt[3]{8} + 6\sqrt[3]{4} = 24 + 6\sqrt[3]{4}$

13. Multiplying: $\sqrt[3]{4}\left(\sqrt[3]{2} + \sqrt[3]{6}\right) = \sqrt[3]{8} + \sqrt[3]{24} = 2 + 2\sqrt[3]{2}$

15. Multiplying: $\sqrt[3]{x}\left(\sqrt[3]{x^2 y^4} + \sqrt[3]{x^5 y}\right) = \sqrt[3]{x^3 y^4} + \sqrt[3]{x^6 y} = xy\sqrt[3]{y} + x^2\sqrt[3]{y}$

17. Multiplying: $\sqrt[4]{2x^3}\left(\sqrt[4]{8x^6} + \sqrt[4]{16x^9}\right) = \sqrt[4]{16x^9} + \sqrt[4]{32x^{12}} = 2x^2\sqrt[4]{x} + 2x^3\sqrt[4]{2}$

19. Multiplying: $\left(\sqrt{3} + \sqrt{2}\right)\left(3\sqrt{3} - \sqrt{2}\right) = 3\sqrt{9} - \sqrt{6} + 3\sqrt{6} - \sqrt{4} = 9 + 2\sqrt{6} - 2 = 7 + 2\sqrt{6}$

21. Multiplying: $\left(\sqrt{x} + 5\right)\left(\sqrt{x} - 3\right) = x - 3\sqrt{x} + 5\sqrt{x} - 15 = x + 2\sqrt{x} - 15$

23. Multiplying: $\left(3\sqrt{6} + 4\sqrt{2}\right)\left(\sqrt{6} + 2\sqrt{2}\right) = 3\sqrt{36} + 4\sqrt{12} + 6\sqrt{12} + 8\sqrt{4} = 18 + 8\sqrt{3} + 12\sqrt{3} + 16 = 34 + 20\sqrt{3}$

25. Multiplying: $\left(\sqrt{3} + 4\right)^2 = \left(\sqrt{3} + 4\right)\left(\sqrt{3} + 4\right) = \sqrt{9} + 4\sqrt{3} + 4\sqrt{3} + 16 = 19 + 8\sqrt{3}$

27. Multiplying: $\left(\sqrt{x} - 3\right)^2 = \left(\sqrt{x} - 3\right)\left(\sqrt{x} - 3\right) = x - 3\sqrt{x} - 3\sqrt{x} + 9 = x - 6\sqrt{x} + 9$

29. Multiplying: $\left(2\sqrt{a} - 3\sqrt{b}\right)^2 = \left(2\sqrt{a} - 3\sqrt{b}\right)\left(2\sqrt{a} - 3\sqrt{b}\right) = 4a - 6\sqrt{ab} - 6\sqrt{ab} + 9b = 4a - 12\sqrt{ab} + 9b$

31. Multiplying: $\left(\sqrt{x-4} + 2\right)^2 = \left(\sqrt{x-4} + 2\right)\left(\sqrt{x-4} + 2\right) = x - 4 + 2\sqrt{x-4} + 2\sqrt{x-4} + 4 = x + 4\sqrt{x-4}$

33. Multiplying: $\left(\sqrt{x-5} - 3\right)^2 = \left(\sqrt{x-5} - 3\right)\left(\sqrt{x-5} - 3\right) = x - 5 - 3\sqrt{x-5} - 3\sqrt{x-5} + 9 = x + 4 - 6\sqrt{x-5}$

35. Multiplying: $\left(\sqrt{3} - \sqrt{2}\right)\left(\sqrt{3} + \sqrt{2}\right) = \left(\sqrt{3}\right)^2 - \left(\sqrt{2}\right)^2 = 3 - 2 = 1$

37. Multiplying: $\left(\sqrt{a} + 7\right)\left(\sqrt{a} - 7\right) = \left(\sqrt{a}\right)^2 - \left(7\right)^2 = a - 49$

39. Multiplying: $\left(5 - \sqrt{x}\right)\left(5 + \sqrt{x}\right) = \left(5\right)^2 - \left(\sqrt{x}\right)^2 = 25 - x$

41. Multiplying: $\left(\sqrt{x-4} + 2\right)\left(\sqrt{x-4} - 2\right) = \left(\sqrt{x-4}\right)^2 - \left(2\right)^2 = x - 4 - 4 = x - 8$

43. Multiplying: $\left(\sqrt{3} + 1\right)^3 = \left(\sqrt{3} + 1\right)\left(3 + 2\sqrt{3} + 1\right) = \left(\sqrt{3} + 1\right)\left(4 + 2\sqrt{3}\right) = 4\sqrt{3} + 4 + 6 + 2\sqrt{3} = 10 + 6\sqrt{3}$

45. Multiplying: $\left(\sqrt[3]{3} + \sqrt[3]{2}\right)\left(\sqrt[3]{9} + \sqrt[3]{4}\right) = \sqrt[3]{27} + \sqrt[3]{12} + \sqrt[3]{18} + \sqrt[3]{8} = 3 + \sqrt[3]{12} + \sqrt[3]{18} + 2 = 5 + \sqrt[3]{12} + \sqrt[3]{18}$

47. Multiplying: $\left(\sqrt[3]{x^5} + \sqrt[3]{y}\right)\left(\sqrt[3]{x} + \sqrt[3]{y^2}\right) = \sqrt[3]{x^6} + \sqrt[3]{x^5 y^2} + \sqrt[3]{xy} + \sqrt[3]{y^3} = x^2 + x\sqrt[3]{x^2 y^2} + \sqrt[3]{xy} + y$

49. Rationalizing the denominator: $\dfrac{1}{\sqrt{2}} = \dfrac{1}{\sqrt{2}} \cdot \dfrac{\sqrt{2}}{\sqrt{2}} = \dfrac{\sqrt{2}}{2}$

51. Rationalizing the denominator: $\dfrac{1}{\sqrt{x}} = \dfrac{1}{\sqrt{x}} \cdot \dfrac{\sqrt{x}}{\sqrt{x}} = \dfrac{\sqrt{x}}{x}$

53. Rationalizing the denominator: $\dfrac{4}{\sqrt{3}} = \dfrac{4}{\sqrt{3}} \cdot \dfrac{\sqrt{3}}{\sqrt{3}} = \dfrac{4\sqrt{3}}{3}$

55. Rationalizing the denominator: $\dfrac{2x}{\sqrt{6}} = \dfrac{2x}{\sqrt{6}} \cdot \dfrac{\sqrt{6}}{\sqrt{6}} = \dfrac{2x\sqrt{6}}{6} = \dfrac{x\sqrt{6}}{3}$

57. Rationalizing the denominator: $\dfrac{4}{\sqrt{10x}} = \dfrac{4}{\sqrt{10x}} \cdot \dfrac{\sqrt{10x}}{\sqrt{10x}} = \dfrac{4\sqrt{10x}}{10x} = \dfrac{2\sqrt{10x}}{5x}$

59. Rationalizing the denominator: $\dfrac{2}{\sqrt{8}} = \dfrac{2}{\sqrt{8}} \cdot \dfrac{\sqrt{2}}{\sqrt{2}} = \dfrac{2\sqrt{2}}{4} = \dfrac{\sqrt{2}}{2}$

61. Rationalizing the denominator: $\sqrt{\dfrac{32x^3y}{4xy^2}} = \sqrt{\dfrac{8x^2}{y}} = \dfrac{2x\sqrt{2}}{\sqrt{y}} \cdot \dfrac{\sqrt{y}}{\sqrt{y}} = \dfrac{2x\sqrt{2y}}{y}$

63. Rationalizing the denominator: $\sqrt{\dfrac{12a^3b^3c}{8ab^5c^2}} = \sqrt{\dfrac{3a^2}{2b^2c}} = \dfrac{a\sqrt{3}}{\sqrt{2b^2c}} \cdot \dfrac{\sqrt{2c}}{\sqrt{2c}} = \dfrac{a\sqrt{6c}}{2bc}$

65. Rationalizing the denominator: $\sqrt[3]{\dfrac{16a^4b^2c^4}{2ab^3c}} = \sqrt[3]{\dfrac{8a^3c^3}{b}} = \dfrac{2ac}{\sqrt[3]{b}} \cdot \dfrac{\sqrt[3]{b^2}}{\sqrt[3]{b^2}} = \dfrac{2ac\sqrt[3]{b^2}}{b}$

67. Rationalizing the denominator: $\dfrac{\sqrt{2}}{\sqrt{6}-\sqrt{2}} = \dfrac{\sqrt{2}}{\sqrt{6}-\sqrt{2}} \cdot \dfrac{\sqrt{6}+\sqrt{2}}{\sqrt{6}+\sqrt{2}} = \dfrac{\sqrt{12}+2}{6-2} = \dfrac{2\sqrt{3}+2}{4} = \dfrac{1+\sqrt{3}}{2}$

69. Rationalizing the denominator: $\dfrac{\sqrt{5}}{\sqrt{5}+1} = \dfrac{\sqrt{5}}{\sqrt{5}+1} \cdot \dfrac{\sqrt{5}-1}{\sqrt{5}-1} = \dfrac{5-\sqrt{5}}{5-1} = \dfrac{5-\sqrt{5}}{4}$

71. Rationalizing the denominator: $\dfrac{\sqrt{x}}{\sqrt{x}-3} = \dfrac{\sqrt{x}}{\sqrt{x}-3} \cdot \dfrac{\sqrt{x}+3}{\sqrt{x}+3} = \dfrac{x+3\sqrt{x}}{x-9}$

73. Rationalizing the denominator: $\dfrac{\sqrt{5}}{2\sqrt{5}-3} = \dfrac{\sqrt{5}}{2\sqrt{5}-3} \cdot \dfrac{2\sqrt{5}+3}{2\sqrt{5}+3} = \dfrac{2\sqrt{25}+3\sqrt{5}}{20-9} = \dfrac{10+3\sqrt{5}}{11}$

75. Rationalizing the denominator: $\dfrac{3}{\sqrt{x}-\sqrt{y}} = \dfrac{3}{\sqrt{x}-\sqrt{y}} \cdot \dfrac{\sqrt{x}+\sqrt{y}}{\sqrt{x}+\sqrt{y}} = \dfrac{3\sqrt{x}+3\sqrt{y}}{x-y}$

77. Rationalizing the denominator: $\dfrac{\sqrt{6}+\sqrt{2}}{\sqrt{6}-\sqrt{2}} = \dfrac{\sqrt{6}+\sqrt{2}}{\sqrt{6}-\sqrt{2}} \cdot \dfrac{\sqrt{6}+\sqrt{2}}{\sqrt{6}+\sqrt{2}} = \dfrac{6+2\sqrt{12}+2}{6-2} = \dfrac{8+4\sqrt{3}}{4} = 2+\sqrt{3}$

79. Rationalizing the denominator: $\dfrac{\sqrt{7}-2}{\sqrt{7}+2} = \dfrac{\sqrt{7}-2}{\sqrt{7}+2} \cdot \dfrac{\sqrt{7}-2}{\sqrt{7}-2} = \dfrac{7-4\sqrt{7}+4}{7-4} = \dfrac{11-4\sqrt{7}}{3}$

81. Rationalizing the denominator: $\dfrac{\sqrt{a}+\sqrt{b}}{\sqrt{a}-\sqrt{b}} = \dfrac{\sqrt{a}+\sqrt{b}}{\sqrt{a}-\sqrt{b}} \cdot \dfrac{\sqrt{a}+\sqrt{b}}{\sqrt{a}+\sqrt{b}} = \dfrac{a+2\sqrt{ab}+b}{a-b}$

83. Rationalizing the denominator: $\dfrac{\sqrt{x}+2}{\sqrt{x}-2} = \dfrac{\sqrt{x}+2}{\sqrt{x}-2} \cdot \dfrac{\sqrt{x}+2}{\sqrt{x}+2} = \dfrac{x+4\sqrt{x}+4}{x-4}$

85. Rationalizing the denominator:

$\dfrac{2\sqrt{3}-\sqrt{7}}{3\sqrt{3}+\sqrt{7}} = \dfrac{2\sqrt{3}-\sqrt{7}}{3\sqrt{3}+\sqrt{7}} \cdot \dfrac{3\sqrt{3}-\sqrt{7}}{3\sqrt{3}-\sqrt{7}} = \dfrac{18-3\sqrt{21}-2\sqrt{21}+7}{27-7} = \dfrac{25-5\sqrt{21}}{20} = \dfrac{5-\sqrt{21}}{4}$

87. Rationalizing the denominator: $\dfrac{3\sqrt{x}+2}{1+\sqrt{x}} = \dfrac{3\sqrt{x}+2}{1+\sqrt{x}} \cdot \dfrac{1-\sqrt{x}}{1-\sqrt{x}} = \dfrac{3\sqrt{x}+2-3x-2\sqrt{x}}{1-x} = \dfrac{\sqrt{x}-3x+2}{1-x}$

89. Simplifying: $\dfrac{2}{\sqrt{3}}+\sqrt{12}=\dfrac{2}{\sqrt{3}}\cdot\dfrac{\sqrt{3}}{\sqrt{3}}+2\sqrt{3}=\dfrac{2\sqrt{3}}{3}+\dfrac{2\sqrt{3}}{1}\cdot\dfrac{3}{3}=\dfrac{2\sqrt{3}}{3}+\dfrac{6\sqrt{3}}{3}=\dfrac{8\sqrt{3}}{3}$

91. Simplifying: $\dfrac{1}{\sqrt{5}}+\sqrt{20}=\dfrac{1}{\sqrt{5}}\cdot\dfrac{\sqrt{5}}{\sqrt{5}}+2\sqrt{5}=\dfrac{\sqrt{5}}{5}+\dfrac{2\sqrt{5}}{1}\cdot\dfrac{5}{5}=\dfrac{\sqrt{5}}{5}+\dfrac{10\sqrt{5}}{5}=\dfrac{11\sqrt{5}}{5}$

93. Simplifying: $\dfrac{6}{\sqrt{12}}-\sqrt{75}=\dfrac{6}{2\sqrt{3}}\cdot\dfrac{\sqrt{3}}{\sqrt{3}}-5\sqrt{3}=\dfrac{6\sqrt{3}}{6}-5\sqrt{3}=\sqrt{3}-5\sqrt{3}=-4\sqrt{3}$

95. Simplifying:

$$\dfrac{1}{\sqrt{3}}+\sqrt{48}+\dfrac{4}{\sqrt{12}}=\dfrac{1}{\sqrt{3}}\cdot\dfrac{\sqrt{3}}{\sqrt{3}}+4\sqrt{3}+\dfrac{4}{2\sqrt{3}}\cdot\dfrac{\sqrt{3}}{\sqrt{3}}$$

$$=\dfrac{\sqrt{3}}{3}+\dfrac{4\sqrt{3}}{1}\cdot\dfrac{3}{3}+\dfrac{4\sqrt{3}}{6}$$

$$=\dfrac{\sqrt{3}}{3}+\dfrac{12\sqrt{3}}{3}+\dfrac{2\sqrt{3}}{3}$$

$$=\dfrac{15\sqrt{3}}{3}$$

$$=5\sqrt{3}$$

97. Simplifying the product: $\left(\sqrt[3]{2}+\sqrt[3]{3}\right)\left(\sqrt[3]{4}-\sqrt[3]{6}+\sqrt[3]{9}\right)=\sqrt[3]{8}-\sqrt[3]{12}+\sqrt[3]{18}+\sqrt[3]{12}-\sqrt[3]{18}+\sqrt[3]{27}=2+3=5$

99. The correct statement is: $5\left(2\sqrt{3}\right)=10\sqrt{3}$

101. The correct statement is: $\left(\sqrt{x}+3\right)^2=\left(\sqrt{x}+3\right)\left(\sqrt{x}+3\right)=x+6\sqrt{x}+9$

103. The correct statement is: $\left(5\sqrt{3}\right)^2=\left(5\sqrt{3}\right)\left(5\sqrt{3}\right)=25\cdot3=75$

105. Substituting $h=50$: $t=\dfrac{\sqrt{100-50}}{4}=\dfrac{\sqrt{50}}{4}=\dfrac{5\sqrt{2}}{4}$ second

Substituting $h=0$: $t=\dfrac{\sqrt{100-0}}{4}=\dfrac{\sqrt{100}}{4}=\dfrac{10}{4}=\dfrac{5}{2}$ second

107. Since the large rectangle is a golden rectangle and $AC=6$, then $CE=6\left(\dfrac{1+\sqrt{5}}{2}\right)=3+3\sqrt{5}$. Since $CD=6$, then

$DE=3+3\sqrt{5}-6=3\sqrt{5}-3$. Now computing the ratio:

$$\dfrac{EF}{DE}=\dfrac{6}{3\sqrt{5}-3}\cdot\dfrac{3\sqrt{5}+3}{3\sqrt{5}+3}=\dfrac{18\left(\sqrt{5}+1\right)}{45-9}=\dfrac{18\left(\sqrt{5}+1\right)}{36}=\dfrac{1+\sqrt{5}}{2}$$

Therefore the smaller rectangle $BDEF$ is also a golden rectangle.

109. Since the large rectangle is a golden rectangle and $AC=2x$, then $CE=2x\left(\dfrac{1+\sqrt{5}}{2}\right)=x\left(1+\sqrt{5}\right)$. Since $CD=2x$, then

$DE=x\left(1+\sqrt{5}\right)-2x=x\left(-1+\sqrt{5}\right)$. Now computing the ratio:

$$\dfrac{EF}{DE}=\dfrac{2x}{x\left(-1+\sqrt{5}\right)}=\dfrac{2}{-1+\sqrt{5}}\cdot\dfrac{-1-\sqrt{5}}{-1-\sqrt{5}}=\dfrac{-2\left(\sqrt{5}+1\right)}{1-5}=\dfrac{-2\left(\sqrt{5}+1\right)}{-4}=\dfrac{1+\sqrt{5}}{2}$$

Therefore the smaller rectangle $BDEF$ is also a golden rectangle.

111. Simplifying the complex fraction: $\dfrac{\dfrac{1}{4}-\dfrac{1}{3}}{\dfrac{1}{2}+\dfrac{1}{6}}=\dfrac{\left(\dfrac{1}{4}-\dfrac{1}{3}\right)\cdot12}{\left(\dfrac{1}{2}+\dfrac{1}{6}\right)\cdot12}=\dfrac{3-4}{6+2}=-\dfrac{1}{8}$

113. Simplifying the complex fraction: $\dfrac{1-\dfrac{2}{y}}{1+\dfrac{2}{y}} = \dfrac{\left(1-\dfrac{2}{y}\right)\bullet y}{\left(1+\dfrac{2}{y}\right)\bullet y} = \dfrac{y-2}{y+2}$

115. Simplifying the complex fraction: $\dfrac{4+\dfrac{4}{x}+\dfrac{1}{x^2}}{4-\dfrac{1}{x^2}} = \dfrac{\left(4+\dfrac{4}{x}+\dfrac{1}{x^2}\right)\bullet x^2}{\left(4-\dfrac{1}{x^2}\right)\bullet x^2} = \dfrac{4x^2+4x+1}{4x^2-1} = \dfrac{(2x+1)^2}{(2x+1)(2x-1)} = \dfrac{2x+1}{2x-1}$

117. Simplifying: $(t+5)^2 = t^2 + 2(5t) + 5^2 = t^2 + 10t + 25$

119. Simplifying: $\sqrt{x} \bullet \sqrt{x} = \sqrt{x^2} = x$

121. Solving the equation:

$$3x + 4 = 5^2$$
$$3x + 4 = 25$$
$$3x = 21$$
$$x = 7$$

123. Solving the equation:

$$t^2 + 7t + 12 = 0$$
$$(t+4)(t+3) = 0$$
$$t = -4, -3$$

125. Solving the equation:

$$t^2 + 10t + 25 = t + 7$$
$$t^2 + 9t + 18 = 0$$
$$(t+6)(t+3) = 0$$
$$t = -6, -3$$

127. Solving the equation:

$$(x+4)^2 = x + 6$$
$$x^2 + 8x + 16 = x + 6$$
$$x^2 + 7x + 10 = 0$$
$$(x+5)(x+2) = 0$$
$$x = -5, -2$$

129. Rationalizing the denominator: $\dfrac{x}{\sqrt{x-2}+4} = \dfrac{x}{\sqrt{x-2}+4} \bullet \dfrac{\sqrt{x-2}-4}{\sqrt{x-2}-4} = \dfrac{x\left(\sqrt{x-2}-4\right)}{x-2-16} = \dfrac{x\left(\sqrt{x-2}-4\right)}{x-18}$

131. Rationalizing the denominator: $\dfrac{x}{\sqrt{x+5}-5} = \dfrac{x}{\sqrt{x+5}-5} \bullet \dfrac{\sqrt{x+5}+5}{\sqrt{x+5}+5} = \dfrac{x\left(\sqrt{x+5}+5\right)}{x+5-25} = \dfrac{x\left(\sqrt{x+5}+5\right)}{x-20}$

133. Rationalizing the denominator: $\dfrac{3x}{\sqrt{5x}+x} = \dfrac{3x}{\sqrt{5x}+x} \bullet \dfrac{\sqrt{5x}-x}{\sqrt{5x}-x} = \dfrac{3x\left(\sqrt{5x}-x\right)}{5x-x^2} = \dfrac{3x\left(\sqrt{5x}-x\right)}{x(5-x)} = \dfrac{3\left(\sqrt{5x}-x\right)}{5-x}$

7.6 Equations with Radicals

1. Solving the equation:
$$\sqrt{2x+1} = 3$$
$$\left(\sqrt{2x+1}\right)^2 = 3^2$$
$$2x+1 = 9$$
$$2x = 8$$
$$x = 4$$

3. Solving the equation:
$$\sqrt{4x+1} = -5$$
$$\left(\sqrt{4x+1}\right)^2 = (-5)^2$$
$$4x+1 = 25$$
$$4x = 24$$
$$x = 6$$
Since this value does not check, the solution set is $\varnothing$.

5. Solving the equation:
$$\sqrt{2y-1} = 3$$
$$\left(\sqrt{2y-1}\right)^2 = 3^2$$
$$2y-1 = 9$$
$$2y = 10$$
$$y = 5$$

7. Solving the equation:
$$\sqrt{5x-7} = -1$$
$$\left(\sqrt{5x-7}\right)^2 = (-1)^2$$
$$5x-7 = 1$$
$$5x = 8$$
$$x = \frac{8}{5}$$
Since this value does not check, the solution set is $\varnothing$.

9. Solving the equation:
$$\sqrt{2x-3} - 2 = 4$$
$$\sqrt{2x-3} = 6$$
$$\left(\sqrt{2x-3}\right)^2 = 6^2$$
$$2x-3 = 36$$
$$2x = 39$$
$$x = \frac{39}{2}$$

11. Solving the equation:
$$\sqrt{4a+1} + 3 = 2$$
$$\sqrt{4a+1} = -1$$
$$\left(\sqrt{4a+1}\right)^2 = (-1)^2$$
$$4a+1 = 1$$
$$4a = 0$$
$$a = 0$$
Since this value does not check, the solution set is $\varnothing$.

13. Solving the equation:
$$\sqrt[4]{3x+1} = 2$$
$$\left(\sqrt[4]{3x+1}\right)^4 = 2^4$$
$$3x+1 = 16$$
$$3x = 15$$
$$x = 5$$

15. Solving the equation:
$$\sqrt[3]{2x-5} = 1$$
$$\left(\sqrt[3]{2x-5}\right)^3 = 1^3$$
$$2x-5 = 1$$
$$2x = 6$$
$$x = 3$$

17. Solving the equation:
$$\sqrt[3]{3a+5} = -3$$
$$\left(\sqrt[3]{3a+5}\right)^3 = (-3)^3$$
$$3a+5 = -27$$
$$3a = -32$$
$$a = -\frac{32}{3}$$

19. Solving the equation:
$$\sqrt{y-3} = y-3$$
$$\left(\sqrt{y-3}\right)^2 = (y-3)^2$$
$$y-3 = y^2 - 6y + 9$$
$$0 = y^2 - 7y + 12$$
$$0 = (y-3)(y-4)$$
$$y = 3, 4$$

21.   Solving the equation:

$$\sqrt{a+2} = a+2$$
$$\left(\sqrt{a+2}\right)^2 = (a+2)^2$$
$$a+2 = a^2 + 4a + 4$$
$$0 = a^2 + 3a + 2$$
$$0 = (a+2)(a+1)$$
$$a = -2, -1$$

23.   Solving the equation:

$$\sqrt{2x+3} = \frac{2x-7}{3}$$
$$\left(\sqrt{2x+3}\right)^2 = \left(\frac{2x-7}{3}\right)^2$$
$$2x+3 = \frac{4x^2 - 28x + 49}{9}$$
$$18x + 27 = 4x^2 - 28x + 49$$
$$0 = 4x^2 - 46x + 22$$
$$0 = 2x^2 - 23x + 11$$
$$0 = (2x-1)(x-11)$$
$$x = \frac{1}{2}, 11$$

The solution is 11 (1/2 does not check).

25.   Solving the equation:

$$\sqrt{4x-3} = \frac{x+3}{2}$$
$$\left(\sqrt{4x-3}\right)^2 = \left(\frac{x+3}{2}\right)^2$$
$$4x-3 = \frac{x^2 + 6x + 9}{4}$$
$$16x - 12 = x^2 + 6x + 9$$
$$0 = x^2 - 10x + 21$$
$$0 = (x-3)(x-7)$$
$$x = 3, 7$$

27.   Solving the equation:

$$\sqrt{7x+2} = \frac{2x+2}{3}$$
$$\left(\sqrt{7x+2}\right)^2 = \left(\frac{2x+2}{3}\right)^2$$
$$7x+2 = \frac{4x^2 + 8x + 4}{9}$$
$$63x + 18 = 4x^2 + 8x + 4$$
$$0 = 4x^2 - 55x - 14$$
$$0 = (4x+1)(x-14)$$
$$x = -\frac{1}{4}, 14$$

29.   Solving the equation:

$$\sqrt{2x+4} = \sqrt{1-x}$$
$$\left(\sqrt{2x+4}\right)^2 = \left(\sqrt{1-x}\right)^2$$
$$2x+4 = 1-x$$
$$3x = -3$$
$$x = -1$$

31.   Solving the equation:

$$\sqrt{4a+7} = -\sqrt{a+2}$$
$$\left(\sqrt{4a+7}\right)^2 = \left(-\sqrt{a+2}\right)^2$$
$$4a+7 = a+2$$
$$3a = -5$$
$$a = -\frac{5}{3}$$

Since this value does not check, the solution set is $\varnothing$.

33.   Solving the equation:

$$\sqrt[4]{5x-8} = \sqrt[4]{4x-1}$$
$$\left(\sqrt[4]{5x-8}\right)^4 = \left(\sqrt[4]{4x-1}\right)^4$$
$$5x-8 = 4x-1$$
$$x = 7$$

35.   Solving the equation:

$$x+1 = \sqrt{5x+1}$$
$$(x+1)^2 = \left(\sqrt{5x+1}\right)^2$$
$$x^2 + 2x + 1 = 5x + 1$$
$$x^2 - 3x = 0$$
$$x(x-3) = 0$$
$$x = 0, 3$$

37. Solving the equation:

$$t + 5 = \sqrt{2t+9}$$
$$(t+5)^2 = \left(\sqrt{2t+9}\right)^2$$
$$t^2 + 10t + 25 = 2t + 9$$
$$t^2 + 8t + 16 = 0$$
$$(t+4)^2 = 0$$
$$t = -4$$

39. Solving the equation:

$$\sqrt{y-8} = \sqrt{8-y}$$
$$\left(\sqrt{y-8}\right)^2 = \left(\sqrt{8-y}\right)^2$$
$$y - 8 = 8 - y$$
$$2y = 16$$
$$y = 8$$

41. Solving the equation:

$$\sqrt[3]{3x+5} = \sqrt[3]{5-2x}$$
$$\left(\sqrt[3]{3x+5}\right)^3 = \left(\sqrt[3]{5-2x}\right)^3$$
$$3x + 5 = 5 - 2x$$
$$5x = 0$$
$$x = 0$$

43. Solving the equation:

$$\sqrt{x-8} = \sqrt{x} - 2$$
$$\left(\sqrt{x-8}\right)^2 = \left(\sqrt{x}-2\right)^2$$
$$x - 8 = x - 4\sqrt{x} + 4$$
$$-12 = -4\sqrt{x}$$
$$\sqrt{x} = 3$$
$$x = 9$$

45. Solving the equation:

$$\sqrt{x+1} = \sqrt{x} + 1$$
$$\left(\sqrt{x+1}\right)^2 = \left(\sqrt{x}+1\right)^2$$
$$x + 1 = x + 2\sqrt{x} + 1$$
$$0 = 2\sqrt{x}$$
$$\sqrt{x} = 0$$
$$x = 0$$

47. Solving the equation:

$$\sqrt{x+8} = \sqrt{x-4} + 2$$
$$\left(\sqrt{x+8}\right)^2 = \left(\sqrt{x-4}+2\right)^2$$
$$x + 8 = x - 4 + 4\sqrt{x-4} + 4$$
$$8 = 4\sqrt{x-4}$$
$$\sqrt{x-4} = 2$$
$$x - 4 = 4$$
$$x = 8$$

49. Solving the equation:

$$\sqrt{x-5} - 3 = \sqrt{x-8}$$
$$\left(\sqrt{x-5}-3\right)^2 = \left(\sqrt{x-8}\right)^2$$
$$x - 5 - 6\sqrt{x-5} + 9 = x - 8$$
$$-6\sqrt{x-5} = -12$$
$$\sqrt{x-5} = 2$$
$$x - 5 = 4$$
$$x = 9$$

Since this value does not check, the solution set is $\varnothing$.

51. Solving the equation:

$$\sqrt{x+4} = 2 - \sqrt{2x}$$
$$\left(\sqrt{x+4}\right)^2 = \left(2-\sqrt{2x}\right)^2$$
$$x + 4 = 4 - 4\sqrt{2x} + 2x$$
$$-x = -4\sqrt{2x}$$
$$(-x)^2 = \left(-4\sqrt{2x}\right)^2$$
$$x^2 = 32x$$
$$x^2 - 32x = 0$$
$$x(x-32) = 0$$
$$x = 0, 32$$

The solution is 0 (32 does not check).

53. Solving the equation:

$$\sqrt{2x+4} = \sqrt{x+3} + 1$$
$$\left(\sqrt{2x+4}\right)^2 = \left(\sqrt{x+3}+1\right)^2$$
$$2x + 4 = x + 3 + 2\sqrt{x+3} + 1$$
$$x = 2\sqrt{x+3}$$
$$x^2 = \left(2\sqrt{x+3}\right)^2$$
$$x^2 = 4x + 12$$
$$x^2 - 4x - 12 = 0$$
$$(x-6)(x+2) = 0$$
$$x = -2, 6$$

The solution is 6 (−2 does not check).

55. Solving the equation:

$$f(x) = 0$$
$$\sqrt{2x-1} = 0$$
$$\left(\sqrt{2x-1}\right)^2 = (0)^2$$
$$2x - 1 = 0$$
$$2x = 1$$
$$x = \frac{1}{2}$$

57. Solving the equation:

$$f(x) = 2x - 1$$
$$\sqrt{2x - 1} = 2x - 1$$
$$\left(\sqrt{2x-1}\right)^2 = (2x-1)^2$$
$$2x - 1 = 4x^2 - 4x + 1$$
$$0 = 4x^2 - 6x + 2$$
$$0 = 2x^2 - 3x + 1$$
$$0 = (2x - 1)(x - 1)$$
$$x = \frac{1}{2}, 1$$

59. Solving the equation:

$$f(x) = \sqrt{x - 4} + 2$$
$$\sqrt{2x - 1} = \sqrt{x - 4} + 2$$
$$\left(\sqrt{2x-1}\right)^2 = \left(\sqrt{x-4} + 2\right)^2$$
$$2x - 1 = x - 4 + 4\sqrt{x - 4} + 4$$
$$x - 1 = 4\sqrt{x - 4}$$
$$(x - 1)^2 = \left(4\sqrt{x-4}\right)^2$$
$$x^2 - 2x + 1 = 16(x - 4)$$
$$x^2 - 2x + 1 = 16x - 64$$
$$x^2 - 18x + 65 = 0$$
$$(x - 5)(x - 13) = 0$$
$$x = 5, 13$$

61. Solving the equation:

$$g(x) = 0$$
$$\sqrt{2x + 3} = 0$$
$$\left(\sqrt{2x+3}\right)^2 = (0)^2$$
$$2x + 3 = 0$$
$$2x = -3$$
$$x = -\frac{3}{2}$$

63. Solving the equation:

$$g(x) = -\sqrt{5x}$$
$$\sqrt{2x + 3} = -\sqrt{5x}$$
$$\left(\sqrt{2x+3}\right)^2 = \left(-\sqrt{5x}\right)^2$$
$$2x + 3 = 5x$$
$$3 = 3x$$
$$x = 1$$

Since this value does not check, the solution set is $\varnothing$.

65. Solving the equation:

$$f(x) = g(x)$$
$$\sqrt{2x} - 1 = 0$$
$$\sqrt{2x} = 1$$
$$\left(\sqrt{2x}\right)^2 = (1)^2$$
$$2x = 1$$
$$x = \frac{1}{2}$$

67. Solving the equation:

$$f(x) = g(x)$$
$$\sqrt{2x} - 1 = \sqrt{2x + 5}$$
$$\left(\sqrt{2x} - 1\right)^2 = \left(\sqrt{2x+5}\right)^2$$
$$2x - 2\sqrt{2x} + 1 = 2x + 5$$
$$-2\sqrt{2x} = 4$$
$$\sqrt{2x} = -2$$
$$\left(\sqrt{2x}\right)^2 = (-2)^2$$
$$2x = 4$$
$$x = 2$$

Since this value does not check, the solution set is $\varnothing$.

69. Solving the equation:

$$h(x) = f(x)$$
$$\sqrt[3]{3x + 5} = 2$$
$$\left(\sqrt[3]{3x+5}\right)^3 = (2)^3$$
$$3x + 5 = 8$$
$$3x = 3$$
$$x = 1$$

71. Solving the equation:

$$h(x) = f(x)$$
$$\sqrt[3]{5 - 2x} = 3$$
$$\left(\sqrt[3]{5-2x}\right)^3 = (3)^3$$
$$5 - 2x = 27$$
$$-2x = 22$$
$$x = -11$$

73. Graphing the equation:

75. Graphing the equation:

77. Graphing the equation:

79. Graphing the equation:

81. Graphing the equation:

83. Graphing the equation:

85. Solving for h:

$$t = \frac{\sqrt{100 - h}}{4}$$
$$4t = \sqrt{100 - h}$$
$$16t^2 = 100 - h$$
$$h = 100 - 16t^2$$

87. Solving for L:

$$2 = 2\left(\frac{22}{7}\right)\sqrt{\frac{L}{32}}$$
$$\frac{7}{22} = \sqrt{\frac{L}{32}}$$
$$\left(\frac{7}{22}\right)^2 = \frac{L}{32}$$
$$L = 32\left(\frac{7}{22}\right)^2 \approx 3.24 \text{ feet}$$

89. The width is $\sqrt{25} = 5$ meters.

91. Solving the equation:

$$\sqrt{x} = 50$$
$$x = 2500$$

The plume is 2,500 meters down river.

93. Multiplying: $\sqrt{2}\left(\sqrt{3} - \sqrt{2}\right) = \sqrt{6} - \sqrt{4} = \sqrt{6} - 2$

95. Multiplying: $\left(\sqrt{x} + 5\right)^2 = \left(\sqrt{x} + 5\right)\left(\sqrt{x} + 5\right) = x + 5\sqrt{x} + 5\sqrt{x} + 25 = x + 10\sqrt{x} + 25$

97. Rationalizing the denominator: $\dfrac{\sqrt{x}}{\sqrt{x} + 3} = \dfrac{\sqrt{x}}{\sqrt{x} + 3} \cdot \dfrac{\sqrt{x} - 3}{\sqrt{x} - 3} = \dfrac{x - 3\sqrt{x}}{x - 9}$

99. Simplifying: $\sqrt{25} = 5$

101. Simplifying: $\sqrt{12} = \sqrt{4 \cdot 3} = 2\sqrt{3}$

103. Simplifying: $(-1)^{15} = -1$

105. Simplifying: $(-1)^{50} = 1$

107. Solving the equation:

$$3x = 12$$
$$x = 4$$

109. Solving the equation:

$$4x - 3 = 5$$
$$4x = 8$$
$$x = 2$$

111. Performing the operations: $(3 + 4x) + (7 - 6x) = 10 - 2x$

113. Performing the operations: $(7 + 3x) - (5 + 6x) = 7 + 3x - 5 - 6x = 2 - 3x$

115. Performing the operations: $(3 - 4x)(2 + 5x) = 6 + 15x - 8x - 20x^2 = 6 + 7x - 20x^2$

117. Performing the operations: $2x(4 - 6x) = 8x - 12x^2$

119. Performing the operations: $(2 + 3x)^2 = 2^2 + 2(2)(3x) + (3x)^2 = 4 + 12x + 9x^2$

121. Performing the operations: $(2 - 3x)(2 + 3x) = 2^2 - (3x)^2 = 4 - 9x^2$

7.7 Complex Numbers

1. Writing in terms of i: $\sqrt{-36} = 6i$

3. Writing in terms of i: $-\sqrt{-25} = -5i$

5. Writing in terms of i: $\sqrt{-72} = 6i\sqrt{2}$

7. Writing in terms of i: $-\sqrt{-12} = -2i\sqrt{3}$

9. Rewriting the expression: $i^{28} = \left(i^4\right)^7 = (1)^7 = 1$

11. Rewriting the expression: $i^{26} = i^{24}i^2 = \left(i^4\right)^6 i^2 = (1)^6(-1) = -1$

13. Rewriting the expression: $i^{75} = i^{72}i^3 = \left(i^4\right)^{18} i^2 i = (1)^{18}(-1)i = -i$

15. Setting real and imaginary parts equal:

$$2x = 6 \qquad 3y = -3$$
$$x = 3 \qquad y = -1$$

17. Setting real and imaginary parts equal:

$$10y = -5$$
$$-x = 2$$
$$x = -2 \qquad y = -\dfrac{1}{2}$$

19. Setting real and imaginary parts equal:

$$2x = -16 \qquad -2y = 10$$
$$x = -8 \qquad y = -5$$

21. Setting real and imaginary parts equal:

$$2x - 4 = 10 \qquad -6y = -3$$
$$2x = 14$$
$$x = 7 \qquad y = \dfrac{1}{2}$$

23. Setting real and imaginary parts equal:

$$7x - 1 = 2 \qquad 5y + 2 = 4$$
$$7x = 3 \qquad 5y = 2$$
$$x = \dfrac{3}{7} \qquad y = \dfrac{2}{5}$$

25. Combining the numbers: $(2 + 3i) + (3 + 6i) = 5 + 9i$

27. Combining the numbers: $(3 - 5i) + (2 + 4i) = 5 - i$

29. Combining the numbers: $(5+2i)-(3+6i)=5+2i-3-6i=2-4i$

31. Combining the numbers: $(3-5i)-(2+i)=3-5i-2-i=1-6i$

33. Combining the numbers: $[(3+2i)-(6+i)]+(5+i)=3+2i-6-i+5+i=2+2i$

35. Combining the numbers: $[(7-i)-(2+4i)]-(6+2i)=7-i-2-4i-6-2i=-1-7i$

37. Combining the numbers:
$$(3+2i)-[(3-4i)-(6+2i)]=(3+2i)-(3-4i-6-2i)=(3+2i)-(-3-6i)=3+2i+3+6i=6+8i$$

39. Combining the numbers: $(4-9i)+[(2-7i)-(4+8i)]=(4-9i)+(2-7i-4-8i)=(4-9i)+(-2-15i)=2-24i$

41. Finding the product: $3i(4+5i)=12i+15i^2=-15+12i$

43. Finding the product: $6i(4-3i)=24i-18i^2=18+24i$

45. Finding the product: $(3+2i)(4+i)=12+8i+3i+2i^2=12+11i-2=10+11i$

47. Finding the product: $(4+9i)(3-i)=12+27i-4i-9i^2=12+23i+9=21+23i$

49. Finding the product: $(1+i)^3=(1+i)(1+i)^2=(1+i)(1+2i-1)=(1+i)(2i)=-2+2i$

51. Finding the product: $(2-i)^3=(2-i)(2-i)^2=(2-i)(4-4i-1)=(2-i)(3-4i)=6-11i-4=2-11i$

53. Finding the product: $(2+5i)^2=(2+5i)(2+5i)=4+10i+10i-25=-21+20i$

55. Finding the product: $(1-i)^2=(1-i)(1-i)=1-i-i-1=-2i$

57. Finding the product: $(3-4i)^2=(3-4i)(3-4i)=9-12i-12i-16=-7-24i$

59. Finding the product: $(2+i)(2-i)=4-i^2=4+1=5$

61. Finding the product: $(6-2i)(6+2i)=36-4i^2=36+4=40$

63. Finding the product: $(2+3i)(2-3i)=4-9i^2=4+9=13$

65. Finding the product: $(10+8i)(10-8i)=100-64i^2=100+64=164$

67. Finding the quotient: $\dfrac{2-3i}{i}=\dfrac{2-3i}{i}\cdot\dfrac{i}{i}=\dfrac{2i+3}{-1}=-3-2i$

69. Finding the quotient: $\dfrac{5+2i}{-i}=\dfrac{5+2i}{-i}\cdot\dfrac{i}{i}=\dfrac{5i-2}{1}=-2+5i$

71. Finding the quotient: $\dfrac{4}{2-3i}=\dfrac{4}{2-3i}\cdot\dfrac{2+3i}{2+3i}=\dfrac{8+12i}{4+9}=\dfrac{8+12i}{13}=\dfrac{8}{13}+\dfrac{12}{13}i$

73. Finding the quotient: $\dfrac{6}{-3+2i}=\dfrac{6}{-3+2i}\cdot\dfrac{-3-2i}{-3-2i}=\dfrac{-18-12i}{9+4}=\dfrac{-18-12i}{13}=-\dfrac{18}{13}-\dfrac{12}{13}i$

75. Finding the quotient: $\dfrac{2+3i}{2-3i}=\dfrac{2+3i}{2-3i}\cdot\dfrac{2+3i}{2+3i}=\dfrac{4+12i-9}{4+9}=\dfrac{-5+12i}{13}=-\dfrac{5}{13}+\dfrac{12}{13}i$

77. Finding the quotient: $\dfrac{5+4i}{3+6i}=\dfrac{5+4i}{3+6i}\cdot\dfrac{3-6i}{3-6i}=\dfrac{15-18i+24}{9+36}=\dfrac{39-18i}{45}=\dfrac{13}{15}-\dfrac{2}{5}i$

79. Dividing to find R: $R=\dfrac{80+20i}{-6+2i}=\dfrac{80+20i}{-6+2i}\cdot\dfrac{-6-2i}{-6-2i}=\dfrac{-480-280i+40}{36+4}=\dfrac{-440-280i}{40}=(-11-7i)$ ohms

81. Solving the equation:

$$\frac{t}{3} - \frac{1}{2} = -1$$

$$6\left(\frac{t}{3} - \frac{1}{2}\right) = 6(-1)$$

$$2t - 3 = -6$$

$$2t = -3$$

$$t = -\frac{3}{2}$$

83. Solving the equation:

$$2 + \frac{5}{y} = \frac{3}{y^2}$$

$$y^2\left(2 + \frac{5}{y}\right) = y^2\left(\frac{3}{y^2}\right)$$

$$2y^2 + 5y = 3$$

$$2y^2 + 5y - 3 = 0$$

$$(2y - 1)(y + 3) = 0$$

$$y = -3, \frac{1}{2}$$

85. Let x represent the number. The equation is:

$$x + \frac{1}{x} = \frac{41}{20}$$

$$20x\left(x + \frac{1}{x}\right) = 20x\left(\frac{41}{20}\right)$$

$$20x^2 + 20 = 41x$$

$$20x^2 - 41x + 20 = 0$$

$$(5x - 4)(4x - 5) = 0$$

$$x = \frac{4}{5}, \frac{5}{4}$$

The number is either $\frac{5}{4}$ or $\frac{4}{5}$.

87. Simplifying: $\frac{1}{i} = \frac{1}{i} \cdot \frac{i}{i} = \frac{i}{i^2} = \frac{i}{-1} = -i$

89. Substituting into the equation: $x^2 - 2x + 2 = (1+i)^2 - 2(1+i) + 2 = 1 + 2i - 1 - 2 - 2i + 2 = 0$

Thus $x = 1 + i$ is a solution to the equation.

91. Substituting into the equation:

$$\begin{aligned} x^3 - 11x + 20 &= (2+i)^3 - 11(2+i) + 20 \\ &= (2+i)(4+4i-1) - 22 - 11i + 20 \\ &= (2+i)(3+4i) - 2 - 11i \\ &= 6 + 11i - 4 - 2 - 11i \\ &= 0 \end{aligned}$$

Thus $x = 2 + i$ is a solution to the equation.

Chapter 7 Review

1. Simplifying: $\sqrt{49} = 7$

3. Simplifying: $16^{1/4} = 2$

5. Simplifying: $\sqrt[5]{32x^{15}y^{10}} = 2x^3y^2$

7. Simplifying: $x^{2/3} \cdot x^{4/3} = x^{2/3+4/3} = x^2$

9. Simplifying: $\frac{a^{3/5}}{a^{1/4}} = a^{3/5-1/4} = a^{12/20-5/20} = a^{7/20}$

11. Multiplying: $\left(3x^{1/2} + 5y^{1/2}\right)\left(4x^{1/2} - 3y^{1/2}\right) = 12x - 9x^{1/2}y^{1/2} + 20x^{1/2}y^{1/2} - 15y = 12x + 11x^{1/2}y^{1/2} - 15y$

13. Dividing: $\frac{28x^{5/6} + 14x^{7/6}}{7x^{1/3}} = \frac{28x^{5/6}}{7x^{1/3}} + \frac{14x^{7/6}}{7x^{1/3}} = 4x^{5/6-1/3} + 2x^{7/6-1/3} = 4x^{1/2} + 2x^{5/6}$

15. Simplifying: $x^{3/4} + \frac{5}{x^{1/4}} = x^{3/4} \cdot \frac{x^{1/4}}{x^{1/4}} + \frac{5}{x^{1/4}} = \frac{x+5}{x^{1/4}}$

17. Simplifying: $\sqrt{50} = \sqrt{25 \cdot 2} = 5\sqrt{2}$

19. Simplifying: $\sqrt{18x^2} = \sqrt{9x^2 \cdot 2} = 3x\sqrt{2}$

21. Simplifying: $\sqrt[4]{32a^4 b^5 c^6} = \sqrt[4]{16a^4 b^4 c^4 \cdot 2bc^2} = 2abc\sqrt[4]{2bc^2}$

23. Rationalizing the denominator: $\dfrac{6}{\sqrt[3]{2}} = \dfrac{6}{\sqrt[3]{2}} \cdot \dfrac{\sqrt[3]{4}}{\sqrt[3]{4}} = \dfrac{6\sqrt[3]{4}}{2} = 3\sqrt[3]{4}$

25. Simplifying: $\sqrt[3]{\dfrac{40x^2 y^3}{3z}} = \dfrac{2y\sqrt[3]{5x^2}}{\sqrt[3]{3z}} \cdot \dfrac{\sqrt[3]{9z^2}}{\sqrt[3]{9z^2}} = \dfrac{2y\sqrt[3]{45x^2 z^2}}{3z}$

27. Combining the expressions: $\sqrt{12} + \sqrt{3} = 2\sqrt{3} + \sqrt{3} = 3\sqrt{3}$

29. Combining the expressions: $3\sqrt{8} - 4\sqrt{72} + 5\sqrt{50} = 6\sqrt{2} - 24\sqrt{2} + 25\sqrt{2} = 7\sqrt{2}$

31. Combining the expressions: $2x\sqrt[3]{xy^3 z^2} - 6y\sqrt[3]{x^4 z^2} = 2xy\sqrt[3]{xz^2} - 6xy\sqrt[3]{xz^2} = -4xy\sqrt[3]{xz^2}$

33. Multiplying: $(\sqrt{x} - 2)(\sqrt{x} - 3) = x - 2\sqrt{x} - 3\sqrt{x} + 6 = x - 5\sqrt{x} + 6$

35. Rationalizing the denominator: $\dfrac{\sqrt{7} + \sqrt{5}}{\sqrt{7} - \sqrt{5}} = \dfrac{\sqrt{7} + \sqrt{5}}{\sqrt{7} - \sqrt{5}} \cdot \dfrac{\sqrt{7} + \sqrt{5}}{\sqrt{7} + \sqrt{5}} = \dfrac{7 + 2\sqrt{35} + 5}{7 - 5} = \dfrac{12 + 2\sqrt{35}}{2} = 6 + \sqrt{35}$

37. Solving the equation:
$$\sqrt{4a+1} = 1$$
$$\left(\sqrt{4a+1}\right)^2 = (1)^2$$
$$4a + 1 = 1$$
$$4a = 0$$
$$a = 0$$

39. Solving the equation:
$$\sqrt{3x+1} - 3 = 1$$
$$\sqrt{3x+1} = 4$$
$$\left(\sqrt{3x+1}\right)^2 = (4)^2$$
$$3x + 1 = 16$$
$$3x = 15$$
$$x = 5$$

41. Graphing the equation:

43. Writing in terms of i: $i^{24} = \left(i^4\right)^6 = (1)^6 = 1$

45. Setting real and imaginary parts equal:
$$-2x = 3 \qquad 8y = -4$$
$$x = -\dfrac{3}{2} \qquad y = -\dfrac{1}{2}$$

47. Combining the numbers: $(3 + 5i) + (6 - 2i) = 9 + 3i$

49. Multiplying: $3i(4 + 2i) = 12i + 6i^2 = -6 + 12i$

51. Multiplying: $(4 + 2i)^2 = (4 + 2i)(4 + 2i) = 16 + 8i + 8i + 4i^2 = 16 + 16i - 4 = 12 + 16i$

53. Dividing: $\dfrac{3+i}{i} = \dfrac{3+i}{i} \cdot \dfrac{i}{i} = \dfrac{3i - 1}{-1} = 1 - 3i$

55. Let l represent the desired length. Using the Pythagorean theorem:

$$18^2 + 13.5^2 = l^2$$
$$l^2 = 506.25$$
$$l = \sqrt{506.25} = 22.5$$

The length of one side of the roof is 22.5 feet.

Chapter 7 Cumulative Review

1. Simplifying: $6 + \left[2 - 5(7 - 2)\right] = 6 + \left[2 - 5(5)\right] = 6 + (2 - 25) = 6 - 23 = -17$

3. Simplifying: $2 + (5 - 7)^2 - (6 - 5)^3 = 2 + (-2)^2 - (1)^3 = 2 + 4 - 1 = 5$

5. Simplifying: $8^{2/3} + 16^{1/4} = \left(8^{1/3}\right)^2 + 16^{1/4} = (2)^2 + 2 = 4 + 2 = 6$

7. Simplifying: $\left(-\dfrac{27}{8}\right)^{-1/3} = \left(-\dfrac{8}{27}\right)^{1/3} = -\dfrac{2}{3}$

9. Reducing to lowest terms: $\dfrac{105}{189} = \dfrac{3 \cdot 5 \cdot 7}{3 \cdot 3 \cdot 3 \cdot 7} = \dfrac{5}{3 \cdot 3} = \dfrac{5}{9}$

11. Reducing to lowest terms: $\dfrac{x^2 - x - 6}{x^2 - 2x - 3} = \dfrac{(x - 3)(x + 2)}{(x - 3)(x + 1)} = \dfrac{x + 2}{x + 1}$

13. Multiplying: $\left(\sqrt{x} - 3\right)\left(2\sqrt{x} - 5\right) = 2x - 5\sqrt{x} - 6\sqrt{x} + 15 = 2x - 11\sqrt{x} + 15$

15. Dividing: $\dfrac{4 + i}{4 + 3i} = \dfrac{4 + i}{4 + 3i} \cdot \dfrac{4 - 3i}{4 - 3i} = \dfrac{16 - 12i + 4i - 3i^2}{16 - 9i^2} = \dfrac{16 - 8i + 3}{16 + 9} = \dfrac{19 - 8i}{25} = \dfrac{19}{25} - \dfrac{8}{25}i$

17. Multiplying: $(2x - 5y)(3x - 2y) = 6x^2 - 4xy - 15xy + 10y^2 = 6x^2 - 19xy + 10y^2$

19. Dividing: $\dfrac{27x^3y^2}{13x^2y^4} \div \dfrac{9xy}{26y} = \dfrac{27x^3y^2}{13x^2y^4} \cdot \dfrac{26y}{9xy} = \dfrac{27 \cdot 26x^3y^3}{9 \cdot 13x^3y^5} = \dfrac{6}{y^2}$

21. Solving the equation:

$$\dfrac{3}{2} + \dfrac{1}{a - 2} = -\dfrac{1}{a - 2}$$
$$2(a - 2)\left(\dfrac{3}{2} + \dfrac{1}{a - 2}\right) = 2(a - 2)\left(-\dfrac{1}{a - 2}\right)$$
$$3(a - 2) + 2 = -2$$
$$3a - 6 + 2 = -2$$
$$3a - 4 = -2$$
$$3a = 2$$
$$a = \dfrac{2}{3}$$

23. Solving the equation:

$$(2x - 5)^2 = 12$$
$$2x - 5 = \pm\sqrt{12}$$
$$2x - 5 = \pm 2\sqrt{3}$$
$$2x = 5 \pm 2\sqrt{3}$$
$$x = \dfrac{5 \pm 2\sqrt{3}}{2}$$

25. Solving the equation:

$$\sqrt{x - 3} = 3 - \sqrt{x}$$
$$\left(\sqrt{x - 3}\right)^2 = \left(3 - \sqrt{x}\right)^2$$
$$x - 3 = 9 - 6\sqrt{x} + x$$
$$-6\sqrt{x} = -12$$
$$\sqrt{x} = 2$$
$$x = 4$$

27. Solving the equation:

$$\dfrac{1}{2}x^2 + \dfrac{3x}{8} = \dfrac{5}{16}$$
$$16\left(\dfrac{1}{2}x^2 + \dfrac{3x}{8}\right) = 16\left(\dfrac{5}{16}\right)$$
$$8x^2 + 6x = 5$$
$$8x^2 + 6x - 5 = 0$$
$$(4x + 5)(2x - 1) = 0$$
$$x = -\dfrac{5}{4}, \dfrac{1}{2}$$

29. Solving the equation:

$$|3x-1|-2=6$$
$$|3x-1|=8$$
$$3x-1=-8, 8$$
$$3x=-7, 9$$
$$x=-\frac{7}{3}, 3$$

31. Solving the inequality:

$$-3 \le \frac{1}{4}x-5 \le -1$$
$$2 \le \frac{1}{4}x \le 4$$
$$8 \le x \le 16$$

Graphing the solution set:

33. Multiply the second equation by 3:

$$2x-3y=-16$$
$$9x+3y=-6$$

Adding yields:

$$11x=-22$$
$$x=-2$$

Substituting into the second equation:

$$3(-2)+y=-2$$
$$-6+y=-2$$
$$y=4$$

The solution is $(-2, 4)$.

35. Multiply the first equation by 2 and add to the second equation:

$$4x+8y+2z=12$$
$$3x-y-2z=4$$

Adding yields the equation $7x+7y=16$. Multiply the first equation by -3 and add it to the third equation:

$$-6x-12y-3z=-18$$
$$x+2y+3z=-10$$

Adding yields the equation $-5x-10y=-28$. So the system of equations becomes:

$$7x+7y=16$$
$$-5x-10y=-28$$

Multiply the first equation by 5 and the second equation by 7:

$$35x+35y=80$$
$$-35x-70y=-196$$

Adding yields:

$$-35y=-116$$
$$y=\frac{116}{35}$$

Substituting to find x:

$$35x+35\left(\frac{116}{35}\right)=80$$
$$35x+116=80$$
$$35x=-36$$
$$x=-\frac{36}{35}$$

Substituting into the original first equation:

$$4\left(-\frac{36}{35}\right) + 8\left(\frac{116}{35}\right) + 2z = 12$$

$$-\frac{144}{35} + \frac{928}{35} + 2z = 12$$

$$\frac{784}{35} + 2z = 12$$

$$2z = -\frac{364}{35}$$

$$z = -\frac{182}{35} = -\frac{26}{5}$$

The solution is $\left(-\frac{36}{35}, \frac{116}{35}, -\frac{26}{5}\right)$.

37. Graphing the line:

39. Finding the slope: $m = \dfrac{-3-4}{-2-(-3)} = \dfrac{-7}{1} = -7$

41. Rationalizing the denominator: $\dfrac{6}{\sqrt[4]{8}} = \dfrac{6}{\sqrt[4]{8}} \cdot \dfrac{\sqrt[4]{2}}{\sqrt[4]{2}} = \dfrac{6\sqrt[4]{2}}{\sqrt[4]{16}} = \dfrac{6\sqrt[4]{2}}{2} = 3\sqrt[4]{2}$

43. Factoring: $625a^4 - 16b^4 = \left(25a^2\right)^2 - \left(4b^2\right)^2 = \left(25a^2 - 4b^2\right)\left(25a^2 + 4b^2\right) = \left(5a - 2b\right)\left(5a + 2b\right)\left(25a^2 + 4b^2\right)$

45. The variation equation is $y = \dfrac{K}{\sqrt{x}}$. Substituting $y = \dfrac{1}{2}$ and $x = 16$:

$$\frac{1}{2} = \frac{K}{\sqrt{16}}$$

$$\frac{1}{2} = \frac{K}{4}$$

$$K = 2$$

So the equation is $y = \dfrac{2}{\sqrt{x}}$. Substituting $x = 4$: $y = \dfrac{2}{\sqrt{4}} = \dfrac{2}{2} = 1$

47. Evaluating the function: $f(b) = b^2 - 3b$

49. Evaluating the function: $(g \circ f)(3) = g[f(3)] = g\left(3^2 - 3 \cdot 3\right) = g(0) = 0 - 1 = -1$

51. Subtracting: $79{,}300{,}000 - 57{,}300{,}000 = 20{,}000{,}000$ more people

Chapter 7 Test

1. Simplifying: $27^{-2/3} = \left(27^{1/3}\right)^{-2} = 3^{-2} = \dfrac{1}{3^2} = \dfrac{1}{9}$ **2.** Simplifying: $\left(\dfrac{25}{49}\right)^{-1/2} = \left(\dfrac{49}{25}\right)^{1/2} = \dfrac{7}{5}$

3. Simplifying: $a^{3/4} \cdot a^{-1/3} = a^{3/4-1/3} = a^{9/12-4/12} = a^{5/12}$

4. Simplifying: $\dfrac{\left(x^{2/3}y^{-3}\right)^{1/2}}{\left(x^{3/4}y^{1/2}\right)^{-1}} = \dfrac{x^{1/3}y^{-3/2}}{x^{-3/4}y^{-1/2}} = x^{1/3+3/4}\,y^{-3/2+1/2} = x^{13/12}y^{-1} = \dfrac{x^{13/12}}{y}$

5. Simplifying: $\sqrt{49x^8y^{10}} = 7x^4y^5$ **6.** Simplifying: $\sqrt[5]{32x^{10}y^{20}} = 2x^2y^4$

7. Simplifying: $\dfrac{\left(36a^8b^4\right)^{1/2}}{\left(27a^9b^6\right)^{1/3}} = \dfrac{6a^4b^2}{3a^3b^2} = 2a$ **8.** Simplifying: $\dfrac{\left(x^n y^{1/n}\right)^n}{\left(x^{1/n} y^n\right)^{n^2}} = \dfrac{x^{n^2} y}{x^n y^{n^3}} = x^{n^2-n}y^{1-n^3}$

9. Multiplying: $2a^{1/2}\left(3a^{3/2} - 5a^{1/2}\right) = 6a^2 - 10a$

10. Multiplying: $\left(4a^{3/2} - 5\right)^2 = \left(4a^{3/2} - 5\right)\left(4a^{3/2} - 5\right) = 16a^3 - 20a^{3/2} - 20a^{3/2} + 25 = 16a^3 - 40a^{3/2} + 25$

11. Factoring: $3x^{2/3} + 5x^{1/3} - 2 = \left(3x^{1/3} - 1\right)\left(x^{1/3} + 2\right)$ **12.** Factoring: $9x^{2/3} - 49 = \left(3x^{1/3} - 7\right)\left(3x^{1/3} + 7\right)$

13. Simplifying: $\sqrt{125x^3y^5} = \sqrt{25x^2y^4 \cdot 5xy} = 5xy^2\sqrt{5xy}$

14. Simplifying: $\sqrt[3]{40x^7y^8} = \sqrt[3]{8x^6y^6 \cdot 5xy^2} = 2x^2y^2\sqrt[3]{5xy^2}$

15. Simplifying: $\sqrt{\dfrac{2}{3}} = \dfrac{\sqrt{2}}{\sqrt{3}} \cdot \dfrac{\sqrt{3}}{\sqrt{3}} = \dfrac{\sqrt{6}}{3}$

16. Simplifying: $\sqrt{\dfrac{12a^4b^3}{5c}} = \dfrac{2a^2b\sqrt{3b}}{\sqrt{5c}} \cdot \dfrac{\sqrt{5c}}{\sqrt{5c}} = \dfrac{2a^2b\sqrt{15bc}}{5c}$

17. Combining: $\dfrac{4}{x^{1/2}} + x^{1/2} = \dfrac{4}{x^{1/2}} + x^{1/2} \cdot \dfrac{x^{1/2}}{x^{1/2}} = \dfrac{x+4}{x^{1/2}}$

18. Combining: $\dfrac{x^2}{\left(x^2-3\right)^{1/2}} - \left(x^2-3\right)^{1/2} = \dfrac{x^2}{\left(x^2-3\right)^{1/2}} - \left(x^2-3\right)^{1/2} \cdot \dfrac{\left(x^2-3\right)^{1/2}}{\left(x^2-3\right)^{1/2}} = \dfrac{x^2 - x^2 + 3}{\left(x^2-3\right)^{1/2}} = \dfrac{3}{\left(x^2-3\right)^{1/2}}$

19. Combining: $3\sqrt{12} - 4\sqrt{27} = 6\sqrt{3} - 12\sqrt{3} = -6\sqrt{3}$

20. Combining: $2\sqrt[3]{24a^3b^3} - 5a\sqrt[3]{3b^3} = 4ab\sqrt[3]{3} - 5ab\sqrt[3]{3} = -ab\sqrt[3]{3}$

21. Multiplying: $\left(\sqrt{x} + 7\right)\left(\sqrt{x} - 4\right) = x - 4\sqrt{x} + 7\sqrt{x} - 28 = x + 3\sqrt{x} - 28$

22. Multiplying: $\left(3\sqrt{2} - \sqrt{3}\right)^2 = \left(3\sqrt{2} - \sqrt{3}\right)\left(3\sqrt{2} - \sqrt{3}\right) = 18 - 3\sqrt{6} - 3\sqrt{6} + 3 = 21 - 6\sqrt{6}$

23. Rationalizing the denominator: $\dfrac{5}{\sqrt{3}-1} = \dfrac{5}{\sqrt{3}-1} \cdot \dfrac{\sqrt{3}+1}{\sqrt{3}+1} = \dfrac{5\sqrt{3}+5}{3-1} = \dfrac{5+5\sqrt{3}}{2}$

24. Rationalizing the denominator: $\dfrac{\sqrt{x}-\sqrt{2}}{\sqrt{x}+\sqrt{2}} = \dfrac{\sqrt{x}-\sqrt{2}}{\sqrt{x}+\sqrt{2}} \cdot \dfrac{\sqrt{x}-\sqrt{2}}{\sqrt{x}-\sqrt{2}} = \dfrac{x - \sqrt{2x} - \sqrt{2x} + 2}{x-2} = \dfrac{x - 2\sqrt{2x} + 2}{x-2}$

25. Solving the equation:

$$\sqrt{3x+1} = x-3$$
$$\left(\sqrt{3x+1}\right)^2 = (x-3)^2$$
$$3x+1 = x^2-6x+9$$
$$0 = x^2-9x+8$$
$$0 = (x-1)(x-8)$$
$$x = 1, 8$$

The solution is 8 (1 does not check).

27. Solving the equation:

$$\sqrt{x+3} = \sqrt{x+4}-1$$
$$\left(\sqrt{x+3}\right)^2 = \left(\sqrt{x+4}-1\right)^2$$
$$x+3 = x+4-2\sqrt{x+4}+1$$
$$-2 = -2\sqrt{x+4}$$
$$\sqrt{x+4} = 1$$
$$x+4 = 1$$
$$x = -3$$

26. Solving the equation:

$$\sqrt[3]{2x+7} = -1$$
$$\left(\sqrt[3]{2x+7}\right)^3 = (-1)^3$$
$$2x+7 = -1$$
$$2x = -8$$
$$x = -4$$

28. Graphing the equation:

$$y = \sqrt{x-2}$$

29. Graphing the equation:

$$y = \sqrt[3]{x} + 3$$

30. Setting the real and imaginary parts equal:

$$2x+5 = 6 \qquad -(y-3) = -4$$
$$2x = 1 \qquad\qquad y-3 = 4$$
$$x = \frac{1}{2} \qquad\qquad y = 7$$

31. Performing the operations:

$$(3+2i)-\left[(7-i)-(4+3i)\right] = (3+2i)-(7-i-4-3i) = (3+2i)-(3-4i) = 3+2i-3+4i = 6i$$

32. Performing the operations: $(2-3i)(4+3i) = 8+6i-12i+9 = 17-6i$

33. Performing the operations: $(5-4i)^2 = (5-4i)(5-4i) = 25-20i-20i-16 = 9-40i$

34. Performing the operations: $\dfrac{2-3i}{2+3i} = \dfrac{2-3i}{2+3i} \cdot \dfrac{2-3i}{2-3i} = \dfrac{4-12i-9}{4+9} = \dfrac{-5-12i}{13} = -\dfrac{5}{13} - \dfrac{12}{13}i$

35. Rewriting the exponent: $i^{38} = \left(i^2\right)^{19} = (-1)^{19} = -1$

36. The correct graph is E.

38. The correct graph is F.

40. The correct graph is D.

37. The correct graph is A.

39. The correct graph is B.

41. The correct graph is C.

Chapter 8
Quadratic Functions

Getting Ready for Chapter 8

1. Simplifying: $\dfrac{3+5}{2} = \dfrac{8}{2} = 4$

2. Simplifying: $\dfrac{3-5}{2} = \dfrac{-2}{2} = -1$

3. Simplifying: $2 + \dfrac{25}{4} = \dfrac{2}{1} \cdot \dfrac{4}{4} + \dfrac{25}{4} = \dfrac{8}{4} + \dfrac{25}{4} = \dfrac{33}{4} = 8\dfrac{1}{4}$

4. Simplifying: $-\dfrac{7}{3} + \dfrac{16}{9} = -\dfrac{7}{3} \cdot \dfrac{3}{3} + \dfrac{16}{9} = -\dfrac{21}{9} + \dfrac{16}{9} = -\dfrac{5}{9}$

5. Solving the equation:
$$x^2 = 25$$
$$x = \pm\sqrt{25} = \pm 5$$

6. Solving the equation:
$$y^2 = 9$$
$$y = \pm\sqrt{9} = \pm 3$$

7. Solving the equation:
$$x^2 = 12$$
$$x = \pm\sqrt{12} = \pm 2\sqrt{3}$$

8. Solving the equation:
$$y^2 = 18$$
$$y = \pm\sqrt{18} = \pm 3\sqrt{2}$$

9. Evaluating when $b = 6$: $\left(\dfrac{b}{2}\right)^2 = \left(\dfrac{6}{2}\right)^2 = (3)^2 = 9$

10. Evaluating when $b = 4$: $\left(\dfrac{b}{2}\right)^2 = \left(\dfrac{4}{2}\right)^2 = (2)^2 = 4$

11. Evaluating when $b = 5$: $\left(\dfrac{b}{2}\right)^2 = \left(\dfrac{5}{2}\right)^2 = \dfrac{25}{4}$

12. Evaluating when $b = 7$: $\left(\dfrac{b}{2}\right)^2 = \left(\dfrac{7}{2}\right)^2 = \dfrac{49}{4}$

13. Evaluating when $b = \dfrac{8}{3}$: $\left(\dfrac{b}{2}\right)^2 = \left(\dfrac{8/3}{2}\right)^2 = \left(\dfrac{4}{3}\right)^2 = \dfrac{16}{9}$

14. Evaluating when $b = \dfrac{2}{5}$: $\left(\dfrac{b}{2}\right)^2 = \left(\dfrac{2/5}{2}\right)^2 = \left(\dfrac{1}{5}\right)^2 = \dfrac{1}{25}$

15. Factoring: $x^2 - 6x + 9 = (x-3)^2$

16. Factoring: $x^2 - 8x + 16 = (x-4)^2$

17. Factoring: $x^2 + 12x + 36 = (x+6)^2$

18. Factoring: $125t^3 + 1 = (5t+1)(25t^2 - 5t + 1)$

19. Simplifying: $-\dfrac{c}{a} + \left(\dfrac{b}{2a}\right)^2 = -\dfrac{c}{a} + \dfrac{b^2}{4a^2} = -\dfrac{c}{a} \cdot \dfrac{4a}{4a} + \dfrac{b^2}{4a^2} = \dfrac{b^2 - 4ac}{4a^2}$

20. Simplifying: $(3a-2)^2 + 2(3a-2) - 3 = 9a^2 - 12a + 4 + 6a - 4 - 3 = 9a^2 - 6a - 3$

259

8.1 Completing the Square

1. Solving the equation:

$$x^2 = 25$$
$$x = \pm\sqrt{25} = \pm 5$$

3. Solving the equation:

$$y^2 = \frac{3}{4}$$
$$y = \pm\sqrt{\frac{3}{4}} = \pm\frac{\sqrt{3}}{2}$$

5. Solving the equation:

$$x^2 + 12 = 0$$
$$x^2 = -12$$
$$x = \pm\sqrt{-12} = \pm 2i\sqrt{3}$$

7. Solving the equation:

$$4a^2 - 45 = 0$$
$$4a^2 = 45$$
$$a^2 = \frac{45}{4}$$
$$a = \pm\sqrt{\frac{45}{4}} = \pm\frac{3\sqrt{5}}{2}$$

9. Solving the equation:

$$(2y-1)^2 = 25$$
$$2y - 1 = \pm\sqrt{25} = \pm 5$$
$$2y - 1 = -5, 5$$
$$2y = -4, 6$$
$$y = -2, 3$$

11. Solving the equation:

$$(2a+3)^2 = -9$$
$$2a + 3 = \pm\sqrt{-9} = \pm 3i$$
$$2a = -3 \pm 3i$$
$$a = \frac{-3 \pm 3i}{2}$$

13. Solving the equation:

$$x^2 + 8x + 16 = -27$$
$$(x+4)^2 = -27$$
$$x + 4 = \pm\sqrt{-27} = \pm 3i\sqrt{3}$$
$$x = -4 \pm 3i\sqrt{3}$$

15. Solving the equation:

$$4a^2 - 12a + 9 = -4$$
$$(2a-3)^2 = -4$$
$$2a - 3 = \pm\sqrt{-4} = \pm 2i$$
$$2a = 3 \pm 2i$$
$$a = \frac{3 \pm 2i}{2}$$

17. Completing the square: $x^2 + 12x + 36 = (x+6)^2$

19. Completing the square: $x^2 - 4x + 4 = (x-2)^2$

21. Completing the square: $a^2 - 10a + 25 = (a-5)^2$

23. Completing the square: $x^2 + 5x + \dfrac{25}{4} = \left(x + \dfrac{5}{2}\right)^2$

25. Completing the square: $y^2 - 7y + \dfrac{49}{4} = \left(y - \dfrac{7}{2}\right)^2$

27. Completing the square: $x^2 + \dfrac{1}{2}x + \dfrac{1}{16} = \left(x + \dfrac{1}{4}\right)^2$

29. Completing the square: $x^2 + \dfrac{2}{3}x + \dfrac{1}{9} = \left(x + \dfrac{1}{3}\right)^2$

31. Solving the equation:

$$x^2 + 12x = -27$$
$$x^2 + 12x + 36 = -27 + 36$$
$$(x+6)^2 = 9$$
$$x + 6 = \pm\sqrt{9} = \pm 3$$
$$x + 6 = -3, 3$$
$$x = -9, -3$$

33. Solving the equation:

$$a^2 - 2a + 5 = 0$$
$$a^2 - 2a + 1 = -5 + 1$$
$$(a-1)^2 = -4$$
$$a - 1 = \pm\sqrt{-4} = \pm 2i$$
$$a = 1 \pm 2i$$

35.   Solving the equation:

$$y^2 - 8y + 1 = 0$$
$$y^2 - 8y + 16 = -1 + 16$$
$$(y-4)^2 = 15$$
$$y - 4 = \pm\sqrt{15}$$
$$y = 4 \pm \sqrt{15}$$

37.   Solving the equation:

$$x^2 - 5x - 3 = 0$$
$$x^2 - 5x + \frac{25}{4} = 3 + \frac{25}{4}$$
$$\left(x - \frac{5}{2}\right)^2 = \frac{37}{4}$$
$$x - \frac{5}{2} = \pm\frac{\sqrt{37}}{2}$$
$$x = \frac{5 \pm \sqrt{37}}{2}$$

39.   Solving the equation:

$$2x^2 - 4x - 8 = 0$$
$$x^2 - 2x - 4 = 0$$
$$x^2 - 2x + 1 = 4 + 1$$
$$(x-1)^2 = 5$$
$$x - 1 = \pm\sqrt{5}$$
$$x = 1 \pm \sqrt{5}$$

41.   Solving the equation:

$$3t^2 - 8t + 1 = 0$$
$$t^2 - \frac{8}{3}t + \frac{1}{3} = 0$$
$$t^2 - \frac{8}{3}t + \frac{16}{9} = -\frac{1}{3} + \frac{16}{9}$$
$$\left(t - \frac{4}{3}\right)^2 = \frac{13}{9}$$
$$t - \frac{4}{3} = +\sqrt{\frac{13}{9}} = \pm\frac{\sqrt{13}}{3}$$
$$t = \frac{4 \pm \sqrt{13}}{3}$$

43.   Solving the equation:

$$4x^2 - 3x + 5 = 0$$
$$x^2 - \frac{3}{4}x + \frac{5}{4} = 0$$
$$x^2 - \frac{3}{4}x + \frac{9}{64} = -\frac{5}{4} + \frac{9}{64}$$
$$\left(x - \frac{3}{8}\right)^2 = -\frac{71}{64}$$
$$x - \frac{3}{8} = \pm\sqrt{-\frac{71}{64}} = \pm\frac{i\sqrt{71}}{8}$$
$$x = \frac{3 \pm i\sqrt{71}}{8}$$

45.   **a.**   No, it cannot be solved by factoring.
   b.   Solving the equation:

$$x^2 = -9$$
$$x = \pm\sqrt{-9}$$
$$x = \pm 3i$$

47. **a.**   Solving by factoring:

$$x^2 - 6x = 0$$
$$x(x - 6) = 0$$
$$x = 0, 6$$

b.   Solving by completing the square:

$$x^2 - 6x = 0$$
$$x^2 - 6x + 9 = 0 + 9$$
$$(x - 3)^2 = 9$$
$$x - 3 = \pm\sqrt{9}$$
$$x - 3 = -3, 3$$
$$x = 0, 6$$

49. **a.**   Solving by factoring:

$$x^2 + 2x = 35$$
$$x^2 + 2x - 35 = 0$$
$$(x + 7)(x - 5) = 0$$
$$x = -7, 5$$

b.   Solving by completing the square:

$$x^2 + 2x = 35$$
$$x^2 + 2x + 1 = 35 + 1$$
$$(x + 1)^2 = 36$$
$$x + 1 = \pm\sqrt{36}$$
$$x + 1 = -6, 6$$
$$x = -7, 5$$

51.   Substituting: $x^2 - 6x - 7 = \left(-3 + \sqrt{2}\right)^2 - 6\left(-3 + \sqrt{2}\right) - 7 = 9 - 6\sqrt{2} + 2 + 18 - 6\sqrt{2} - 7 = 22 - 12\sqrt{2}$

No, $x = -3 + \sqrt{2}$ is not a solution to the equation.

53. **a.**   Solving the equation:

$$5x - 7 = 0$$
$$5x = 7$$
$$x = \frac{7}{5}$$

b.   Solving the equation:

$$5x - 7 = 8$$
$$5x = 15$$
$$x = 3$$

c.   Solving the equation:

$$(5x - 7)^2 = 8$$
$$5x - 7 = \pm\sqrt{8}$$
$$5x - 7 = \pm 2\sqrt{2}$$
$$5x = 7 \pm 2\sqrt{2}$$
$$x = \frac{7 \pm 2\sqrt{2}}{5}$$

d.   Solving the equation:

$$\sqrt{5x - 7} = 0$$
$$\left(\sqrt{5x - 7}\right)^2 = (0)^2$$
$$5x - 7 = 0$$
$$5x = 7$$
$$x = \frac{7}{5}$$

e.   Solving the equation:

$$\frac{5}{2} - \frac{7}{2x} = \frac{4}{x}$$
$$2x\left(\frac{5}{2} - \frac{7}{2x}\right) = 2x\left(\frac{4}{x}\right)$$
$$5x - 7 = 8$$
$$5x = 15$$
$$x = 3$$

55.   Solving the equation:

$$(x + 5)^2 + (x - 5)^2 = 52$$
$$x^2 + 10x + 25 + x^2 - 10x + 25 = 52$$
$$2x^2 + 50 = 52$$
$$2x^2 = 2$$
$$x^2 = 1$$
$$x = \pm\sqrt{1} = \pm 1$$

57.   Solving the equation:

$$(2x + 3)^2 + (2x - 3)^2 = 26$$
$$4x^2 + 12x + 9 + 4x^2 - 12x + 9 = 26$$
$$8x^2 + 18 = 26$$
$$8x^2 = 8$$
$$x^2 = 1$$
$$x = \pm\sqrt{1} = \pm 1$$

59. Solving the equation:
$$(3x+4)(3x-4)-(x+2)(x-2) = -4$$
$$9x^2 - 16 - x^2 + 4 = -4$$
$$8x^2 - 12 = -4$$
$$8x^2 = 8$$
$$x^2 = 1$$
$$x = \pm\sqrt{1} = \pm 1$$

61. Completing the table:

x	$f(x)$	$g(x)$	$h(x)$
-2	49	49	25
-1	25	25	13
0	9	9	9
1	1	1	13
2	1	1	25

63. Solving the equation:
$$f(x) = 0$$
$$(x-3)^2 = 0$$
$$x - 3 = 0$$
$$x = 3$$

65. a. Finding the x-intercepts:
$$x^2 - 5x - 6 = 0$$
$$(x+1)(x-6) = 0$$
$$x = -1, 6$$

b. Setting $f(x) = 0$:
$$f(x) = 0$$
$$x^2 - 5x - 6 = 0$$
$$(x+1)(x-6) = 0$$
$$x = -1, 6$$

c. Finding the value: $f(0) = (0)^2 - 5(0) - 6 = -6$

d. Finding the value: $f(1) = (1)^2 - 5(1) - 6 = -10$

67. The other two sides are $\dfrac{\sqrt{3}}{2}$ inch, 1 inch .

69. The hypotenuse is $\sqrt{2}$ inches.

71. Let x represent the horizontal distance. Using the Pythagorean theorem:
$$x^2 + 120^2 = 790^2$$
$$x^2 + 14400 = 624100$$
$$x^2 = 609700$$
$$x = \sqrt{609700} \approx 781 \text{ feet}$$

73. Solving for r:
$$3456 = 3000(1+r)^2$$
$$(1+r)^2 = 1.152$$
$$1 + r = \sqrt{1.152}$$
$$r = \sqrt{1.152} - 1 \approx 0.073$$
The annual interest rate is 7.3%.

75. Its length is $20\sqrt{2} \approx 28$ feet.

77. Simplifying: $\sqrt{45} = \sqrt{9 \cdot 5} = 3\sqrt{5}$

79. Simplifying: $\sqrt{27y^5} = \sqrt{9y^4 \cdot 3y} = 3y^2\sqrt{3y}$

81. Simplifying: $\sqrt[3]{54x^6y^5} = \sqrt[3]{27x^6y^3 \cdot 2y^2} = 3x^2y\sqrt[3]{2y^2}$

83. Rationalizing the denominator: $\dfrac{3}{\sqrt{2}} = \dfrac{3}{\sqrt{2}} \cdot \dfrac{\sqrt{2}}{\sqrt{2}} = \dfrac{3\sqrt{2}}{2}$

85. Rationalizing the denominator: $\dfrac{2}{\sqrt[3]{4}} = \dfrac{2}{\sqrt[3]{4}} \cdot \dfrac{\sqrt[3]{2}}{\sqrt[3]{2}} = \dfrac{2\sqrt[3]{2}}{2} = \sqrt[3]{2}$

87. Simplifying: $\sqrt{49 - 4(6)(-5)} = \sqrt{49 + 120} = \sqrt{169} = 13$

89. Simplifying: $\sqrt{(-27)^2 - 4(0.1)(1,700)} = \sqrt{729 - 680} = \sqrt{49} = 7$

91. Simplifying: $\dfrac{-7 + \sqrt{169}}{12} = \dfrac{-7 + 13}{12} = \dfrac{6}{12} = \dfrac{1}{2}$

93. Substituting values: $\sqrt{b^2 - 4ac} = \sqrt{7^2 - 4(6)(-5)} = \sqrt{49 + 120} = \sqrt{169} = 13$

95. Factoring: $27t^3 - 8 = (3t - 2)(9t^2 + 6t + 4)$

97. Solving for x:

$$(x + a)^2 + (x - a)^2 = 10a^2$$
$$x^2 + 2ax + a^2 + x^2 - 2ax + a^2 = 10a^2$$
$$2x^2 + 2a^2 = 10a^2$$
$$2x^2 = 8a^2$$
$$x^2 = 4a^2$$
$$x = \pm 2a$$

99. Solving for x:

$$x^2 + px + q = 0$$
$$x^2 + px + \frac{p^2}{4} = -q + \frac{p^2}{4}$$
$$\left(x + \frac{p}{2}\right)^2 = \frac{p^2 - 4q}{4}$$
$$x + \frac{p}{2} = \pm\sqrt{\frac{p^2 - 4q}{4}} = \pm\frac{\sqrt{p^2 - 4q}}{2}$$
$$x = \frac{-p \pm \sqrt{p^2 - 4q}}{2}$$

101. Solving for x:

$$3x^2 + px + q = 0$$
$$x^2 + \frac{p}{3}x + \frac{q}{3} = 0$$
$$x^2 + \frac{p}{3}x + \frac{p^2}{36} = -\frac{q}{3} + \frac{p^2}{36}$$
$$\left(x + \frac{p}{6}\right)^2 = \frac{p^2 - 12q}{36}$$
$$x + \frac{p}{6} = \pm\sqrt{\frac{p^2 - 12q}{36}} = \pm\frac{\sqrt{p^2 - 12q}}{6}$$
$$x = \frac{-p \pm \sqrt{p^2 - 12q}}{6}$$

103. Completing the square:

$$x^2 - 10x + y^2 - 6y = -30$$
$$\left(x^2 - 10x + 25\right) + \left(y^2 - 6y + 9\right) = -30 + 25 + 9$$
$$(x - 5)^2 + (y - 3)^2 = 4$$
$$(x - 5)^2 + (y - 3)^2 = 2^2$$

8.2 The Quadratic Formula

1. **a.** Using the quadratic formula:
    $$x = \frac{-4 \pm \sqrt{4^2 - 4(3)(-2)}}{2(3)} = \frac{-4 \pm \sqrt{16 + 24}}{6} = \frac{-4 \pm \sqrt{40}}{6} = \frac{-4 \pm 2\sqrt{10}}{6} = \frac{-2 \pm \sqrt{10}}{3}$$

    b. Using the quadratic formula: $x = \dfrac{4 \pm \sqrt{(-4)^2 - 4(3)(-2)}}{2(3)} = \dfrac{4 \pm \sqrt{16 + 24}}{6} = \dfrac{4 \pm \sqrt{40}}{6} = \dfrac{4 \pm 2\sqrt{10}}{6} = \dfrac{2 \pm \sqrt{10}}{3}$

    c. Using the quadratic formula: $x = \dfrac{-4 \pm \sqrt{4^2 - 4(3)(2)}}{2(3)} = \dfrac{-4 \pm \sqrt{16 - 24}}{6} = \dfrac{-4 \pm \sqrt{-8}}{6} = \dfrac{-4 \pm 2i\sqrt{2}}{6} = \dfrac{-2 \pm i\sqrt{2}}{3}$

    d. Using the quadratic formula:
    $$x = \frac{-4 \pm \sqrt{4^2 - 4(2)(-3)}}{2(2)} = \frac{-4 \pm \sqrt{16 + 24}}{4} = \frac{-4 \pm \sqrt{40}}{4} = \frac{-4 \pm 2\sqrt{10}}{4} = \frac{-2 \pm \sqrt{10}}{2}$$

    e. Using the quadratic formula: $x = \dfrac{4 \pm \sqrt{(-4)^2 - 4(2)(3)}}{2(2)} = \dfrac{4 \pm \sqrt{16 - 24}}{4} = \dfrac{4 \pm \sqrt{-8}}{4} = \dfrac{4 \pm 2i\sqrt{2}}{4} = \dfrac{2 \pm i\sqrt{2}}{2}$

3. **a.** Using the quadratic formula: $x = \dfrac{2 \pm \sqrt{(-2)^2 - 4(1)(2)}}{2(1)} = \dfrac{2 \pm \sqrt{4 - 8}}{2} = \dfrac{2 \pm \sqrt{-4}}{2} = \dfrac{2 \pm 2i}{2} = 1 \pm i$

    b. Using the quadratic formula: $x = \dfrac{2 \pm \sqrt{(-2)^2 - 4(1)(5)}}{2(1)} = \dfrac{2 \pm \sqrt{4 - 20}}{2} = \dfrac{2 \pm \sqrt{-16}}{2} = \dfrac{2 \pm 4i}{2} = 1 \pm 2i$

    c. Using the quadratic formula: $x = \dfrac{-2 \pm \sqrt{2^2 - 4(1)(2)}}{2(1)} = \dfrac{-2 \pm \sqrt{4 - 8}}{2} = \dfrac{-2 \pm \sqrt{-4}}{2} = \dfrac{-2 \pm 2i}{2} = -1 \pm i$

5. Solving the equation:
    $$\frac{1}{6}x^2 - \frac{1}{2}x + \frac{1}{3} = 0$$
    $$x^2 - 3x + 2 = 0$$
    $$(x - 1)(x - 2) = 0$$
    $$x = 1, 2$$

7. Solving the equation:
    $$\frac{x^2}{2} + 1 = \frac{2x}{3}$$
    $$3x^2 + 6 = 4x$$
    $$3x^2 - 4x + 6 = 0$$
    $$x = \frac{4 \pm \sqrt{16 - 72}}{6} = \frac{4 \pm \sqrt{-56}}{6} = \frac{4 \pm 2i\sqrt{14}}{6} = \frac{2 \pm i\sqrt{14}}{3}$$

9. Solving the equation:
    $$y^2 - 5y = 0$$
    $$y(y - 5) = 0$$
    $$y = 0, 5$$

11. Solving the equation:
    $$30x^2 + 40x = 0$$
    $$10x(3x + 4) = 0$$
    $$x = -\frac{4}{3}, 0$$

13. Solving the equation:

$$\frac{2t^2}{3} - t = -\frac{1}{6}$$

$$4t^2 - 6t = -1$$

$$4t^2 - 6t + 1 = 0$$

$$t = \frac{6 \pm \sqrt{36-16}}{8} = \frac{6 \pm \sqrt{20}}{8} = \frac{6 \pm 2\sqrt{5}}{8} = \frac{3 \pm \sqrt{5}}{4}$$

15. Solving the equation:

$$0.01x^2 + 0.06x - 0.08 = 0$$

$$x^2 + 6x - 8 = 0$$

$$x = \frac{-6 \pm \sqrt{36+32}}{2} = \frac{-6 \pm \sqrt{68}}{2} = \frac{-6 \pm 2\sqrt{17}}{2} = -3 \pm \sqrt{17}$$

17. Solving the equation:

$$2x + 3 = -2x^2$$

$$2x^2 + 2x + 3 = 0$$

$$x = \frac{-2 \pm \sqrt{4-24}}{4} = \frac{-2 \pm \sqrt{-20}}{4} = \frac{-2 \pm 2i\sqrt{5}}{4} = \frac{-1 \pm i\sqrt{5}}{2}$$

19. Solving the equation:

$$100x^2 - 200x + 100 = 0$$

$$100\left(x^2 - 2x + 1\right) = 0$$

$$100(x-1)^2 = 0$$

$$x = 1$$

21. Solving the equation:

$$\frac{1}{2}r^2 = \frac{1}{6}r - \frac{2}{3}$$

$$3r^2 = r - 4$$

$$3r^2 - r + 4 = 0$$

$$r = \frac{1 \pm \sqrt{1-48}}{6} = \frac{1 \pm \sqrt{-47}}{6} = \frac{1 \pm i\sqrt{47}}{6}$$

23. Solving the equation:

$$(x-3)(x-5) = 1$$

$$x^2 - 8x + 15 = 1$$

$$x^2 - 8x + 14 = 0$$

$$x = \frac{8 \pm \sqrt{64-56}}{2} = \frac{8 \pm \sqrt{8}}{2} = \frac{8 \pm 2\sqrt{2}}{2} = 4 \pm \sqrt{2}$$

25. Solving the equation:

$$(x+3)^2 + (x-8)(x-1) = 16$$

$$x^2 + 6x + 9 + x^2 - 9x + 8 = 16$$

$$2x^2 - 3x + 1 = 0$$

$$(2x-1)(x-1) = 0$$

$$x = \frac{1}{2}, 1$$

27.   Solving the equation:

$$\frac{x^2}{3} - \frac{5x}{6} = \frac{1}{2}$$
$$2x^2 - 5x = 3$$
$$2x^2 - 5x - 3 = 0$$
$$(2x+1)(x-3) = 0$$
$$x = -\frac{1}{2}, 3$$

29.   Solving the equation:

$$\frac{1}{x+1} - \frac{1}{x} = \frac{1}{2}$$
$$2x(x+1)\left(\frac{1}{x+1} - \frac{1}{x}\right) = 2x(x+1) \cdot \frac{1}{2}$$
$$2x - (2x+2) = x^2 + x$$
$$2x - 2x - 2 = x^2 + x$$
$$x^2 + x + 2 = 0$$
$$x = \frac{-1 \pm \sqrt{1-8}}{2} = \frac{-1 \pm \sqrt{-7}}{2} = \frac{-1 \pm i\sqrt{7}}{2}$$

31.   Solving the equation:

$$\frac{1}{y-1} + \frac{1}{y+1} = 1$$
$$(y+1)(y-1)\left(\frac{1}{y-1} + \frac{1}{y+1}\right) = (y+1)(y-1) \cdot 1$$
$$y+1+y-1 = y^2 - 1$$
$$2y = y^2 - 1$$
$$y^2 - 2y - 1 = 0$$
$$y = \frac{2 \pm \sqrt{4+4}}{2} = \frac{2 \pm \sqrt{8}}{2} = \frac{2 \pm 2\sqrt{2}}{2} = 1 \pm \sqrt{2}$$

33.   Solving the equation:

$$\frac{1}{x+2} + \frac{1}{x+3} = 1$$
$$(x+2)(x+3)\left(\frac{1}{x+2} + \frac{1}{x+3}\right) = (x+2)(x+3) \cdot 1$$
$$x+3+x+2 = x^2 + 5x + 6$$
$$2x+5 = x^2 + 5x + 6$$
$$x^2 + 3x + 1 = 0$$
$$x = \frac{-3 \pm \sqrt{9-4}}{2} = \frac{-3 \pm \sqrt{5}}{2}$$

35.   Solving the equation:

$$\frac{6}{r^2-1}-\frac{1}{2}=\frac{1}{r+1}$$

$$2(r+1)(r-1)\left(\frac{6}{(r+1)(r-1)}-\frac{1}{2}\right)=2(r+1)(r-1)\cdot\frac{1}{r+1}$$

$$12-\left(r^2-1\right)=2r-2$$

$$12-r^2+1=2r-2$$

$$r^2+2r-15=0$$

$$(r+5)(r-3)=0$$

$$r=-5,3$$

37.   Solving the equation:

$$x^3-8=0$$

$$(x-2)\left(x^2+2x+4\right)=0$$

$$x=2\quad\text{or}\quad x=\frac{-2\pm\sqrt{4-16}}{2}=\frac{-2\pm\sqrt{-12}}{2}=\frac{-2\pm2i\sqrt{3}}{2}=-1\pm i\sqrt{3}$$

$$x=2,-1\pm i\sqrt{3}$$

39.   Solving the equation:

$$8a^3+27=0$$

$$(2a+3)\left(4a^2-6a+9\right)=0$$

$$a=-\frac{3}{2}\quad\text{or}\quad a=\frac{6\pm\sqrt{36-144}}{8}=\frac{6\pm\sqrt{-108}}{8}=\frac{6\pm6i\sqrt{3}}{8}=\frac{3\pm3i\sqrt{3}}{4}$$

$$a=-\frac{3}{2},\frac{3\pm3i\sqrt{3}}{4}$$

41.   Solving the equation:

$$125t^3-1=0$$

$$(5t-1)\left(25t^2+5t+1\right)=0$$

$$t=\frac{1}{5}\quad\text{or}\quad t=\frac{-5\pm\sqrt{25-100}}{50}=\frac{-5\pm\sqrt{-75}}{50}=\frac{-5\pm5i\sqrt{3}}{50}=\frac{-1\pm i\sqrt{3}}{10}$$

$$t=\frac{1}{5},\frac{-1\pm i\sqrt{3}}{10}$$

43.   Solving the equation:

$$2x^3+2x^2+3x=0$$

$$x\left(2x^2+2x+3\right)=0$$

$$x=0\quad\text{or}\quad x=\frac{-2\pm\sqrt{4-24}}{4}=\frac{-2\pm\sqrt{-20}}{4}=\frac{-2\pm2i\sqrt{5}}{4}=\frac{-1\pm i\sqrt{5}}{2}$$

$$x=0,\frac{-1\pm i\sqrt{5}}{2}$$

45.   Solving the equation:

$$3y^4=6y^3-6y^2$$

$$3y^4-6y^3+6y^2=0$$

$$3y^2\left(y^2-2y+2\right)=0$$

$$y=0\quad\text{or}\quad y=\frac{2\pm\sqrt{4-8}}{2}=\frac{2\pm\sqrt{-4}}{2}=\frac{2\pm2i}{2}=1\pm i$$

$$y=0,1\pm i$$

47. Solving the equation:

$$6t^5 + 4t^4 = -2t^3$$
$$6t^5 + 4t^4 + 2t^3 = 0$$
$$2t^3\left(3t^2 + 2t + 1\right) = 0$$

$$t = 0 \quad \text{or} \quad t = \frac{-2 \pm \sqrt{4 - 12}}{6} = \frac{-2 \pm \sqrt{-8}}{6} = \frac{-2 \pm 2i\sqrt{2}}{6} = \frac{-1 \pm i\sqrt{2}}{3}$$

$$t = 0, \frac{-1 \pm i\sqrt{2}}{3}$$

49. The expressions from **a** and **b** are equivalent, since: $\frac{6 + 2\sqrt{3}}{4} = \frac{2\left(3 + \sqrt{3}\right)}{4} = \frac{3 + \sqrt{3}}{2}$

51. **a.** Solving by factoring:

$$3x^2 - 5x = 0$$
$$x\left(3x - 5\right) = 0$$
$$x = 0, \frac{5}{3}$$

b. Using the quadratic formula: $x = \frac{5 \pm \sqrt{(-5)^2 - 4(3)(0)}}{2(3)} = \frac{5 \pm \sqrt{25 - 0}}{6} = \frac{5 \pm 5}{6} = 0, \frac{5}{3}$

53. No, it cannot be solved by factoring. Using the quadratic formula:

$$x = \frac{4 \pm \sqrt{(-4)^2 - 4(1)(7)}}{2(1)} = \frac{4 \pm \sqrt{16 - 28}}{2} = \frac{4 \pm \sqrt{-12}}{2} = \frac{4 \pm 2i\sqrt{3}}{2} = 2 \pm i\sqrt{3}$$

55. Substituting: $x^2 + 2x = (-1 + i)^2 + 2(-1 + i) = 1 - 2i + i^2 - 2 + 2i = 1 - 2i - 1 - 2 + 2i = -2$

Yes, $x = -1 + i$ is a solution to the equation.

57. **a.** Solving the equation:

$$f(x) = 0$$
$$x^2 - 2x - 3 = 0$$
$$(x + 1)(x - 3) = 0$$
$$x = -1, 3$$

b. Solving the equation:

$$f(x) = -11$$
$$x^2 - 2x - 3 = -11$$
$$x^2 - 2x = -8$$
$$x^2 - 2x + 1 = -8 + 1$$
$$(x - 1)^2 = -7$$
$$x - 1 = \pm i\sqrt{7}$$
$$x = 1 \pm i\sqrt{7}$$

c. Solving the equation:

$$f(x) = -2x + 1$$
$$x^2 - 2x - 3 = -2x + 1$$
$$x^2 - 4 = 0$$
$$(x + 2)(x - 2) = 0$$
$$x = -2, 2$$

d. Solving the equation:

$$f(x) = 2x + 1$$
$$x^2 - 2x - 3 = 2x + 1$$
$$x^2 - 4x = 4$$
$$x^2 - 4x + 4 = 4 + 4$$
$$(x - 2)^2 = 8$$
$$x - 2 = \pm 2\sqrt{2}$$
$$x = 2 \pm 2\sqrt{2}$$

59.   a.   Solving the equation:

$$f(x) = g(x)$$
$$\frac{10}{x^2} = 3 + \frac{1}{x}$$
$$x^2\left(\frac{10}{x^2}\right) = x^2\left(3 + \frac{1}{x}\right)$$
$$10 = 3x^2 + x$$
$$0 = 3x^2 + x - 10$$
$$0 = (3x - 5)(x + 2)$$
$$x = -2, \frac{5}{3}$$

b.   Solving the equation:

$$f(x) = g(x)$$
$$\frac{10}{x^2} = 8x - \frac{17}{x^2}$$
$$x^2\left(\frac{10}{x^2}\right) = x^2\left(8x - \frac{17}{x^2}\right)$$
$$10 = 8x^3 - 17$$
$$0 = 8x^3 - 27$$
$$0 = (2x - 3)\left(4x^2 + 6x + 9\right)$$
$$x = \frac{3}{2}, x = \frac{-6 \pm \sqrt{6^2 - 4(4)(9)}}{2(4)} = \frac{-6 \pm \sqrt{-108}}{8} = \frac{-6 \pm 6i\sqrt{3}}{8} = \frac{-3 \pm 3i\sqrt{3}}{4}$$

c.   Solving the equation:

$$f(x) = g(x)$$
$$\frac{10}{x^2} = 0$$
$$10 = 0$$
$$(\text{impossible})$$

There is no solution ($\varnothing$).

d.   Solving the equation:

$$f(x) = g(x)$$
$$\frac{10}{x^2} = 10$$
$$x^2\left(\frac{10}{x^2}\right) = x^2(10)$$
$$10 = 10x^2$$
$$x^2 = 1$$
$$x = \pm 1$$

61.   Substituting $s = 74$:

$$5t + 16t^2 = 74$$
$$16t^2 + 5t - 74 = 0$$
$$(16t + 37)(t - 2) = 0$$
$$t = 2 \quad \left(t = -\frac{37}{16} \text{ is impossible}\right)$$

It will take 2 seconds for the object to fall 74 feet.

63.   Since profit is revenue minus the cost, the equation is:

$$100x - 0.5x^2 - (60x + 300) = 300$$
$$100x - 0.5x^2 - 60x - 300 = 300$$
$$-0.5x^2 + 40x - 600 = 0$$
$$x^2 - 80x + 1,200 = 0$$
$$(x - 20)(x - 60) = 0$$
$$x = 20, 60$$

The weekly profit is $300 if 20 items or 60 items are sold.

65. Let x represent the width of strip being cut off. After removing the strip, the overall area is 80% of its original area. The equation is:

$$(10.5 - 2x)(8.2 - 2x) = 0.80(10.5 \times 8.2)$$
$$86.1 - 37.4x + 4x^2 = 68.88$$
$$4x^2 - 37.4x + 17.22 = 0$$

$$x = \frac{37.4 \pm \sqrt{(-37.4)^2 - 4(4)(17.22)}}{8} = \frac{37.4 \pm \sqrt{1123.24}}{8} \approx \frac{37.4 \pm 33.5}{8} \approx 0.49, 8.86$$

The width of strip is 0.49 centimeter (8.86 cm is impossible).

67. **a.** The two equations are: $l + w = 10, lw = 15$

b. Since $l = 10 - w$, the equation is:

$$w(10 - w) = 15$$
$$10w - w^2 = 15$$
$$w^2 - 10w + 15 = 0$$

$$w = \frac{10 \pm \sqrt{100 - 60}}{2} = \frac{10 \pm \sqrt{40}}{2} = \frac{10 \pm 2\sqrt{10}}{2} = 5 \pm \sqrt{10} \approx 1.84, 8.16$$

The length and width are 8.16 yards and 1.84 yards.

c. Two answers are possible because either dimension (long or short) may be considered the length.

69. Dividing using long division:

$$
\begin{array}{r}
4y + 1 \\
2y - 7 \overline{\smash{\big)}\ 8y^2 - 26y - 9} \\
\underline{8y^2 - 28y} \\
2y - 9 \\
\underline{2y - 7} \\
-2
\end{array}
$$

The quotient is $4y + 1 - \dfrac{2}{2y - 7}$.

71. Dividing using long division:

$$
\begin{array}{r}
x^2 + 7x + 12 \\
x + 2 \overline{\smash{\big)}\ x^3 + 9x^2 + 26x + 24} \\
\underline{x^3 + 2x^2} \\
7x^2 + 26x \\
\underline{7x^2 + 14x} \\
12x + 24 \\
\underline{12x + 24} \\
0
\end{array}
$$

The quotient is $x^2 + 7x + 12$.

73. Simplifying: $25^{1/2} = \sqrt{25} = 5$

75. Simplifying: $\left(\dfrac{9}{25}\right)^{3/2} = \left(\dfrac{3}{5}\right)^3 = \dfrac{27}{125}$

77. Simplifying: $8^{-2/3} = \left(8^{-1/3}\right)^2 = \left(\dfrac{1}{2}\right)^2 = \dfrac{1}{4}$

79. Simplifying: $\dfrac{\left(49x^8y^{-4}\right)^{1/2}}{\left(27x^{-3}y^9\right)^{-1/3}} = \dfrac{7x^4y^{-2}}{\frac{1}{3}xy^{-3}} = 21x^{4-1}y^{-2+3} = 21x^3y$

81. Evaluating $b^2 - 4ac$: $b^2 - 4ac = (-3)^2 - 4(1)(-40) = 9 + 160 = 169$

83. Evaluating $b^2 - 4ac$: $b^2 - 4ac = 12^2 - 4(4)(9) = 144 - 144 = 0$

85. Solving the equation:

$$k^2 - 144 = 0$$
$$(k + 12)(k - 12) = 0$$
$$k = -12, 12$$

87. Multiplying: $(x - 3)(x + 2) = x^2 + 2x - 3x - 6 = x^2 - x - 6$

89. Multiplying:
$$(x-3)(x-3)(x+2) = (x^2 - 6x + 9)(x+2)$$
$$= x^3 - 6x^2 + 9x + 2x^2 - 12x + 18$$
$$= x^3 - 4x^2 - 3x + 18$$

91. Solving the equation:
$$x^2 + \sqrt{3}x - 6 = 0$$
$$(x + 2\sqrt{3})(x - \sqrt{3}) = 0$$
$$x = -2\sqrt{3}, \sqrt{3}$$

93. Solving the equation:
$$\sqrt{2}x^2 + 2x - \sqrt{2} = 0$$
$$x = \frac{-2 \pm \sqrt{4 + 4(2)}}{2\sqrt{2}} = \frac{-2 \pm \sqrt{12}}{2\sqrt{2}} = \frac{-2 \pm 2\sqrt{3}}{2\sqrt{2}} \cdot \frac{\sqrt{2}}{\sqrt{2}} = \frac{-2\sqrt{2} \pm 2\sqrt{6}}{4} = \frac{-\sqrt{2} \pm \sqrt{6}}{2}$$

95. Solving the equation:
$$x^2 + ix + 2 = 0$$
$$x = \frac{-i \pm \sqrt{-1 - 8}}{2} = \frac{-i \pm 3i}{2} = -2i, i$$

8.3 Additional Items Involving Solutions to Equations

1. Computing the discriminant: $D = (-6)^2 - 4(1)(5) = 36 - 20 = 16$. The equation will have two rational solutions.

3. First write the equation as $4x^2 - 4x + 1 = 0$. Computing the discriminant: $D = (-4)^2 - 4(4)(1) = 16 - 16 = 0$ The equation will have one rational solution.

5. Computing the discriminant: $D = 1^2 - 4(1)(-1) = 1 + 4 = 5$. The equation will have two irrational solutions.

7. First write the equation as $2y^2 - 3y - 1 = 0$. Computing the discriminant: $D = (-3)^2 - 4(2)(-1) = 9 + 8 = 17$ The equation will have two irrational solutions.

9. Computing the discriminant: $D = 0^2 - 4(1)(-9) = 36$. The equation will have two rational solutions.

11. First write the equation as $5a^2 - 4a - 5 = 0$. Computing the discriminant: $D = (-4)^2 - 4(5)(-5) = 16 + 100 = 116$ The equation will have two irrational solutions.

13. Setting the discriminant equal to 0:
$$(-k)^2 - 4(1)(25) = 0$$
$$k^2 - 100 = 0$$
$$k^2 = 100$$
$$k = \pm 10$$

15. First write the equation as $x^2 - kx + 36 = 0$. Setting the discriminant equal to 0:
$$(-k)^2 - 4(1)(36) = 0$$
$$k^2 - 144 = 0$$
$$k^2 = 144$$
$$k = \pm 12$$

17. Setting the discriminant equal to 0:
$$(-12)^2 - 4(4)(k) = 0$$
$$144 - 16k = 0$$
$$16k = 144$$
$$k = 9$$

19. First write the equation as $kx^2 - 40x - 25 = 0$. Setting the discriminant equal to 0:
$$(-40)^2 - 4(k)(-25) = 0$$
$$1600 + 100k = 0$$
$$100k = -1600$$
$$k = -16$$

21. Setting the discriminant equal to 0:
$$(-k)^2 - 4(3)(2) = 0$$
$$k^2 - 24 = 0$$
$$k^2 = 24$$
$$k = \pm\sqrt{24} = \pm 2\sqrt{6}$$

23. Writing the equation:
$$(x-5)(x-2) = 0$$
$$x^2 - 7x + 10 = 0$$

25. Writing the equation:
$$(t+3)(t-6) = 0$$
$$t^2 - 3t - 18 = 0$$

27. Writing the equation:
$$(y-2)(y+2)(y-4) = 0$$
$$(y^2 - 4)(y-4) = 0$$
$$y^3 - 4y^2 - 4y + 16 = 0$$

29. Writing the equation:
$$(2x-1)(x-3) = 0$$
$$2x^2 - 7x + 3 = 0$$

31. Writing the equation:
$$(4t+3)(t-3) = 0$$
$$4t^2 - 9t - 9 = 0$$

33. Writing the equation:
$$(x-3)(x+3)(6x-5) = 0$$
$$(x^2 - 9)(6x-5) = 0$$
$$6x^3 - 5x^2 - 54x + 45 = 0$$

35. Writing the equation:
$$(2a+1)(5a-3) = 0$$
$$10a^2 - a - 3 = 0$$

37. Writing the equation:
$$(3x+2)(3x-2)(x-1) = 0$$
$$(9x^2 - 4)(x-1) = 0$$
$$9x^3 - 9x^2 - 4x + 4 = 0$$

39. Writing the equation:
$$(x-2)(x+2)(x-3)(x+3) = 0$$
$$(x^2 - 4)(x^2 - 9) = 0$$
$$x^4 - 13x^2 + 36 = 0$$

41. Writing the equation:
$$(x-1)(x+2) = 0$$
$$x^2 + x - 2 = 0$$

So $f(x) = x^2 + x - 2$.

43. Starting with the solutions:
$$x = 3 \pm \sqrt{2}$$
$$x - 3 = \pm\sqrt{2}$$
$$(x-3)^2 = \left(\pm\sqrt{2}\right)^2$$
$$x^2 - 6x + 9 = 2$$
$$x^2 - 6x + 7 = 0$$
So $f(x) = x^2 - 6x + 7$.

45. Starting with the solutions:
$$x = 2 \pm i$$
$$x - 2 = \pm i$$
$$(x-2)^2 = (\pm i)^2$$
$$x^2 - 4x + 4 = -1$$
$$x^2 - 4x + 5 = 0$$
So $f(x) = x^2 - 4x + 5$.

47. Starting with the solutions:
$$x = \frac{5 \pm \sqrt{7}}{2}$$
$$2x = 5 \pm \sqrt{7}$$
$$(2x-5)^2 = \left(\pm\sqrt{7}\right)^2$$
$$4x^2 - 20x + 25 = 7$$
$$4x^2 - 20x + 18 = 0$$
$$2x^2 - 10x + 9 = 0$$
So $f(x) = 2x^2 - 10x + 9$.

49. Starting with the solutions:
$$x = \frac{3 \pm i\sqrt{5}}{2}$$
$$2x = 3 \pm i\sqrt{5}$$
$$(2x-3)^2 = \left(\pm i\sqrt{5}\right)^2$$
$$4x^2 - 12x + 9 = -5$$
$$4x^2 - 12x + 14 = 0$$
$$2x^2 - 6x + 7 = 0$$
So $f(x) = 2x^2 - 6x + 7$.

51. Starting with the solutions:
$$x = 2 \pm i\sqrt{3}$$
$$(x-2)^2 = \left(\pm i\sqrt{3}\right)^2$$
$$x^2 - 4x + 4 = -3$$
$$x^2 - 4x + 7 = 0$$
So $f(x) = x\left(x^2 - 4x + 7\right) = x^3 - 4x^2 + 7x$.

53. **a.** This graph has one x-intercept, which is $y = g(x)$.

b. This graph has no x-intercepts, which is $y = h(x)$.

c. This graph has two x-intercepts, which is $y = f(x)$.

55. Multiplying: $a^4\left(a^{3/2} - a^{1/2}\right) = a^{11/2} - a^{9/2}$

57. Multiplying: $\left(x^{3/2} - 3\right)^2 = \left(x^{3/2} - 3\right)\left(x^{3/2} - 3\right) = x^3 - 6x^{3/2} + 9$

59. Dividing: $\dfrac{30x^{3/4} - 25x^{5/4}}{5x^{1/4}} = \dfrac{30x^{3/4}}{5x^{1/4}} - \dfrac{25x^{5/4}}{5x^{1/4}} = 6x^{1/2} - 5x$

61. Factoring: $10(x-3)^{3/2} - 15(x-3)^{1/2} = 5(x-3)^{1/2}\left[2(x-3) - 3\right] = 5(x-3)^{1/2}(2x-9)$

63. Factoring: $2x^{2/3} - 11x^{1/3} + 12 = \left(2x^{1/3} - 3\right)\left(x^{1/3} - 4\right)$

65. Simplifying: $(x+3)^2 - 2(x+3) - 8 = x^2 + 6x + 9 - 2x - 6 - 8 = x^2 + 4x - 5$

67. Simplifying: $(2a-3)^2 - 9(2a-3) + 20 = 4a^2 - 12a + 9 - 18a + 27 + 20 = 4a^2 - 30a + 56$

69. Simplifying:
$$2(4a+2)^2 - 3(4a+2) - 20 = 2\left(16a^2 + 16a + 4\right) - 3(4a+2) - 20$$
$$= 32a^2 + 32a + 8 - 12a - 6 - 20$$
$$= 32a^2 + 20a - 18$$

71. Solving the equation:
$$x^2 = \frac{1}{4}$$
$$x = \pm\sqrt{\frac{1}{4}} = \pm\frac{1}{2}$$

73. Since $\sqrt{x} \ge 0$, this equation has no solution.

75. Solving the equation:
$$x + 3 = 4$$
$$x = 1$$

77. Solving the equation:
$$y^2 - 2y - 8 = 0$$
$$(y+2)(y-4) = 0$$
$$y = -2, 4$$

79. Solving the equation:

$$4y^2 + 7y - 2 = 0$$
$$(4y-1)(y+2) = 0$$
$$y = -2, \frac{1}{4}$$

81. Solving the equation:

$$x^2 = 4x + 5$$
$$x^2 - 4x - 5 = 0$$
$$(x-5)(x+1) = 0$$
$$x = -1, 5$$

83. Solving the equation:

$$x^2 - 1 = 2x$$
$$x^2 - 2x - 1 = 0$$
$$x = \frac{2 \pm \sqrt{4+4}}{2} = \frac{2 \pm \sqrt{8}}{2} = \frac{2 \pm 2\sqrt{2}}{2} = 1 \pm \sqrt{2} \approx -0.41, 2.41$$

85. Solving the equation:

$$2x^3 - x^2 - 2x + 1 = 0$$
$$x^2(2x-1) - 1(2x-1) = 0$$
$$(2x-1)(x^2-1) = 0$$
$$(2x-1)(x+1)(x-1) = 0$$
$$x = -1, \frac{1}{2}, 1$$

87. Solving the equation:

$$2x^3 + 2 = x^2 + 4x$$
$$2x^3 - x^2 - 4x + 2 = 0$$
$$x^2(2x-1) - 2(2x-1) = 0$$
$$(2x-1)(x^2-2) = 0$$
$$x = \frac{1}{2}, \pm\sqrt{2}$$
$$x \approx -1.41, \frac{1}{2}, 1.41$$

8.4 Equations Quadratic in Form

1. Solving the equation:

$$(x-3)^2 + 3(x-3) + 2 = 0$$
$$(x-3+2)(x-3+1) = 0$$
$$(x-1)(x-2) = 0$$
$$x = 1, 2$$

3. Solving the equation:

$$2(x+4)^2 + 5(x+4) - 12 = 0$$
$$[2(x+4)-3][(x+4)+4] = 0$$
$$(2x+5)(x+8) = 0$$
$$x = -8, -\frac{5}{2}$$

5. Solving the equation:

$$x^4 - 10x^2 + 9 = 0$$
$$(x^2-1)(x^2-9) = 0$$
$$x^2 = 1, 9$$
$$x = \pm1, \pm3$$

7. Solving the equation:

$$x^4 - 7x^2 + 12 = 0$$
$$(x^2-4)(x^2-3) = 0$$
$$x^2 = 4, 3$$
$$x = \pm2, \pm\sqrt{3}$$

9. Solving the equation:

$$x^4 - 6x^2 - 27 = 0$$
$$(x^2-9)(x^2+3) = 0$$
$$x^2 = 9, -3$$
$$x = \pm3, \pm i\sqrt{3}$$

11. Solving the equation:

$$x^4 + 9x^2 = -20$$
$$x^4 + 9x^2 + 20 = 0$$
$$(x^2+4)(x^2+5) = 0$$
$$x^2 = -4, -5$$
$$x = \pm2i, \pm i\sqrt{5}$$

13. Solving the equation:

$$(2a-3)^2 - 9(2a-3) = -20$$
$$(2a-3)^2 - 9(2a-3) + 20 = 0$$
$$(2a-3-4)(2a-3-5) = 0$$
$$(2a-7)(2a-8) = 0$$
$$a = \frac{7}{2}, 4$$

15. Solving the equation:

$$2(4a+2)^2 = 3(4a+2) + 20$$
$$2(4a+2)^2 - 3(4a+2) - 20 = 0$$
$$[2(4a+2)+5][(4a+2)-4] = 0$$
$$(8a+9)(4a-2) = 0$$
$$a = -\frac{9}{8}, \frac{1}{2}$$

17. Solving the equation:

$$6t^4 = -t^2 + 5$$
$$6t^4 + t^2 - 5 = 0$$
$$(6t^2 - 5)(t^2 + 1) = 0$$
$$t^2 = \frac{5}{6}, -1$$
$$t = \pm\sqrt{\frac{5}{6}} = \pm\frac{\sqrt{30}}{6}, \pm i$$

19. Solving the equation:

$$9x^4 - 49 = 0$$
$$(3x^2 - 7)(3x^2 + 7) = 0$$
$$x^2 = \frac{7}{3}, -\frac{7}{3}$$
$$x = \pm\sqrt{\frac{7}{3}}, \pm\sqrt{-\frac{7}{3}}$$
$$t = \pm\frac{\sqrt{21}}{3}, \pm\frac{i\sqrt{21}}{3}$$

21. Solving the equation:

$$x - 7\sqrt{x} + 10 = 0$$
$$(\sqrt{x} - 5)(\sqrt{x} - 2) = 0$$
$$\sqrt{x} = 2, 5$$
$$x = 4, 25$$

Both values check in the original equation.

23. Solving the equation:

$$t - 2\sqrt{t} - 15 = 0$$
$$(\sqrt{t} - 5)(\sqrt{t} + 3) = 0$$
$$\sqrt{t} = -3, 5$$
$$t = 9, 25$$

Only $t = 25$ checks in the original equation.

25. Solving the equation:

$$6x + 11\sqrt{x} = 35$$
$$6x + 11\sqrt{x} - 35 = 0$$
$$(3\sqrt{x} - 5)(2\sqrt{x} + 7) = 0$$
$$\sqrt{x} = \frac{5}{3}, -\frac{7}{2}$$
$$x = \frac{25}{9}, \frac{49}{4}$$

Only $x = \frac{25}{9}$ checks in the original equation.

27. Solving the equation:

$$x - 2\sqrt{x} - 8 = 0$$
$$(\sqrt{x} - 4)(\sqrt{x} + 2) = 0$$
$$\sqrt{x} = -2, 4$$
$$x = 4, 16$$

Only $x = 16$ checks in the original equation.

29. Solving the equation:

$$x + 3\sqrt{x} - 18 = 0$$
$$(\sqrt{x} - 3)(\sqrt{x} + 6) = 0$$
$$\sqrt{x} = -6, 3$$
$$x = 36, 9$$

Only $x = 9$ checks in the original equation.

31. Solving the equation:

$$2x + 9\sqrt{x} - 5 = 0$$
$$(2\sqrt{x} - 1)(\sqrt{x} + 5) = 0$$
$$\sqrt{x} = -5, \frac{1}{2}$$
$$x = 25, \frac{1}{4}$$

Only $x = \frac{1}{4}$ checks in the original equation.

33. Solving the equation:

$$(a-2)-11\sqrt{a-2}+30=0$$
$$\left(\sqrt{a-2}-6\right)\left(\sqrt{a-2}-5\right)=0$$
$$\sqrt{a-2}=5,6$$
$$a-2=25,36$$
$$a=27,38$$

35. Solving the equation:

$$(2x+1)-8\sqrt{2x+1}+15=0$$
$$\left(\sqrt{2x+1}-3\right)\left(\sqrt{2x+1}-5\right)=0$$
$$\sqrt{2x+1}=3,5$$
$$2x+1=9,25$$
$$2x=8,24$$
$$x=4,12$$

37. Solving the equation:

$$\left(x^2+1\right)^2-2\left(x^2+1\right)-15=0$$
$$\left(x^2+1-5\right)\left(x^2+1+3\right)=0$$
$$\left(x^2-4\right)\left(x^2+4\right)=0$$
$$x^2=4,-4$$
$$x=\pm 2,\pm 2i$$

39. Solving the equation:

$$\left(x^2+5\right)^2-6\left(x^2+5\right)-27=0$$
$$\left(x^2+5-9\right)\left(x^2+5+3\right)=0$$
$$\left(x^2-4\right)\left(x^2+8\right)=0$$
$$x^2=4,-8$$
$$x=\pm 2,\pm 2i\sqrt{2}$$

41. Solving the equation:

$$x^{-2}-3x^{-1}+2=0$$
$$\left(x^{-1}-2\right)\left(x^{-1}-1\right)=0$$
$$x^{-1}=2,1$$
$$x=\frac{1}{2},1$$

43. Solving the equation:

$$y^{-4}-5y^{-2}=0$$
$$y^{-2}\left(y^{-2}-5\right)=0$$
$$y^{-2}=0,5$$
$$y^2=\frac{1}{5}$$
$$y=\pm\sqrt{\frac{1}{5}}=\pm\frac{\sqrt{5}}{5}$$

45. Solving the equation:

$$x^{2/3}+4x^{1/3}-12=0$$
$$\left(x^{1/3}+6\right)\left(x^{1/3}-2\right)=0$$
$$x^{1/3}=-6,2$$
$$x=(-6)^3,2^3$$
$$x=-216,8$$

47. Solving the equation:

$$x^{4/3}+6x^{2/3}+9=0$$
$$\left(x^{2/3}+3\right)^2=0$$
$$x^{2/3}=-3$$
$$x^2=(-3)^3=-27$$
$$x=\pm\sqrt{-27}=\pm 3i\sqrt{3}$$

49. Solving for t:

$$16t^2-vt-h=0$$
$$t=\frac{v\pm\sqrt{v^2-4(16)(-h)}}{32}=\frac{v\pm\sqrt{v^2+64h}}{32}$$

51. Solving for x:

$$kx^2+8x+4=0$$
$$x=\frac{-8\pm\sqrt{64-16k}}{2k}=\frac{-8\pm 4\sqrt{4-k}}{2k}=\frac{-4\pm 2\sqrt{4-k}}{k}$$

53. Solving for x:

$$x^2+2xy+y^2=0$$
$$x=\frac{-2y\pm\sqrt{4y^2-4y^2}}{2}=\frac{-2y}{2}=-y$$

55. Solving for t (note that $t>0$):

$$16t^2-8t-h=0$$
$$t=\frac{8+\sqrt{64+64h}}{32}=\frac{8+8\sqrt{1+h}}{32}=\frac{1+\sqrt{1+h}}{4}$$

57.   a.   Sketching the graph:

$y = -\frac{1}{150}x^2 + \frac{21}{5}x$

b.   Finding the x-intercepts:

$$-\frac{1}{150}x^2 + \frac{21}{5}x = 0$$

$$-\frac{1}{150}x(x - 630) = 0$$

$$x = 0, 630$$

The width is 630 feet.

59.   Let $x = BC$. Solving the proportion:

$$\frac{AB}{BC} = \frac{BC}{AC}$$

$$\frac{4}{x} = \frac{x}{4 + x}$$

$$16 + 4x = x^2$$

$$0 = x^2 - 4x - 16$$

$$x = \frac{4 \pm \sqrt{16 + 64}}{2} = \frac{4 \pm \sqrt{80}}{2} = \frac{4 \pm 4\sqrt{5}}{2} = 2 \pm 2\sqrt{5}$$

Thus $BC = 2 + 2\sqrt{5} = 4\left(\frac{1 + \sqrt{5}}{2}\right)$.

61.   Combining: $5\sqrt{7} - 2\sqrt{7} = 3\sqrt{7}$

63.   Combining: $\sqrt{18} - \sqrt{8} + \sqrt{32} = 3\sqrt{2} - 2\sqrt{2} + 4\sqrt{2} = 5\sqrt{2}$

65.   Combining: $9x\sqrt{20x^3y^2} + 7y\sqrt{45x^5} = 9x \cdot 2xy\sqrt{5x} + 7y \cdot 3x^2\sqrt{5x} = 18x^2y\sqrt{5x} + 21x^2y\sqrt{5x} = 39x^2y\sqrt{5x}$

67.   Multiplying: $(\sqrt{5} - 2)(\sqrt{5} + 8) = 5 - 2\sqrt{5} + 8\sqrt{5} - 16 = -11 + 6\sqrt{5}$

69.   Multiplying: $(\sqrt{x} + 2)^2 = (\sqrt{x})^2 + 4\sqrt{x} + 4 = x + 4\sqrt{x} + 4$

71.   Rationalizing the denominator: $\dfrac{\sqrt{7}}{\sqrt{7} - 2} \cdot \dfrac{\sqrt{7} + 2}{\sqrt{7} + 2} = \dfrac{7 + 2\sqrt{7}}{7 - 4} = \dfrac{7 + 2\sqrt{7}}{3}$

73.   Evaluating when $x = 1$: $y = 3(1)^2 - 6(1) + 1 = 3 - 6 + 1 = -2$

75.   Evaluating: $P(135) = -0.1(135)^2 + 27(135) - 500 = -1,822.5 + 3,645 - 500 = 1,322.5$

77.   Solving the equation:

$$0 = a(80)^2 + 70$$

$$0 = 6400a + 70$$

$$6400a = -70$$

$$a = -\frac{7}{640}$$

79.   Solving the equation:

$$x^2 - 6x + 5 = 0$$

$$(x - 1)(x - 5) = 0$$

$$x = 1, 5$$

81. Solving the equation:
$$-x^2 - 2x + 3 = 0$$
$$x^2 + 2x - 3 = 0$$
$$(x+3)(x-1) = 0$$
$$x = -3, 1$$

83. Solving the equation:
$$2x^2 - 6x + 5 = 0$$

$$x = \frac{6 \pm \sqrt{(-6)^2 - 4(2)(5)}}{2(2)} = \frac{6 \pm \sqrt{36-40}}{4} = \frac{6 \pm \sqrt{-4}}{4} = \frac{6 \pm 2i}{4} = \frac{3 \pm i}{2} = \frac{3}{2} \pm \frac{1}{2}i$$

85. Completing the square: $x^2 - 6x + 9 = (x-3)^2$ **87.** Completing the square: $y^2 + 2y + 1 = (y+1)^2$

89. To find the x-intercepts, set $y = 0$:
$$x^3 - 4x = 0$$
$$x(x^2 - 4) = 0$$
$$x(x+2)(x-2) = 0$$
$$x = -2, 0, 2$$

To find the y-intercept, set $x = 0$: $y = 0^3 - 4(0) = 0$. The x-intercepts are $-2, 0, 2$ and the y-intercept is 0.

91. To find the x-intercepts, set $y = 0$:
$$3x^3 + x^2 - 27x - 9 = 0$$
$$x^2(3x+1) - 9(3x+1) = 0$$
$$(3x+1)(x^2 - 9) = 0$$
$$(3x+1)(x+3)(x-3) = 0$$
$$x = -3, -\frac{1}{3}, 3$$

To find the y-intercept, set $x = 0$: $y = 3(0)^3 + (0)^2 - 27(0) - 9 = -9$

The x-intercepts are $-3, -\frac{1}{3}, 3$ and the y-intercept is -9.

93. Using long division:

$$\begin{array}{r}
2x^2 + x - 1 \\
x-4 \overline{)\, 2x^3 - 7x^2 - 5x + 4} \\
\underline{2x^3 - 8x^2} \\
x^2 - 5x \\
\underline{x^2 - 4x} \\
-x + 4 \\
\underline{-x + 4} \\
0
\end{array}$$

Now solve the equation:
$$2x^2 + x - 1 = 0$$
$$(2x-1)(x+1) = 0$$
$$x = -1, \frac{1}{2}$$

It also crosses the x-axis at $\frac{1}{2}, -1$.

8.5 Graphing Parabolas

1. First complete the square: $y = x^2 + 2x - 3 = \left(x^2 + 2x + 1\right) - 1 - 3 = (x+1)^2 - 4$

The x-intercepts are $-3, 1$ and the vertex is $(-1, -4)$. Graphing the parabola:

3. First complete the square: $y = -x^2 - 4x + 5 = -\left(x^2 + 4x + 4\right) + 4 + 5 = -(x+2)^2 + 9$

The x-intercepts are $-5, 1$ and the vertex is $(-2, 9)$. Graphing the parabola:

5. The x-intercepts are $-1, 1$ and the vertex is $(0, -1)$. Graphing the parabola:

7. The *x*-intercepts are –3,3 and the vertex is (0,9). Graphing the parabola:

$y = -x^2 + 9$

9. First complete the square: $f(x) = 2x^2 - 4x - 6 = 2(x^2 - 2x + 1) - 2 - 6 = 2(x-1)^2 - 8$

    The *x*-intercepts are –1,3 and the vertex is (1,–8). Graphing the parabola:

$f(x) = 2x^2 - 4x - 6$

11. First complete the square: $f(x) = x^2 - 2x - 4 = (x^2 - 2x + 1) - 1 - 4 = (x-1)^2 - 5$

    The *x*-intercepts are $1 \pm \sqrt{5}$ and the vertex is (1,–5). Graphing the parabola:

$f(x) = x^2 - 2x - 4$

13.   First complete the square: $y = x^2 - 4x - 4 = \left(x^2 - 4x + 4\right) - 4 - 4 = (x-2)^2 - 8$

The vertex is (2,–8). Graphing the parabola:

15.   First complete the square: $y = -x^2 + 2x - 5 = -\left(x^2 - 2x + 1\right) + 1 - 5 = -(x-1)^2 - 4$

The vertex is (1,–4). Graphing the parabola:

17.   The vertex is (0,1). Graphing the parabola:

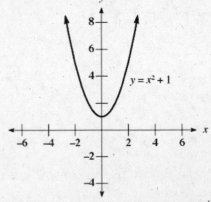

19.   The vertex is (0,–3). Graphing the parabola:

21.   Completing the square: $y = x^2 - 6x + 5 = \left(x^2 - 6x + 9\right) - 9 + 5 = (x-3)^2 - 4$

The vertex is (3,–4), which is the lowest point on the graph.

23.   Completing the square: $y = -x^2 + 2x + 8 = -\left(x^2 - 2x + 1\right) + 1 + 8 = -(x-1)^2 + 9$

The vertex is (1,9), which is the highest point on the graph.

25.   Completing the square: $y = -x^2 + 4x + 12 = -\left(x^2 - 4x + 4\right) + 4 + 12 = -(x-2)^2 + 16$

The vertex is (2,16), which is the highest point on the graph.

27. Completing the square: $y = -x^2 - 8x = -\left(x^2 + 8x + 16\right) + 16 = -\left(x + 4\right)^2 + 16$

The vertex is $(-4,16)$, which is the highest point on the graph.

29. Completing the square: $x = y^2 + 2y - 3 = \left(y^2 + 2y + 1\right) - 3 - 1 = \left(y + 1\right)^2 - 4$

The vertex is $(-4,-1)$. Graphing the parabola:

31. Completing the square: $x = -y^2 - 4y + 5 = -\left(y^2 + 4y + 4\right) + 5 + 4 = -\left(y + 2\right)^2 + 9$

The vertex is $(9,-2)$. Graphing the parabola:

33. The vertex is $(-1,0)$. Graphing the parabola:

35. The vertex is $(9,0)$. Graphing the parabola:

37. Completing the square: $x = 2y^2 - 4y - 6 = 2\left(y^2 - 2y + 1\right) - 2 - 6 = 2\left(y - 1\right)^2 - 8$

The vertex is $(-8,1)$. Graphing the parabola:

39.   Completing the square: $x = y^2 - 2y - 4 = \left(y^2 - 2y + 1\right) - 1 - 4 = \left(y - 1\right)^2 - 5$

The vertex is (–5,1). Graphing the parabola:

41.   First complete the square:

$$P(x) = -0.002x^2 + 3.5x - 800 = -0.002\left(x^2 - 1750x + 765,625\right) + 1,531.25 - 800 = -0.002\left(x - 875\right)^2 + 731.25$$

It must sell 875 patterns to obtain a maximum profit of $731.25.

43.   The ball is in her hand at times 0 sec and 2 sec.

Completing the square: $h(t) = -16t^2 + 32t = -16\left(t^2 - 2t + 1\right) + 16 = -16\left(t - 1\right)^2 + 16$

The maximum height of the ball is 16 feet.

45.   Completing the square: $R = xp = 1200p - 100p^2 = -100\left(p^2 - 12p + 36\right) + 3600 = -100\left(p - 6\right)^2 + 3600$

The price is $6.00 and the maximum revenue is $3,600. Sketching the graph:

47.   Completing the square: $R = xp = 1700p - 100p^2 = -100\left(p^2 - 17p + 72.25\right) + 7225 = -100\left(p - 8.5\right)^2 + 7225$

The price is $8.50 and the maximum revenue is $7,225. Sketching the graph:

49.   The equation is given on the graph:

$$y = -\frac{1}{135}(x-90)^2 + 60$$

51.   Solving the equation:

$$x^2 - 2x - 8 = 0$$
$$(x-4)(x+2) = 0$$
$$x = -2, 4$$

53.   Solving the equation:

$$6x^2 - x = 2$$
$$6x^2 - x - 2 = 0$$
$$(2x+1)(3x-2) = 0$$
$$x = -\frac{1}{2}, \frac{2}{3}$$

55.   Solving the equation:

$$x^2 - 6x + 9 = 0$$
$$(x-3)^2 = 0$$
$$x = 3$$

57.   Performing the operations: $(3-5i)-(2-4i) = 3-5i-2+4i = 1-i$

59.   Performing the operations: $(3+2i)(7-3i) = 21+5i-6i^2 = 27+5i$

61.   Performing the operations: $\dfrac{i}{3+i} = \dfrac{i}{3+i} \cdot \dfrac{3-i}{3-i} = \dfrac{3i-i^2}{9-i^2} = \dfrac{3i+1}{9+1} = \dfrac{1}{10} + \dfrac{3}{10}i$

63.   The equation is $y = (x-2)^2 - 4$.

65.   a.   The angle should be either 20° or 70°.
       b.   If the cannon was inclined at 45°, he would land approximately 170 feet horizontally.
       c.   If the cannon was inclined at 45°, he would reach a height of approximately 42 or 43 feet.
       d.   Yes, an angle of 10° should result in the same horizontal distance.
       e.   It would imply a quadratic equation.

8.6 Quadratic Inequalities

1. Factoring the inequality:
$$x^2 + x - 6 > 0$$
$$(x+3)(x-2) > 0$$
Forming the sign chart:

The solution set is $x < -3$ or $x > 2$. Graphing the solution set:

3. Factoring the inequality:
$$x^2 - x - 12 \le 0$$
$$(x+3)(x-4) \le 0$$
Forming the sign chart:

The solution set is $-3 \le x \le 4$. Graphing the solution set:

5. Factoring the inequality:
$$x^2 + 5x \ge -6$$
$$x^2 + 5x + 6 \ge 0$$
$$(x+2)(x+3) \ge 0$$
Forming the sign chart:

The solution set is $x \le -3$ or $x \ge -2$. Graphing the solution set:

7. Factoring the inequality:
$$6x^2 < 5x - 1$$
$$6x^2 - 5x + 1 < 0$$
$$(3x-1)(2x-1) < 0$$
Forming the sign chart:

The solution set is $\dfrac{1}{3} < x < \dfrac{1}{2}$. Graphing the solution set:

9.   Factoring the inequality:
$$x^2 - 9 < 0$$
$$(x+3)(x-3) < 0$$
Forming the sign chart:

The solution set is $-3 < x < 3$. Graphing the solution set:

11.   Factoring the inequality:
$$4x^2 - 9 \geq 0$$
$$(2x+3)(2x-3) \geq 0$$
Forming the sign chart:

The solution set is $x \leq -\dfrac{3}{2}$ or $x \geq \dfrac{3}{2}$. Graphing the solution set:

13.   Factoring the inequality:
$$2x^2 - x - 3 < 0$$
$$(2x-3)(x+1) < 0$$
Forming the sign chart:

The solution set is $-1 < x < \dfrac{3}{2}$. Graphing the solution set:

15.   Factoring the inequality:
$$x^2 - 4x + 4 \geq 0$$
$$(x-2)^2 \geq 0$$
Since this inequality is always true, the solution set is all real numbers. Graphing the solution set:

17.   Factoring the inequality:
$$x^2 - 10x + 25 < 0$$
$$(x-5)^2 < 0$$
Since this inequality is never true, there is no solution.

19.   Forming the sign chart:

The solution set is $2 < x < 3$ or $x > 4$. Graphing the solution set:

21.   Forming the sign chart:

The solution set is $x \leq -3$ or $-2 \leq x \leq -1$. Graphing the solution set:

23.   Forming the sign chart:

The solution set is $-4 < x \leq 1$. Graphing the solution set:

25.   Forming the sign chart:

The solution set is $x < -6$ or $x > \dfrac{8}{3}$. Graphing the solution set:

27.   Forming the sign chart:

The solution set is $x < 2$ or $x > 6$. Graphing the solution set:

29.   Forming the sign chart:

The solution set is $x < -3$ or $2 < x < 4$. Graphing the solution set:

31.   Forming the sign chart:

The solution set is $2 < x < 3$ or $x > 4$. Graphing the solution set:

33.   **a.**   Writing the statement: $x - 1 > 0$ **b.**   Writing the statement: $x - 1 \geq 0$
          c.   Writing the statement: $x - 1 \geq 0$

35.   This statement is never true, so there is no solution ($\varnothing$).

37.   This statement is only true if $x = 1$.   **39.**   This statement is true for all real numbers.

41.   This statement is true if $x \neq 1$, or $x < 1$ or $x > 1$.

43.   Since $x^2 - 6x + 9 = (x - 3)^2$, this statement is never true, so there is no solution ($\varnothing$).

45.   Since $x^2 - 6x + 9 = (x - 3)^2$, this statement is true if $x \neq 3$, or $x < 3$ or $x > 3$.

47.   **a.**   The solution set is $-2 < x < 2$.   **49.**   **a.**   The solution set is $-2 < x < 5$.
          b.   The solution set is $x < -2$ or $x > 2$.   **b.**   The solution set is $x < -2$ or $x > 5$.
          c.   The solution set is $x = -2, 2$.   **c.**   The solution set is $x = -2, 5$.

51.   **a.**   The solution set is $x < -1$ or $1 < x < 3$.   **b.**   The solution set is $-1 < x < 1$ or $x > 3$.
          c.   The solution set is $x = -1, 1, 3$.

53.   Let w represent the width and $2w + 3$ represent the length. Using the area formula:
$$w(2w + 3) \geq 44$$
$$2w^2 + 3w > 44$$
$$2w^2 + 3w - 44 \geq 0$$
$$(2w + 11)(w - 4) \geq 0$$

Forming the sign chart:

The width is at least 4 inches.

55.   Solving the inequality:

$$1300p - 100p^2 \geq 4000$$
$$-100p^2 + 1300p - 4000 \geq 0$$
$$p^2 - 13p + 40 \leq 0$$
$$(p - 8)(p - 5) \leq 0$$

Forming the sign chart:

She should charge at least \$5 but no more than \$8 per radio.

57.   Using a calculator: $\dfrac{50,000}{32,000} = 1.5625$

59.   Using a calculator: $\dfrac{1}{2}\left(\dfrac{4.5926}{1.3876} - 2 \right) \approx 0.6549$

61.   Solving the equation:

$$\sqrt{3t - 1} = 2$$
$$\left(\sqrt{3t - 1} \right)^2 = (2)^2$$
$$3t - 1 = 4$$
$$3t = 5$$
$$t = \frac{5}{3}$$

The solution is $\dfrac{5}{3}$.

63.   Solving the equation:

$$\sqrt{x + 3} = x - 3$$
$$\left(\sqrt{x + 3} \right)^2 = (x - 3)^2$$
$$x + 3 = x^2 - 6x + 9$$
$$0 = x^2 - 7x + 6$$
$$0 = (x - 6)(x - 1)$$
$$x = 1, 6 \qquad (x = 1 \text{ does not check})$$

The solution is 6.

65.   Graphing the equation:

67.   Using the quadratic formula: $x = \dfrac{2 \pm \sqrt{4 - 4(-1)}}{2(1)} = \dfrac{2 \pm \sqrt{8}}{2} = \dfrac{2 \pm 2\sqrt{2}}{2} = 1 \pm \sqrt{2}$

The inequality is satisfied when $1 - \sqrt{2} < x < 1 + \sqrt{2}$.

69.   Using the quadratic formula: $x = \dfrac{8 \pm \sqrt{64 - 4(13)}}{2(1)} = \dfrac{8 \pm \sqrt{12}}{2} = \dfrac{8 \pm 2\sqrt{3}}{2} = 4 \pm \sqrt{3}$

The inequality is satisfied when $x < 4 - \sqrt{3}$ or $x > 4 + \sqrt{3}$.

Chapter 8 Review

1. Solving the equation:

$$(2t-5)^2 = 25$$
$$2t-5 = \pm 5$$
$$2t-5 = -5, 5$$
$$2t = 0, 10$$
$$t = 0, 5$$

3. Solving the equation:

$$(3y-4)^2 = -49$$
$$3y-4 = \pm\sqrt{-49}$$
$$3y-4 = \pm 7i$$
$$3y = 4 \pm 7i$$
$$y = \frac{4 \pm 7i}{3}$$

5. Solving by completing the square:

$$2x^2 + 6x - 20 = 0$$
$$x^2 + 3x = 10$$
$$x^2 + 3x + \frac{9}{4} = 10 + \frac{9}{4}$$
$$\left(x + \frac{3}{2}\right)^2 = \frac{49}{4}$$
$$x + \frac{3}{2} = -\frac{7}{2}, \frac{7}{2}$$
$$x = -5, 2$$

7. Solving by completing the square:

$$a^2 + 9 = 6a$$
$$a^2 - 6a = -9$$
$$a^2 - 6a + 9 = -9 + 9$$
$$(a-3)^2 = 0$$
$$a - 3 = 0$$
$$a = 3$$

9. Solving by completing the square:

$$2y^2 + 6y = -3$$
$$y^2 + 3y = -\frac{3}{2}$$
$$y^2 + 3y + \frac{9}{4} = -\frac{3}{2} + \frac{9}{4}$$
$$\left(y + \frac{3}{2}\right)^2 = \frac{3}{4}$$
$$y + \frac{3}{2} = \pm\frac{\sqrt{3}}{2}$$
$$y = \frac{-3 \pm \sqrt{3}}{2}$$

11. Solving by completing the square:

$$\frac{1}{6}x^2 + \frac{1}{2}x - \frac{5}{3} = 0$$
$$x^2 + 3x - 10 = 0$$
$$(x+5)(x-2) = 0$$
$$x = -5, 2$$

13. Solving the equation:

$$4t^2 - 8t + 19 = 0$$
$$t = \frac{8 \pm \sqrt{64 - 304}}{8} = \frac{8 \pm \sqrt{-240}}{8} = \frac{8 \pm 4i\sqrt{15}}{8} = \frac{2 \pm i\sqrt{15}}{2}$$

15. Solving the equation:

$$0.06a^2 + 0.05a = 0.04$$
$$0.06a^2 + 0.05a - 0.04 = 0$$
$$6a^2 + 5a - 4 = 0$$
$$(2a-1)(3a+4) = 0$$
$$a = -\frac{4}{3}, \frac{1}{2}$$

17.   Solving the equation:
$$(2x+1)(x-5)-(x+3)(x-2)=-17$$
$$2x^2-9x-5-x^2-x+6=-17$$
$$x^2-10x+1=-17$$
$$x^2-10x+18=0$$
$$x=\frac{10\pm\sqrt{100-72}}{2}=\frac{10\pm\sqrt{28}}{2}=\frac{10\pm2\sqrt{7}}{2}=5\pm\sqrt{7}$$

19.   Solving the equation:
$$5x^2=-2x+3$$
$$5x^2+2x-3=0$$
$$(x+1)(5x-3)=0$$
$$x=-1,\frac{3}{5}$$

21.   Solving the equation:
$$3-\frac{2}{x}+\frac{1}{x^2}=0$$
$$3x^2-2x+1=0$$
$$x=\frac{2\pm\sqrt{4-12}}{6}=\frac{2\pm\sqrt{-8}}{6}=\frac{2\pm2i\sqrt{2}}{6}=\frac{1\pm i\sqrt{2}}{3}$$

23.   The profit equation is given by:
$$34x-0.1x^2-7x-400=1300$$
$$-0.1x^2+27x-1700=0$$
$$x^2-270x+17000=0$$
$$(x-100)(x-170)=0$$
$$x=100,170$$
The company must sell either 100 or 170 items for its weekly profit to be $1,300.

25.   First write the equation as $2x^2-8x+8=0$. Computing the discriminant: $D=(-8)^2-4(2)(8)=64-64=0$
The equation will have one rational solution.

27.   Computing the discriminant: $D=(1)^2-4(2)(-3)=1+24=25$. The equation will have two rational solutions.

29.   First write the equation as $x^2-x-1=0$. Computing the discriminant: $D=(-1)^2-4(1)(-1)=1+4=5$
The equation will have two irrational solutions.

31.   First write the equation as $3x^2+5x+4=0$. Computing the discriminant: $D=(5)^2-4(3)(4)=25-48=-23$
The equation will have two non-real complex solutions.

33.   Setting the discriminant equal to 0:
$$(-k)^2-4(25)(4)=0$$
$$k^2-400=0$$
$$k^2=400$$
$$k=\pm20$$

35.   Setting the discriminant equal to 0:
$$(12)^2-4(k)(9)=0$$
$$144-36k=0$$
$$36k=144$$
$$k=4$$

37.   Setting the discriminant equal to 0:
$$30^2-4(9)(k)=0$$
$$900-36k=0$$
$$36k=900$$
$$k=25$$

39.   Writing the equation:
$$(x-3)(x-5)=0$$
$$x^2-8x+15=0$$

41. Writing the equation:

$$(2y-1)(y+4)=0$$
$$2y^2+7y-4=0$$

43. Writing the equation:
$$(x-2)^2-4(x-2)-60=0$$
$$(x-2-10)(x-2+6)=0$$
$$(x-12)(x+4)=0$$
$$x=-4,12$$

45. Solving the equation:
$$x^4-x^2=12$$
$$x^4-x^2-12=0$$
$$(x^2-4)(x^2+3)=0$$
$$x^2=4,-3$$
$$x=\pm2,\pm i\sqrt{3}$$

47. Solving the equation:
$$2x-11\sqrt{x}=-12$$
$$2x-11\sqrt{x}+12=0$$
$$(2\sqrt{x}-3)(\sqrt{x}-4)=0$$
$$\sqrt{x}=\frac{3}{2},4$$
$$x=\frac{9}{4},16$$

49. Solving the equation:
$$\sqrt{y+21}+\sqrt{y}=7$$
$$\left(\sqrt{y+21}\right)^2=\left(7-\sqrt{y}\right)^2$$
$$y+21=49-14\sqrt{y}+y$$
$$14\sqrt{y}=28$$
$$\sqrt{y}=2$$
$$y=4$$

51. Solving for t:
$$16t^2-10t-h=0$$
$$t=\frac{10\pm\sqrt{100+64h}}{32}=\frac{10\pm2\sqrt{25+16h}}{32}=\frac{5\pm\sqrt{25+16h}}{16}$$

53. First complete the square: $y=x^2-6x+8=\left(x^2-6x+9\right)+8-9=(x-3)^2-1$

The x-intercepts are 2,4, and the vertex is (3,–1). Graphing the parabola:

55. Factoring the inequality:

$$x^2 - x - 2 < 0$$
$$(x-2)(x+1) < 0$$

Forming a sign chart:

The solution set is $-1 < x < 2$. Graphing the solution set:

57. Factoring the inequality:

$$2x^2 + 5x - 12 \geq 0$$
$$(2x-3)(x+4) \geq 0$$

Forming a sign chart:

The solution set is $x \leq -4$ or $x \geq \dfrac{3}{2}$. Graphing the solution set:

Chapter 8 Cumulative Review

1. Simplifying: $11 + 20 \div 5 - 3 \cdot 5 = 11 + 4 - 15 = 0$

3. Simplifying: $\left(-\dfrac{2}{3}\right)^3 = \left(-\dfrac{2}{3}\right)\left(-\dfrac{2}{3}\right)\left(-\dfrac{2}{3}\right) = -\dfrac{8}{27}$

5. Simplifying: $4 + 8x - 3(5x - 2) = 4 + 8x - 15x + 6 = -7x + 10$

7. Simplifying: $\left(\dfrac{x^{-5}y^4}{x^{-2}y^{-3}}\right)^{-1} = \dfrac{x^5 y^{-4}}{x^2 y^3} = x^{5-2} y^{-4-3} = x^3 y^{-7} = \dfrac{x^3}{y^7}$

9. Simplifying: $\sqrt[3]{32} = \sqrt[3]{8 \cdot 4} = 2\sqrt[3]{4}$

11. Simplifying: $8^{-2/3} + 25^{-1/2} = \left(8^{1/3}\right)^{-2} + \left(25^{1/2}\right)^{-1} = 2^{-2} + 5^{-1} = \dfrac{1}{4} + \dfrac{1}{5} = \dfrac{5}{20} + \dfrac{4}{20} = \dfrac{9}{20}$

13. Simplifying: $\dfrac{1 - \dfrac{3}{4}}{1 + \dfrac{3}{4}} = \dfrac{1 - \dfrac{3}{4}}{1 + \dfrac{3}{4}} \cdot \dfrac{4}{4} = \dfrac{4-3}{4+3} = \dfrac{1}{7}$

15. Reducing the fraction: $\dfrac{5x^2 - 26xy - 24y^2}{5x + 4y} = \dfrac{(5x+4y)(x-6y)}{5x+4y} = x - 6y$

17. Dividing: $\dfrac{3+i}{i} = \dfrac{3+i}{i} \cdot \dfrac{i}{i} = \dfrac{3i + i^2}{i^2} = \dfrac{3i - 1}{-1} = 1 - 3i$

19. Solving the equation:

$$\frac{7}{5}a - 6 = 15$$
$$\frac{7}{5}a = 21$$
$$7a = 105$$
$$a = 15$$

21. Solving the equation:

$$|a| - 6 = 3$$
$$|a| = 9$$
$$a = \pm 9$$

23. Solving the equation:

$$\frac{y}{3} + \frac{5}{y} = -\frac{8}{3}$$
$$3y\left(\frac{y}{3} + \frac{5}{y}\right) = 3y\left(-\frac{8}{3}\right)$$
$$y^2 + 15 = -8y$$
$$y^2 + 8y + 15 = 0$$
$$(y + 5)(y + 3) = 0$$
$$y = -5, -3$$

25. Solving the equation:

$$(3x - 4)^2 = 18$$
$$3x - 4 = \pm\sqrt{18}$$
$$3x - 4 = \pm 3\sqrt{2}$$
$$3x = 4 \pm 3\sqrt{2}$$
$$x = \frac{4 \pm 3\sqrt{2}}{3}$$

27. Solving the equation:

$$\frac{2}{15}x^2 + \frac{1}{3}x + \frac{1}{5} = 0$$
$$2x^2 + 5x + 3 = 0$$
$$(2x + 3)(x + 1) = 0$$
$$x = -\frac{3}{2}, -1$$

29. Solving the equation:

$$\sqrt{y + 3} = y + 3$$
$$\left(\sqrt{y + 3}\right)^2 = (y + 3)^2$$
$$y + 3 = y^2 + 6y + 9$$
$$0 = y^2 + 5y + 6$$
$$0 = (y + 2)(y + 3)$$
$$y = -3, -2$$

31. Solving the equation:

$$x - 3\sqrt{x} - 54 = 0$$
$$\left(\sqrt{x} - 9\right)\left(\sqrt{x} + 6\right) = 0$$
$$\sqrt{x} = -6, 9$$
$$x = 36, 81$$

Only $x = 81$ checks in the original equation.

33. Solving the inequality:

$$5 \le \frac{1}{4}x + 3 \le 8$$
$$2 \le \frac{1}{4}x \le 5$$
$$8 \le x \le 20$$

Graphing the solution set:

35. Factoring the inequality:

$$x^2 + 2x < 8$$
$$x^2 + 2x - 8 < 0$$
$$(x + 4)(x - 2) < 0$$

Forming the sign chart:

The solution set is $-4 < x < 2$. Graphing the solution set:

37. Multiply the first equation by 2:

$$6x - 2y = 44$$
$$-6x + 2y = -4$$

Adding yields $0 = 40$, which is false. There is no solution (lines are parallel).

39.   Graphing the line:

41.   First find the slope: $m = \dfrac{-\dfrac{1}{3} - \dfrac{4}{3}}{\dfrac{1}{4} - \dfrac{3}{2}} = \dfrac{-\dfrac{5}{3}}{-\dfrac{5}{4}} = \dfrac{4}{3}$. Using the point-slope formula:

$$y - \frac{4}{3} = \frac{4}{3}\left(x - \frac{3}{2}\right)$$

$$y - \frac{4}{3} = \frac{4}{3}x - 2$$

$$y = \frac{4}{3}x - \frac{2}{3}$$

43.   Factoring completely: $x^2 + 8x + 16 - y^2 = (x+4)^2 - y^2 = (x+4+y)(x+4-y)$

45.   Rationalizing the denominator: $\dfrac{7}{\sqrt[3]{9}} = \dfrac{7}{\sqrt[3]{9}} \cdot \dfrac{\sqrt[3]{3}}{\sqrt[3]{3}} = \dfrac{7\sqrt[3]{3}}{3}$

47.   The set is $\{2,4\}$.

49.   Let x represent the largest angle, $\dfrac{1}{4}x$ represent the smallest angle, and $\dfrac{1}{4}x + 30$ represent the remaining angle.

The equation is:

$$x + \frac{1}{4}x + \frac{1}{4}x + 30 = 180$$

$$\frac{3}{2}x + 30 = 180$$

$$\frac{3}{2}x = 150$$

$$3x = 300$$

$$x = 100$$

$$\frac{1}{4}x = 25$$

$$\frac{1}{4}x + 30 = 55$$

The angles are $25°, 55°$, and $100°$.

51. The variation equation is $y = \dfrac{K}{x^2}$. Substituting $y = 4$ and $x = \dfrac{5}{3}$:

$$4 = \frac{K}{\left(\dfrac{5}{3}\right)^2}$$

$$K = 4 \cdot \frac{25}{9} = \frac{100}{9}$$

So the equation is $y = \dfrac{100}{9x^2}$. Substituting $x = \dfrac{8}{3}$: $y = \dfrac{100}{9\left(\dfrac{8}{3}\right)^2} = \dfrac{100}{9 \cdot \dfrac{64}{9}} = \dfrac{100}{64} = \dfrac{25}{16}$

53. Finding the increase: $0.205(2,500) = 512.50$

The total sales is: $\$2,500 + \$512.50 = \$3,012.50$

Chapter 8 Test

1. Solving the equation:

$$(2x+4)^2 = 25$$
$$2x+4 = \pm 5$$
$$2x+4 = -5, 5$$
$$2x = -9, 1$$
$$x = -\frac{9}{2}, \frac{1}{2}$$

2. Solving the equation:

$$(2x-6)^2 = -8$$
$$2x-6 = \pm\sqrt{-8}$$
$$2x-6 = \pm 2i\sqrt{2}$$
$$2x = 6 \pm 2i\sqrt{2}$$
$$x = 3 \pm i\sqrt{2}$$

3. Solving the equation:

$$y^2 - 10y + 25 = -4$$
$$(y-5)^2 = -4$$
$$y-5 = \pm 2i$$
$$y = 5 \pm 2i$$

4. Solving the equation:

$$(y+1)(y-3) = -6$$
$$y^2 - 2y - 3 = -6$$
$$y^2 - 2y + 3 = 0$$
$$y = \frac{2 \pm \sqrt{4-12}}{2} = \frac{2 \pm \sqrt{-8}}{2} = \frac{2 \pm 2i\sqrt{2}}{2} = 1 \pm i\sqrt{2}$$

5. Solving the equation:

$$8t^3 - 125 = 0$$
$$(2t-5)(4t^2 + 10t + 25) = 0$$
$$t = \frac{5}{2}, \frac{-10 \pm \sqrt{100-400}}{8} = \frac{-10 \pm \sqrt{-300}}{8} = \frac{-10 \pm 10i\sqrt{3}}{8} = \frac{-5 \pm 5i\sqrt{3}}{4}$$

6.   Solving the equation:

$$\frac{1}{a+2} - \frac{1}{3} = \frac{1}{a}$$

$$3a(a+2)\left(\frac{1}{a+2} - \frac{1}{3}\right) = 3a(a+2)\left(\frac{1}{a}\right)$$

$$3a - a(a+2) = 3(a+2)$$

$$3a - a^2 - 2a = 3a + 6$$

$$a^2 + 2a + 6 = 0$$

$$a = \frac{-2 \pm \sqrt{4-24}}{2} = \frac{-2 \pm \sqrt{-20}}{2} = \frac{-2 \pm 2i\sqrt{5}}{2} = -1 \pm i\sqrt{5}$$

7.   Solving for r:

$$64(1+r)^2 = A$$

$$(1+r)^2 = \frac{A}{64}$$

$$1 + r = \pm\frac{\sqrt{A}}{8}$$

$$r = -1 \pm \frac{\sqrt{A}}{8}$$

8.   Solving by completing the square:

$$x^2 - 4x = -2$$

$$x^2 - 4x + 4 = -2 + 4$$

$$(x-2)^2 = 2$$

$$x - 2 = \pm\sqrt{2}$$

$$x = 2 \pm \sqrt{2}$$

9.   The shorter side is $\frac{4}{2} = 2$ inches and the middle side is $2\sqrt{3}$ inches.

10.   Solving the equation:

$$32t - 16t^2 = 12$$

$$-16t^2 + 32t - 12 = 0$$

$$4t^2 - 8t + 3 = 0$$

$$(2t-1)(2t-3) = 0$$

$$t = \frac{1}{2}, \frac{3}{2}$$

The object will be 12 feet above the ground after $\frac{1}{2}$ or $\frac{3}{2}$ sec.

11.   Setting the profit equal to $200:

$$25x - 0.2x^2 - 2x - 100 = 200$$

$$-0.2x^2 + 23x - 300 = 0$$

$$x^2 - 115x + 1500 = 0$$

$$(x-15)(x-100) = 0$$

$$x = 15, 100$$

The company must sell 15 or 100 cups to make a weekly profit of $200.

12.   First write the equation as $kx^2 - 12x + 4 = 0$. Setting the discriminant equal to 0:

$$(-12)^2 - 4(k)(4) = 0$$

$$144 - 16k = 0$$

$$16k = 144$$

$$k = 9$$

13.   First write the equation as $2x^2 - 5x - 7 = 0$. Finding the discriminant: $D = (-5)^2 - 4(2)(-7) = 25 + 56 = 81$

The equation has two rational solutions.

14.   Finding the equation:

$$(x-5)(3x+2) = 0$$

$$3x^2 - 13x - 10 = 0$$

15.   Finding the equation:

$$(x-2)(x+2)(x-7) = 0$$

$$(x^2-4)(x-7) = 0$$

$$x^3 - 7x^2 - 4x + 28 = 0$$

16. Solving the equation:

$$4x^4 - 7x^2 - 2 = 0$$
$$(x^2 - 2)(4x^2 + 1) = 0$$
$$x^2 = 2, -\frac{1}{4}$$
$$x = \pm\sqrt{2}, \pm\frac{1}{2}i$$

17. Solving the equation:

$$(2t+1)^2 - 5(2t+1) + 6 = 0$$
$$(2t+1-3)(2t+1-2) = 0$$
$$(2t-2)(2t-1) = 0$$
$$t = \frac{1}{2}, 1$$

18. Solving the equation:

$$2t - 7\sqrt{t} + 3 = 0$$
$$(2\sqrt{t} - 1)(\sqrt{t} - 3) = 0$$
$$\sqrt{t} = \frac{1}{2}, 3$$
$$t = \frac{1}{4}, 9$$

19. Solving for t:

$$16t^2 - 14t - h = 0$$

$$t = \frac{14 + \sqrt{196 - 4(16)(-h)}}{32} = \frac{14 + \sqrt{196 + 64h}}{32} = \frac{14 + 4\sqrt{49 + 16h}}{32} = \frac{7 + \sqrt{16h + 49}}{16}$$

20. Completing the square: $y = x^2 - 2x - 3 = (x^2 - 2x + 1) - 1 - 3 = (x-1)^2 - 4$

The vertex is $(1, -4)$. Graphing the parabola:

21. Completing the square: $y = -x^2 + 2x + 8 = -(x^2 - 2x + 1) + 1 + 8 = -(x-1)^2 + 9$

The vertex is $(1, 9)$. Graphing the parabola:

22.   The vertex is $(-1,4)$ and the parabola is pointed down, so the equation is: $y = -(x+1)^2 + 4$

23.   Factoring the inequality:

$$x^2 - x - 6 \le 0$$
$$(x-3)(x+2) \le 0$$

Forming a sign chart:

The solution set is $-2 \le x \le 3$. Graphing the solution set:

24.   Factoring the inequality:

$$2x^2 + 5x > 3$$
$$2x^2 + 5x - 3 > 0$$
$$(x+3)(2x-1) > 0$$

Forming a sign chart:

The solution set is $x < -3$ or $x > \dfrac{1}{2}$. Graphing the solution set:

25.   Finding the profit and completing the square:

$$P = 25x - 0.1x^2 - 5x - 100$$
$$= -0.1x^2 + 20x - 100$$
$$= -0.1\left(x^2 - 200x + 10000\right) + 1000 - 100$$
$$= -0.1(x-100)^2 + 900$$

The maximum weekly profit is $900, obtained by selling 100 items per week.

26.   **a.**   The solution is $x < -2$ or $-1 < x < 1$.

    b.   The solution is $-2 < x < -1$ or $x > 1$.

    c.   The solution is $x = -2, -1, 1$.

Chapter 9
Exponential and Logarithmic Functions

Getting Ready for Chapter 9

1. Evaluating: $2^0 = 1$

2. Evaluating: $2^2 = 2 \cdot 2 = 4$

3. Evaluating: $2^{-2} = \dfrac{1}{2^2} = \dfrac{1}{2 \cdot 2} = \dfrac{1}{4}$

4. Evaluating: $27^{2/3} = \left(27^{1/3}\right)^2 = (3)^2 = 9$

5. Evaluating: $\left(\dfrac{1}{3}\right)^0 = 1$

6. Evaluating: $\left(\dfrac{1}{3}\right)^2 = \dfrac{1}{3} \cdot \dfrac{1}{3} = \dfrac{1}{9}$

7. Evaluating: $\left(\dfrac{1}{3}\right)^{-2} = 3^2 = 9$

8. Calculating: $2^{-1.25} \approx 0.4204$

9. Calculating: $2^{-2.5} \approx 0.1768$

10. Calculating: $1,200 \cdot 2^{-1.25} \approx 504.5378$

11. Calculating: $1,500 \cdot 2^{-2.5} \approx 265.1650$

12. Calculating: $500 \cdot (1.02)^{20} \approx 742.9737$

13. Calculating: $500 \cdot (1.02)^{30} \approx 905.6808$

14. Calculating: $1 + \dfrac{0.09}{12} = 1.0075$

15. Calculating: $1 + \dfrac{0.08}{4} = 1.02$

16. Solving the equation:
$$(x+3)(x) = 2^2$$
$$x^2 + 3x = 4$$
$$x^2 + 3x - 4 = 0$$
$$(x+4)(x-1) = 0$$
$$x = -4, 1$$

17. Solving the equation:
$$\dfrac{x+1}{x-4} = 25$$
$$x+1 = 25(x-4)$$
$$x+1 = 25x - 100$$
$$101 = 24x$$
$$r = \dfrac{101}{24}$$

18. Solving for y:
$$x = 2y - 3$$
$$x + 3 = 2y$$
$$y = \dfrac{x+3}{2}$$

19. Solving for y:
$$x = y^2 - 2$$
$$x + 2 = y^2$$
$$y = \pm\sqrt{x+2}$$

20.   Solving for y:

$$(2y+4)^2 + y^2 = 4$$
$$4y^2 + 16y + 16 + y^2 = 4$$
$$5y^2 + 16y + 12 = 0$$
$$(5y+6)(y+2) = 0$$
$$y = -2, -\frac{6}{5}$$

9.1 Exponential Functions

1.   Evaluating: $g(0) = \left(\frac{1}{2}\right)^0 = 1$

3.   Evaluating: $g(-1) = \left(\frac{1}{2}\right)^{-1} = 2$

5.   Evaluating: $f(-3) = 3^{-3} = \frac{1}{27}$

7.   Evaluating: $f(2) + g(-2) = 3^2 + \left(\frac{1}{2}\right)^{-2} = 9 + 4 = 13$

9.   Graphing the function:

11.   Graphing the function:

13.   Graphing the function:

15.   Graphing the function:

17.   Graphing the functions:

19.   Graphing the functions:

21.   The equation is: $h(n) = 6\left(\dfrac{2}{3}\right)^n$

Substituting $n = 5$: $h(5) = 6\left(\dfrac{2}{3}\right)^5 \approx 0.79$ feet

23.   After 8 days, there will be: $1400 \cdot 2^{-8/8} = 1400 \cdot \dfrac{1}{2} = 700$ micrograms

After 11 days, there will be: $1400 \cdot 2^{-11/8} \approx 539.8$ micrograms

25.   The equation is $I(t) = 600 \cdot 2^{-t/2}$. Evaluating the function:

$$I(4) = 600 \cdot 2^{-4/2} = 600 \cdot 2^{-2} = 150 \text{ mg}$$

$$I(8) = 600 \cdot 2^{-8/2} = 600 \cdot 2^{-4} = 37.5 \text{ mg}$$

After 4 hours there are 150 mg of ibuprofin, and after 8 hours there are 37.5 mg of ibuprofin.

27.   **a.**   The equation is $A(t) = 1200\left(1 + \dfrac{0.06}{4}\right)^{4t}$.   **b.**   Substitute $t = 8$: $A(8) = 1200\left(1 + \dfrac{0.06}{4}\right)^{32} \approx \$1{,}932.39$

c.   Using a graphing calculator, the time is approximately 11.6 years.

d.   Substitute $t = 8$ into the compound interest formula: $A(8) = 1200e^{0.06 \times 8} \approx \$1{,}939.29$

29.   **a.**   Substitute $t = 20$: $E(20) = 78.16(1.11)^{20} \approx \$630$ billion. The estimate is $69 billion too low.

b.   For 2008, substitute $t = 38$: $E(38) = 78.16(1.11)^{38} \approx \$4{,}123$ billion

For 2009, substitute $t = 39$: $E(39) = 78.16(1.11)^{39} \approx \$4{,}577$ billion

For 2010, substitute $t = 40$: $E(40) = 78.16(1.11)^{40} \approx \$5{,}080$ billion

31.   Graphing the function:

33.   The painting will be worth $600 after approximately 6 years.

35.   **a.**   Substitute $t = 3.5$: $V(5) = 450,000(1 - 0.30)^5 \approx \$129,138.48$

    b.   The domain is $\{t \mid 0 \le t \le 6\}$.

    c.   Sketching the graph:

$V(t) = 450,000(1 - 0.30)^t$

    d.   The range is $\{V(t) \mid 52,942.05 \le V(t) \le 450,000\}$.

    e.   From the graph, the crane will be worth $85,000 after approximately 4.7 years, or 4 years 8 months.

37.   The domain is $\{1, 3, 4\}$ and the range is $\{1, 2, 4\}$. This is a function.

39.   Find where the quantity inside the radical is non-negative:

$$3x + 1 \ge 0$$
$$3x \ge -1$$
$$x \ge -\frac{1}{3}$$

The domain is $\left\{x \mid x \ge -\frac{1}{3}\right\}$.

41.   Evaluating the function: $f(0) = 2(0)^2 - 18 = 0 - 18 = -18$

43.   Simplifying the function: $\dfrac{g(x+h) - g(x)}{h} = \dfrac{2(x+h) - 6 - (2x - 6)}{h} = \dfrac{2x + 2h - 6 - 2x + 6}{h} = \dfrac{2h}{h} = 2$

45.   Factoring: $2x^2 + 5x - 12 = (2x - 3)(x + 4)$

47.   Factoring: $5x^3 - 22x^2 + 8x = x(5x^2 - 22x + 8) = x(5x - 2)(x - 4)$

49.   Factoring: $x^4 - 64x = x(x^3 - 64) = x(x - 4)(x^2 + 4x + 16)$

51.   Solving for y:

$$x = 2y - 3$$
$$2y = x + 3$$
$$y = \frac{x + 3}{2}$$

53.   Solving for y:

$$x = y^2 - 2$$
$$y^2 = x + 2$$
$$y = \pm\sqrt{x + 2}$$

55.   Solving for y:

$$x = \frac{y - 4}{y - 2}$$
$$x(y - 2) = y - 4$$
$$xy - 2x = y - 4$$
$$xy - y = 2x - 4$$
$$y(x - 1) = 2x - 4$$
$$y = \frac{2x - 4}{x - 1}$$

57.   Solving for y:

$$x = \sqrt{y - 3}$$
$$x^2 = y - 3$$
$$y = x^2 + 3$$

59. **a.** Completing the table:

x	y
0	−6
3	0
5	4
8	10

b. Completing the table:

x	y
0	3
4	5
10	8
−6	0

61. Graphing the function:

$s(t) = 250(1 - 1.5^{-t})$

9.2 The Inverse of a Function

1. Let $y = f(x)$. Switch x and y and solve for y:

$$3y - 1 = x$$
$$3y = x + 1$$
$$y = \frac{x+1}{3}$$

The inverse is $f^{-1}(x) = \frac{x+1}{3}$.

3. Let $y = f(x)$. Switch x and y and solve for y:

$$y^3 = x$$
$$y = \sqrt[3]{x}$$

The inverse is $f^{-1}(x) = \sqrt[3]{x}$.

5. Let $y = f(x)$. Switch x and y and solve for y:

$$\frac{y-3}{y-1} = x$$
$$y - 3 = xy - x$$
$$y - xy = 3 - x$$
$$y(1 - x) = 3 - x$$
$$y = \frac{3-x}{1-x} = \frac{x-3}{x-1}$$

The inverse is $f^{-1}(x) = \frac{x-3}{x-1}$.

7. Let $y = f(x)$. Switch x and y and solve for y:

$$\frac{y-3}{4} = x$$
$$y - 3 = 4x$$
$$y = 4x + 3$$

The inverse is $f^{-1}(x) = 4x + 3$.

9. Let $y = f(x)$. Switch x and y and solve for y:

$$\frac{1}{2}y - 3 = x$$
$$y - 6 = 2x$$
$$y = 2x + 6$$

The inverse is $f^{-1}(x) = 2x + 6$.

11. Let $y = f(x)$. Switch x and y and solve for y:

$$\frac{2y+1}{3y+1} = x$$
$$2y + 1 = 3xy + x$$
$$2y - 3xy = x - 1$$
$$y(2 - 3x) = x - 1$$
$$y = \frac{x-1}{2-3x} = \frac{1-x}{3x-2}$$

The inverse is $f^{-1}(x) = \frac{1-x}{3x-2}$.

13. Finding the inverse:

$$2y - 1 = x$$
$$2y = x + 1$$
$$y = \frac{x+1}{2}$$

The inverse is $y^{-1} = \dfrac{x+1}{2}$. Graphing each curve:

15. Finding the inverse:

$$y^2 - 3 = x$$
$$y^2 = x + 3$$
$$y = \pm\sqrt{x+3}$$

The inverse is $y^{-1} = \pm\sqrt{x+3}$. Graphing each curve:

17. Finding the inverse:

$$y^2 - 2y - 3 = x$$
$$y^2 - 2y + 1 = x + 3 + 1$$
$$(y-1)^2 = x + 4$$
$$y - 1 = \pm\sqrt{x+4}$$
$$y = 1 \pm\sqrt{x+4}$$

The inverse is $y^{-1} = 1 \pm\sqrt{x+4}$. Graphing each curve:

19. The inverse is $x = 3^y$. Graphing each curve:

21. The inverse is $x = 4$. Graphing each curve:

23. Finding the inverse:

$$\frac{1}{2}y^3 = x$$
$$y^3 = 2x$$
$$y = \sqrt[3]{2x}$$

The inverse is $y^{-1} = \sqrt[3]{2x}$. Graphing each curve:

25. Finding the inverse:

$$\frac{1}{2}y + 2 = x$$
$$y + 4 = 2x$$
$$y = 2x - 4$$

The inverse is $y^{-1} = 2x - 4$. Graphing each curve:

27. Finding the inverse:

$$\sqrt{y + 2} = x$$
$$y + 2 = x^2$$
$$y = x^2 - 2$$

The inverse is $y^{-1} = x^2 - 2, x \geq 0$. Graphing each curve:

29. **a.** Yes, this function is one-to-one.   **b.** No, this function is not one-to-one.
     c. Yes, this function is one-to-one.

31. **a.** Evaluating the function: $f(2) = 3(2) - 2 = 6 - 2 = 4$

     b. Evaluating the function: $f^{-1}(2) = \dfrac{2+2}{3} = \dfrac{4}{3}$

     c. Evaluating the function: $f\left(f^{-1}(2)\right) = f\left(\dfrac{4}{3}\right) = 3\left(\dfrac{4}{3}\right) - 2 = 4 - 2 = 2$

     d. Evaluating the function: $f^{-1}\left(f(2)\right) = f^{-1}(4) = \dfrac{4+2}{3} = \dfrac{6}{3} = 2$

33. Let $y = f(x)$. Switch x and y and solve for y:

$$\frac{1}{y} = x$$
$$y = \frac{1}{x}$$

The inverse is $f^{-1}(x) = \dfrac{1}{x}$.

35. **a.** The value is -3.   **b.** The value is -6.
     c. The value is 2.   **d.** The value is 3.
     e. The value is -2.   **f.** The value is 3.
     g. Each is an inverse of the other.

37. Solving the equation:

$$(2x-1)^2 = 25$$
$$2x - 1 = \pm\sqrt{25}$$
$$2x - 1 = -5, 5$$
$$2x = -4, 6$$
$$x = -2, 3$$

39. The number is 25, since $x^2 - 10x + 25 = (x-5)^2$.

41. Solving the equation:

$$x^2 - 10x + 8 = 0$$
$$x^2 - 10x + 25 = -8 + 25$$
$$(x-5)^2 = 17$$
$$x - 5 = \pm\sqrt{17}$$
$$x = 5 \pm \sqrt{17}$$

43. Solving the equation:

$$3x^2 - 6x + 6 = 0$$
$$x^2 - 2x + 2 = 0$$
$$x^2 - 2x + 1 = -2 + 1$$
$$(x-1)^2 = -1$$
$$x - 1 = \pm i$$
$$x = 1 \pm i$$

45. Simplifying: $3^{-2} = \dfrac{1}{3^2} = \dfrac{1}{9}$

47. Solving the equation:

$$2 = 3x$$
$$x = \frac{2}{3}$$

49. Solving the equation:

$$4 = x^3$$
$$x = \sqrt[3]{4}$$

51. Completing the statement: $8 = 2^3$

53. Completing the statement: $10,000 = 10^4$

55. Completing the statement: $81 = 3^4$

57. Completing the statement: $6 = 6^1$

59. Finding the inverse:

$$3y + 5 = x$$
$$3y = x - 5$$
$$y = \frac{x - 5}{3}$$

So $f^{-1}(x) = \frac{x-5}{3}$. Now verifying the inverse: $f\left[f^{-1}(x)\right] = f\left(\frac{x-5}{3}\right) = 3\left(\frac{x-5}{3}\right) + 5 = x - 5 + 5 = x$

61. Finding the inverse:

$$y^3 + 1 = x$$
$$y^3 = x - 1$$
$$y = \sqrt[3]{x - 1}$$

So $f^{-1}(x) = \sqrt[3]{x-1}$. Now verifying the inverse: $f\left[f^{-1}(x)\right] = f\left(\sqrt[3]{x-1}\right) = \left(\sqrt[3]{x-1}\right)^3 + 1 = x - 1 + 1 = x$

63. Finding the inverse:

$$\frac{y-4}{y-2} = x$$
$$y - 4 = xy - 2x$$
$$y - xy = 4 - 2x$$
$$y(1 - x) = 4 - 2x$$
$$y = \frac{4 - 2x}{1 - x} = \frac{2x - 4}{x - 1}$$

So $f^{-1}(x) = \frac{2x-4}{x-1}$. Now verifying the inverse:

$$f\left[f^{-1}(x)\right] = f\left(\frac{2x-4}{x-1}\right) = \frac{\frac{2x-4}{x-1} - 4}{\frac{2x-4}{x-1} - 2} = \frac{2x-4-4(x-1)}{2x-4-2(x-1)} = \frac{2x-4-4x+4}{2x-4-2x+2} = \frac{-2x}{-2} = x$$

65.
  a. From the graph: $f(0) = 1$
  b. From the graph: $f(1) = 2$
  c. From the graph: $f(2) = 5$
  d. From the graph: $f^{-1}(1) = 0$
  e. From the graph: $f^{-1}(2) = 1$
  f. From the graph: $f^{-1}(5) = 2$
  g. From the graph: $f^{-1}(f(2)) = 2$
  h. From the graph: $f(f^{-1}(5)) = 5$

9.3 Logarithms Are Exponents

1. Writing in logarithmic form: $\log_2 16 = 4$

3. Writing in logarithmic form: $\log_5 125 = 3$

5. Writing in logarithmic form: $\log_{10} 0.01 = -2$

7. Writing in logarithmic form: $\log_2 \frac{1}{32} = -5$

9. Writing in logarithmic form: $\log_{1/2} 8 = -3$

11. Writing in logarithmic form: $\log_3 27 = 3$

13. Writing in exponential form: $10^2 = 100$

15. Writing in exponential form: $2^6 = 64$

17. Writing in exponential form: $8^0 = 1$

19. Writing in exponential form: $10^{-3} = 0.001$

21. Writing in exponential form: $6^2 = 36$

23. Writing in exponential form: $5^{-2} = \frac{1}{25}$

25. Solving the equation:

$$\log_3 x = 2$$
$$x = 3^2 = 9$$

27. Solving the equation:

$$\log_5 x = -3$$
$$x = 5^{-3} = \frac{1}{125}$$

29.   Solving the equation:
$$\log_2 16 = x$$
$$2^x = 16$$
$$x = 4$$

31.   Solving the equation:
$$\log_8 2 = x$$
$$8^x = 2$$
$$x = \frac{1}{3}$$

33.   Solving the equation:
$$\log_x 4 = 2$$
$$x^2 = 4$$
$$x = 2$$

35.   Solving the equation:
$$\log_x 5 = 3$$
$$x^3 = 5$$
$$x = \sqrt[3]{5}$$

37.   Sketching the graph:

39.   Sketching the graph:

41.   Sketching the graph:

43.   Sketching the graph:

45.   Simplifying the logarithm:
$$x = \log_2 16$$
$$2^x = 16$$
$$x = 4$$

47.   Simplifying the logarithm:
$$x = \log_{25} 125$$
$$25^x = 125$$
$$5^{2x} = 5^3$$
$$2x = 3$$
$$x = \frac{3}{2}$$

49.   Simplifying the logarithm:
$$x = \log_{10} 1000$$
$$10^x = 1000$$
$$x = 3$$

51.   Simplifying the logarithm:
$$x = \log_3 3$$
$$3^x = 3$$
$$x = 1$$

53.   Simplifying the logarithm:
$$x = \log_5 1$$
$$5^x = 1$$
$$x = 0$$

55. First find $\log_6 6$:

$$x = \log_6 6$$
$$6^x = 6$$
$$x = 1$$

Now find $\log_3 1$:

$$x = \log_3 1$$
$$3^x = 1$$
$$x = 0$$

57. First find $\log_2 16$:

$$x = \log_2 16$$
$$2^x = 16$$
$$x = 4$$

Now find $\log_4 2$:

$$x = \log_4 2$$
$$4^x = 2$$
$$2^{2x} = 2$$
$$2x = 1$$
$$x = \frac{1}{2}$$

Now find $\log_2 4$:

$$x = \log_2 4$$
$$2^x = 4$$
$$x = 2$$

59. The pH is given by: $\text{pH} = -\log_{10}\left(10^{-7}\right) = -(-7) = 7$

61. The $\left[\text{H}^+\right]$ is given by:

$$-\log_{10}\left[\text{H}^+\right] = 6$$
$$\log_{10}\left[\text{H}^+\right] = -6$$
$$\left[\text{H}^+\right] = 10^{-6}$$

63. Using the relationship $M = \log_{10} T$:

$$M = \log_{10} 100$$
$$10^M = 100$$
$$M = 2$$

65. It is 10^8 times as large.

67. Completing the square: $x^2 + 10x + 25 = (x+5)^2$

69. Completing the square: $y^2 - 2y + 1 = (y-1)^2$

71. Solving the equation:

$$-y^2 = 9$$
$$y^2 = -9$$
$$y = \pm\sqrt{-9} = \pm 3i$$

73. Solving the equation:

$$-x^2 - 8 = -4$$
$$-x^2 = 4$$
$$x^2 = -4$$
$$x = \pm\sqrt{-4} = \pm 2i$$

75. Solving the equation:

$$2x^2 + 4x - 3 = 0$$
$$x = \frac{-4 \pm \sqrt{16 + 24}}{4} = \frac{-4 \pm \sqrt{40}}{4} = \frac{-4 \pm 2\sqrt{10}}{4} = \frac{-2 \pm \sqrt{10}}{2}$$

77. Solving the equation:

$$(2y - 3)(2y - 1) = -4$$
$$4y^2 - 8y + 3 = -4$$
$$4y^2 - 8y + 7 = 0$$
$$y = \frac{8 \pm \sqrt{64 - 112}}{8} = \frac{8 \pm \sqrt{-48}}{8} = \frac{8 \pm 4i\sqrt{3}}{8} = \frac{2 \pm i\sqrt{3}}{2}$$

79. Solving the equation:

$$t^3 - 125 = 0$$
$$(t - 5)\left(t^2 + 5t + 25\right) = 0$$
$$t = 5, \frac{-5 \pm \sqrt{25 - 100}}{2} = \frac{-5 \pm \sqrt{-75}}{2} = \frac{-5 \pm 5i\sqrt{3}}{2}$$

81.   Solving the equation:

$$4x^5 - 16x^4 = 20x^3$$
$$4x^5 - 16x^4 - 20x^3 = 0$$
$$4x^3\left(x^2 - 4x - 5\right) = 0$$
$$4x^3\left(x - 5\right)\left(x + 1\right) = 0$$
$$x = -1, 0, 5$$

83.   Solving the equation:

$$\frac{1}{x-3} + \frac{1}{x+2} = 1$$
$$x + 2 + x - 3 = (x-3)(x+2)$$
$$2x - 1 = x^2 - x - 6$$
$$x^2 - 3x - 5 = 0$$
$$x = \frac{3 \pm \sqrt{9+20}}{2} = \frac{3 \pm \sqrt{29}}{2}$$

85.   Simplifying; $8^{2/3} = \left(8^{1/3}\right)^2 = \left(\sqrt[3]{8}\right)^2 = 2^2 = 4$

87.   Solving the equation:

$$(x+2)(x) = 2^3$$
$$x^2 + 2x = 8$$
$$x^2 + 2x - 8 = 0$$
$$(x+4)(x-2) = 0$$
$$x = -4, 2$$

89.   Solving the equation:

$$\frac{x-2}{x+1} = 9$$
$$x - 2 = 9(x+1)$$
$$x - 2 = 9x + 9$$
$$-8x = 11$$
$$x = -\frac{11}{8}$$

91.   Writing in exponential form: $2^3 = (x+2)(x)$

93.   Writing in exponential form: $3^4 = \dfrac{x-2}{x+1}$

95.   **a.**   Completing the table:

x	-1	0	1	2
$f(x)$	$\frac{1}{8}$	1	8	64

b.   Completing the table:

x	$\frac{1}{8}$	1	8	64
$f^{-1}(x)$	-1	0	1	2

c.   The equation is $f(x) = 8^x$.

d.   The equation is $f^{-1}(x) = \log_8 x$.

9.4 Properties of Logarithms

1.   Using properties of logarithms: $\log_3 4x = \log_3 4 + \log_3 x$

3.   Using properties of logarithms: $\log_6 \dfrac{5}{x} = \log_6 5 - \log_6 x$

5.   Using properties of logarithms: $\log_2 y^5 = 5\log_2 y$

7.   Using properties of logarithms: $\log_9 \sqrt[3]{z} = \log_9 z^{1/3} = \dfrac{1}{3}\log_9 z$

9.   Using properties of logarithms: $\log_6 x^2 y^4 = \log_6 x^2 + \log_6 y^4 = 2\log_6 x + 4\log_6 y$

11.   Using properties of logarithms: $\log_5 \sqrt{x} \bullet y^4 = \log_5 x^{1/2} + \log_5 y^4 = \dfrac{1}{2}\log_5 x + 4\log_5 y$

13.   Using properties of logarithms: $\log_b \dfrac{xy}{z} = \log_b xy - \log_b z = \log_b x + \log_b y - \log_b z$

15.   Using properties of logarithms: $\log_{10} \dfrac{4}{xy} = \log_{10} 4 - \log_{10} xy = \log_{10} 4 - \log_{10} x - \log_{10} y$

17.   Using properties of logarithms: $\log_{10} \dfrac{x^2 y}{\sqrt{z}} = \log_{10} x^2 + \log_{10} y - \log_{10} z^{1/2} = 2\log_{10} x + \log_{10} y - \dfrac{1}{2}\log_{10} z$

19.   Using properties of logarithms: $\log_{10} \dfrac{x^3 \sqrt{y}}{z^4} = \log_{10} x^3 + \log_{10} y^{1/2} - \log_{10} z^4 = 3\log_{10} x + \dfrac{1}{2}\log_{10} y - 4\log_{10} z$

21.   Using properties of logarithms:

$$\log_b \sqrt[3]{\dfrac{x^2 y}{z^4}} = \log_b \dfrac{x^{2/3} y^{1/3}}{z^{4/3}} = \log_b x^{2/3} + \log_b y^{1/3} - \log_b z^{4/3} = \dfrac{2}{3}\log_b x + \dfrac{1}{3}\log_b y - \dfrac{4}{3}\log_b z$$

23.   Writing as a single logarithm: $\log_b x + \log_b z = \log_b xz$

25.   Writing as a single logarithm: $2\log_3 x - 3\log_3 y = \log_3 x^2 - \log_3 y^3 = \log_3 \dfrac{x^2}{y^3}$

27.   Writing as a single logarithm: $\dfrac{1}{2}\log_{10} x + \dfrac{1}{3}\log_{10} y = \log_{10} x^{1/2} + \log_{10} y^{1/3} = \log_{10} \sqrt{x}\,\sqrt[3]{y}$

29.   Writing as a single logarithm: $3\log_2 x + \dfrac{1}{2}\log_2 y - \log_2 z = \log_2 x^3 + \log_2 y^{1/2} - \log_2 z = \log_2 \dfrac{x^3 \sqrt{y}}{z}$

31.   Writing as a single logarithm: $\dfrac{1}{2}\log_2 x - 3\log_2 y - 4\log_2 z = \log_2 x^{1/2} - \log_2 y^3 - \log_2 z^4 = \log_2 \dfrac{\sqrt{x}}{y^3 z^4}$

33.   Writing as a single logarithm:

$$\dfrac{3}{2}\log_{10} x - \dfrac{3}{4}\log_{10} y - \dfrac{4}{5}\log_{10} z = \log_{10} x^{3/2} - \log_{10} y^{3/4} - \log_{10} z^{4/5} = \log_{10} \dfrac{x^{3/2}}{y^{3/4} z^{4/5}}$$

35.   Solving the equation:

$$\log_2 x + \log_2 3 = 1$$
$$\log_2 3x = 1$$
$$3x = 2^1$$
$$3x = 2$$
$$x = \dfrac{2}{3}$$

37.   Solving the equation:

$$\log_3 x - \log_3 2 = 2$$
$$\log_3 \dfrac{x}{2} = 2$$
$$\dfrac{x}{2} = 3^2$$
$$\dfrac{x}{2} = 9$$
$$x = 18$$

39.   Solving the equation:

$$\log_3 x + \log_3 (x-2) = 1$$
$$\log_3 (x^2 - 2x) = 1$$
$$x^2 - 2x = 3^1$$
$$x^2 - 2x - 3 = 0$$
$$(x-3)(x+1) = 0$$
$$x = 3, -1$$

The solution is 3 (–1 does not check).

41.   Solving the equation:

$$\log_3 (x+3) - \log_3 (x-1) = 1$$
$$\log_3 \dfrac{x+3}{x-1} = 1$$
$$\dfrac{x+3}{x-1} = 3^1$$
$$x + 3 = 3x - 3$$
$$-2x = -6$$
$$x = 3$$

43.   Solving the equation:

$$\log_2 x + \log_2 (x-2) = 3$$
$$\log_2 (x^2 - 2x) = 3$$
$$x^2 - 2x = 2^3$$
$$x^2 - 2x - 8 = 0$$
$$(x-4)(x+2) = 0$$
$$x = 4, -2$$

The solution is 4 (–2 does not check).

45.   Solving the equation:

$$\log_8 x + \log_8 (x-3) = \dfrac{2}{3}$$
$$\log_8 (x^2 - 3x) = \dfrac{2}{3}$$
$$x^2 - 3x = 8^{2/3}$$
$$x^2 - 3x - 4 = 0$$
$$(x-4)(x+1) = 0$$
$$x = 4, -1$$

The solution is 4 (–1 does not check).

47. Solving the equation:

$$\log_5 \sqrt{x} + \log_5 \sqrt{6x+5} = 1$$
$$\log_5 \sqrt{6x^2 + 5x} = 1$$
$$\frac{1}{2}\log_5 \left(6x^2 + 5x\right) = 1$$
$$\log_5 \left(6x^2 + 5x\right) = 2$$
$$6x^2 + 5x = 5^2$$
$$6x^2 + 5x - 25 = 0$$
$$(3x - 5)(2x + 5) = 0$$
$$x = \frac{5}{3}, -\frac{5}{2}$$

The solution is $\frac{5}{3}$ ($-\frac{5}{2}$ does not check).

49. Rewriting the formula:

$$D = 10\log_{10}\left(\frac{I}{I_0}\right)$$
$$D = 10\left(\log_{10} I - \log_{10} I_0\right)$$

51. Solving for N:

$$N = \log_{10}\frac{100}{1} = \log_{10}10^2 = 2 \qquad N = \log_{10}100 - \log_{10}1 = \log_{10}10^2 = 2$$

So $N = 2$ in both cases.

53. Dividing: $\frac{12x^2 + y^2}{36} = \frac{12x^2}{36} + \frac{y^2}{36} = \frac{x^2}{3} + \frac{y^2}{36}$ **55.** Dividing: $\frac{25x^2 + 4y^2}{100} = \frac{25x^2}{100} + \frac{4y^2}{100} = \frac{x^2}{4} + \frac{y^2}{25}$

57. Computing the discriminant: $D = (-5)^2 - 4(2)(4) = 25 - 32 = -7$. There are two complex solutions.

59. Writing the equation:
$$(x+3)(x-5) = 0$$
$$x^2 - 2x - 15 = 0$$

61. Writing the equation:
$$(3y-2)(y-3) = 0$$
$$3y^2 - 11y + 6 = 0$$

63. Simplifying: $5^0 = 1$

65. Simplifying: $\log_3 3 = \log_3 3^1 = 1$

67. Simplifying: $\log_b b^4 = 4$

69. Using a calculator: $10^{-5.6} \approx 2.5 \times 10^{-6}$

71. Using a calculator: $\frac{2.00 \times 10^8}{3.96 \times 10^6} \approx 51$

9.5 Common Logarithms and Natural Logarithms

1. Evaluating the logarithm: $\log 378 \approx 2.5775$

3. Evaluating the logarithm: $\log 37.8 \approx 1.5775$

5. Evaluating the logarithm: $\log 3,780 \approx 3.5775$

7. Evaluating the logarithm: $\log 0.0378 \approx -1.4225$

9. Evaluating the logarithm: $\log 37,800 \approx 4.5775$

11. Evaluating the logarithm: $\log 600 \approx 2.7782$

13. Evaluating the logarithm: $\log 2,010 \approx 3.3032$

15. Evaluating the logarithm: $\log 0.00971 \approx -2.0128$

17. Evaluating the logarithm: $\log 0.0314 \approx -1.5031$

19. Evaluating the logarithm: $\log 0.399 \approx -0.3990$

21. Solving for x:
$$\log x = 2.8802$$
$$x = 10^{2.8802} \approx 759$$

23. Solving for x:
$$\log x = -2.1198$$
$$x = 10^{-2.1198} \approx 0.00759$$

25. Solving for x:
$$\log x = 3.1553$$
$$x = 10^{3.1553} \approx 1,430$$

27. Solving for x:
$$\log x = -5.3497$$
$$x = 10^{-5.3497} \approx 0.00000447$$

29.   Solving for x:
$$\log x = -7.0372$$
$$x = 10^{-7.0372} \approx 0.0000000918$$

31.   Solving for x:
$$\log x = 10$$
$$x = 10^{10}$$

33.   Solving for x:
$$\log x = -10$$
$$x = 10^{-10}$$

35.   Solving for x:
$$\log x = 20$$
$$x = 10^{20}$$

37.   Solving for x:
$$\log x = -2$$
$$x = 10^{-2} = \frac{1}{100}$$

39.   Solving for x:
$$\log x = \log_2 8$$
$$\log x = 3$$
$$x = 10^3 = 1,000$$

41.   Simplifying the logarithm: $\ln e = \ln e^1 = 1$

43.   Simplifying the logarithm: $\ln e^5 = 5$

45.   Simplifying the logarithm: $\ln e^x = x$

47.   Using properties of logarithms: $\ln 10 e^{3t} = \ln 10 + \ln e^{3t} = \ln 10 + 3t$

49.   Using properties of logarithms: $\ln A e^{-2t} = \ln A + \ln e^{-2t} = \ln A - 2t$

51.   Evaluating the logarithm: $\ln 15 = \ln(3 \bullet 5) = \ln 3 + \ln 5 = 1.0986 + 1.6094 = 2.7080$

53.   Evaluating the logarithm: $\ln \dfrac{1}{3} = \ln 3^{-1} = -\ln 3 = -1.0986$

55.   Evaluating the logarithm: $\ln 9 = \ln 3^2 = 2 \ln 3 = 2(1.0986) = 2.1972$

57.   Evaluating the logarithm: $\ln 16 = \ln 2^4 = 4 \ln 2 = 4(0.6931) = 2.7724$

59.   **a.**   Using a calculator: $\dfrac{\log 25}{\log 15} \approx 1.1886$

  b.   Using a calculator: $\log \dfrac{25}{15} \approx 0.2218$

61.   **a.**   Using a calculator: $\dfrac{\log 4}{\log 8} \approx 0.6667$

  b.   Using a calculator: $\log \dfrac{4}{8} \approx -0.3010$

63.   Computing the pH: $\text{pH} = -\log\left(6.50 \times 10^{-4}\right) \approx 3.19$

65.   Finding the concentration:
$$4.75 = -\log\left[\text{H}^+\right]$$
$$-4.75 = \log\left[\text{H}^+\right]$$
$$\left[\text{H}^+\right] = 10^{-4.75} \approx 1.78 \times 10^{-5}$$

67.   Finding the magnitude:
$$5.5 = \log T$$
$$T = 10^{5.5} \approx 3.16 \times 10^5$$

69.   Finding the magnitude:
$$8.3 = \log T$$
$$T = 10^{8.3} \approx 2.00 \times 10^8$$

71.   For the first earthquake:
$$\log T_1 = 6.5$$
$$T_1 = 10^{6.5}$$

For the second earthquake:
$$\log T_2 = 5.5$$
$$T_2 = 10^{5.5}$$

The ratio is $\dfrac{T_1}{T_2} = \dfrac{10^{6.5}}{10^{5.5}} = 10$ times stronger.

73.   Completing the table:

Location	Date	Magnitude (M)	Shockwave (T)
Moresby Island	January 23	4.0	1.00×10^4
Vancouver Island	April 30	5.3	1.99×10^5
Quebec City	June 29	3.2	1.58×10^3
Mould Bay	November 13	5.2	1.58×10^5
St. Lawrence	December 14	3.7	5.01×10^3

75. Finding the rate of depreciation:

$$\log(1-r) = \frac{1}{5}\log\frac{4500}{9000}$$
$$\log(1-r) \approx -0.0602$$
$$1-r \approx 10^{-0.0602}$$
$$r = 1-10^{-0.0602}$$
$$r \approx 0.129 = 12.9\%$$

77. Finding the rate of depreciation:

$$\log(1-r) = \frac{1}{5}\log\frac{5750}{7550}$$
$$\log(1-r) \approx -0.0237$$
$$1-r \approx 10^{-0.0237}$$
$$r = 1-10^{-0.0237}$$
$$r \approx 0.053 = 5.3\%$$

x	$(1+x)^{1/x}$
1	2
0.5	2.25
0.1	2.5937
0.01	2.7048
0.001	2.7169
0.0001	2.7181
0.00001	2.7183

79. It appears to approach e. Completing the table:

81. Solving the equation:

$$(y+3)^2 + y^2 = 9$$
$$y^2 + 6y + 9 + y^2 = 9$$
$$2y^2 + 6y = 0$$
$$2y(y+3) = 0$$
$$y = -3, 0$$

83. Solving the equation:

$$(x+3)^2 + 1^2 = 2$$
$$x^2 + 6x + 9 + 1 = 2$$
$$x^2 + 6x + 8 = 0$$
$$(x+4)(x+2) = 0$$
$$x = -4, -2$$

85. Solving the equation:

$$x^4 - 2x^2 - 8 = 0$$
$$(x^2 - 4)(x^2 + 2) = 0$$
$$x^2 = 4, -2$$
$$x = \pm 2, \pm i\sqrt{2}$$

87. Solving the equation:

$$2x - 5\sqrt{x} + 3 = 0$$
$$(2\sqrt{x} - 3)(\sqrt{x} - 1) = 0$$
$$\sqrt{x} = 1, \frac{3}{2}$$
$$x = 1, \frac{9}{4}$$

89. Solving the equation:

$$5(2x+1) = 12$$
$$10x + 5 = 12$$
$$10x = 7$$
$$x = \frac{7}{10} = 0.7$$

91. Using a calculator: $\dfrac{100,000}{32,000} = 3.125$

93. Using a calculator: $\dfrac{1}{2}\left(\dfrac{-0.6931}{1.4289} + 3\right) \approx 1.2575$

95. Rewriting the logarithm: $\log 1.05^t = t\log 1.05$

97. Simplifying: $\ln e^{0.05t} = 0.05t$

9.6 Exponential Equations and Change of Base

1. Solving the equation:
    $$3^x = 5$$
    $$\ln 3^x = \ln 5$$
    $$x \ln 3 = \ln 5$$
    $$x = \frac{\ln 5}{\ln 3} \approx 1.4650$$

3. Solving the equation:
    $$5^x = 3$$
    $$\ln 5^x = \ln 3$$
    $$x \ln 5 = \ln 3$$
    $$x = \frac{\ln 3}{\ln 5} \approx 0.6826$$

5. Solving the equation:
    $$5^{-x} = 12$$
    $$\ln 5^{-x} = \ln 12$$
    $$-x \ln 5 = \ln 12$$
    $$x = -\frac{\ln 12}{\ln 5} \approx -1.5440$$

7. Solving the equation:
    $$12^{-x} = 5$$
    $$\ln 12^{-x} = \ln 5$$
    $$-x \ln 12 = \ln 5$$
    $$x = -\frac{\ln 5}{\ln 12} \approx -0.6477$$

9. Solving the equation:
    $$8^{x+1} = 4$$
    $$2^{3x+3} = 2^2$$
    $$3x + 3 = 2$$
    $$3x = -1$$
    $$x = -\frac{1}{3} \approx -0.3333$$

11. Solving the equation:

    $$4^{x-1} = 4$$
    $$4^{x-1} = 4^1$$
    $$x - 1 = 1$$
    $$x = 2$$

13. Solving the equation:
    $$3^{2x+1} = 2$$
    $$\ln 3^{2x+1} = \ln 2$$
    $$(2x+1)\ln 3 = \ln 2$$
    $$2x + 1 = \frac{\ln 2}{\ln 3}$$
    $$2x = \frac{\ln 2}{\ln 3} - 1$$
    $$x = \frac{1}{2}\left(\frac{\ln 2}{\ln 3} - 1\right) \approx -0.1845$$

15. Solving the equation:
    $$3^{1-2x} = 2$$
    $$\ln 3^{1-2x} = \ln 2$$
    $$(1-2x)\ln 3 = \ln 2$$
    $$1 - 2x = \frac{\ln 2}{\ln 3}$$
    $$-2x = \frac{\ln 2}{\ln 3} - 1$$
    $$x = \frac{1}{2}\left(1 - \frac{\ln 2}{\ln 3}\right) \approx 0.1845$$

17. Solving the equation:
    $$15^{3x-4} = 10$$
    $$\ln 15^{3x-4} = \ln 10$$
    $$(3x-4)\ln 15 = \ln 10$$
    $$3x - 4 = \frac{\ln 10}{\ln 15}$$
    $$3x = \frac{\ln 10}{\ln 15} + 4$$
    $$x = \frac{1}{3}\left(\frac{\ln 10}{\ln 15} + 4\right) \approx 1.6168$$

19. Solving the equation:
    $$6^{5-2x} = 4$$
    $$\ln 6^{5-2x} = \ln 4$$
    $$(5-2x)\ln 6 = \ln 4$$
    $$5 - 2x = \frac{\ln 4}{\ln 6}$$
    $$-2x = \frac{\ln 4}{\ln 6} - 5$$
    $$x = \frac{1}{2}\left(5 - \frac{\ln 4}{\ln 6}\right) \approx 2.1131$$

21. Evaluating the logarithm: $\log_8 16 = \dfrac{\log 16}{\log 8} \approx 1.3333$

23. Evaluating the logarithm: $\log_{16} 8 = \dfrac{\log 8}{\log 16} = 0.75$

25. Evaluating the logarithm: $\log_7 15 = \dfrac{\log 15}{\log 7} \approx 1.3917$

27. Evaluating the logarithm: $\log_{15} 7 = \dfrac{\log 7}{\log 15} \approx 0.7186$

29. Evaluating the logarithm: $\log_8 240 = \dfrac{\log 240}{\log 8} \approx 2.6356$

31. Evaluating the logarithm: $\log_4 321 = \dfrac{\log 321}{\log 4} \approx 4.1632$

33. Evaluating the logarithm: $\ln 345 \approx 5.8435$

35. Evaluating the logarithm: $\ln 0.345 \approx -1.0642$

37. Evaluating the logarithm: $\ln 10 \approx 2.3026$

39. Evaluating the logarithm: $\ln 45,000 \approx 10.7144$

41. Using the compound interest formula:

$$500\left(1+\frac{0.06}{2}\right)^{2t}=1000$$

$$\left(1+\frac{0.06}{2}\right)^{2t}=2$$

$$\ln\left(1+\frac{0.06}{2}\right)^{2t}=\ln 2$$

$$2t\ln\left(1+\frac{0.06}{2}\right)=\ln 2$$

$$t=\frac{\ln 2}{2\ln\left(1+\frac{0.06}{2}\right)}\approx 11.7$$

It will take 11.7 years.

43. Using the compound interest formula:

$$1000\left(1+\frac{0.12}{6}\right)^{6t}=3000$$

$$\left(1+\frac{0.12}{6}\right)^{6t}=3$$

$$\ln\left(1+\frac{0.12}{6}\right)^{6t}=\ln 3$$

$$6t\ln\left(1+\frac{0.12}{6}\right)=\ln 3$$

$$t=\frac{\ln 3}{6\ln\left(1+\frac{0.12}{6}\right)}\approx 9.25$$

It will take 9.25 years.

45. Using the compound interest formula:

$$P\left(1+\frac{0.08}{4}\right)^{4t}=2P$$

$$\left(1+\frac{0.08}{4}\right)^{4t}=2$$

$$\ln\left(1+\frac{0.08}{4}\right)^{4t}=\ln 2$$

$$4t\ln\left(1+\frac{0.08}{4}\right)=\ln 2$$

$$t=\frac{\ln 2}{4\ln\left(1+\frac{0.08}{4}\right)}\approx 8.75$$

It will take 8.75 years.

47. Using the compound interest formula:

$$25\left(1+\frac{0.06}{2}\right)^{2t}=75$$

$$\left(1+\frac{0.06}{2}\right)^{2t}=3$$

$$\ln\left(1+\frac{0.06}{2}\right)^{2t}=\ln 3$$

$$2t\ln\left(1+\frac{0.06}{2}\right)=\ln 3$$

$$t=\frac{\ln 3}{2\ln\left(1+\frac{0.06}{2}\right)}\approx 18.58$$

It was invested 18.58 years ago.

49. Using the continuous interest formula:

$$500e^{0.06t}=1000$$

$$e^{0.06t}=2$$

$$0.06t=\ln 2$$

$$t=\frac{\ln 2}{0.06}\approx 11.55$$

It will take 11.55 years.

51. Using the continuous interest formula:

$$500e^{0.06t}=1500$$

$$e^{0.06t}=3$$

$$0.06t=\ln 3$$

$$t=\frac{\ln 3}{0.06}\approx 18.31$$

It will take 18.31 years.

53. Completing the square: $y=2x^2+8x-15=2\left(x^2+4x+4\right)-8-15=2\left(x+2\right)^2-23$. The lowest point is $(-2,-23)$.

55. Completing the square: $y=12x-4x^2=-4\left(x^2-3x+\frac{9}{4}\right)+9=-4\left(x-\frac{3}{2}\right)^2+9$. The highest point is $\left(\frac{3}{2},9\right)$.

57. Completing the square: $y=64t-16t^2=-16\left(t^2-4t+4\right)+64=-16\left(t-2\right)^2+64$

The object reaches a maximum height after 2 seconds, and the maximum height is 64 feet.

59. Using the population model:

$$32000e^{0.05t} = 64000$$
$$e^{0.05t} = 2$$
$$0.05t = \ln 2$$
$$t = \frac{\ln 2}{0.05} \approx 13.9$$

The city will reach 64,000 toward the end of the year 2023.

61. Finding when $P(t) = 45,000$:

$$15,000e^{0.04t} = 45,000$$
$$e^{0.04t} = 3$$
$$0.04t = \ln 3$$
$$t = \frac{\ln 3}{0.04} \approx 27.5$$

It will take approximately 27.5 years.

63. Solving for t:

$$A = Pe^{rt}$$
$$e^{rt} = \frac{A}{P}$$
$$rt = \ln\frac{A}{P}$$
$$t = \frac{\ln A - \ln P}{r} = \frac{1}{r}\ln\frac{A}{P}$$

65. Solving for t:

$$A = P2^{-kt}$$
$$2^{-kt} = \frac{A}{P}$$
$$\log 2^{-kt} = \log\frac{A}{P}$$
$$-kt\log 2 = \log A - \log P$$
$$t = \frac{\log A - \log P}{-k\log 2} = \frac{\log P - \log A}{k\log 2}$$

67. Solving for t:

$$A = P(1-r)^t$$
$$(1-r)^t = \frac{A}{P}$$
$$\log(1-r)^t = \log\frac{A}{P}$$
$$t\log(1-r) = \log A - \log P$$
$$t = \frac{\log A - \log P}{\log(1-r)}$$

Chapter 9 Review

1.   Evaluating the function: $f(4) = 2^4 = 16$

3.   Evaluating the function: $g(2) = \left(\dfrac{1}{3}\right)^2 = \dfrac{1}{9}$

5.   Evaluating the function: $f(-1) + g(1) = 2^{-1} + \left(\dfrac{1}{3}\right)^1 = \dfrac{1}{2} + \dfrac{1}{3} = \dfrac{5}{6}$

7.   Graphing the function:

9.   Finding the inverse:
$$2y + 1 = x$$
$$2y = x - 1$$
$$y = \frac{x-1}{2}$$

The inverse is $y^{-1} = \dfrac{x-1}{2}$. Sketching the graph:

11.   Finding the inverse:
$$2y + 3 = x$$
$$2y = x - 3$$
$$y = \frac{x-3}{2}$$

The inverse is $f^{-1}(x) = \dfrac{x-3}{2}$.

13.   Finding the inverse:
$$\frac{1}{2}y + 2 = x$$
$$y + 4 = 2x$$
$$y = 2x - 4$$

The inverse is $f^{-1}(x) = 2x - 4$.

15.   Writing in logarithmic form: $\log_3 81 = 4$

17.   Writing in logarithmic form: $\log_{10} 0.01 = -2$

19.   Writing in exponential form: $2^3 = 8$

21.   Writing in exponential form: $4^{1/2} = 2$

23. Solving for x:

$$\log_5 x = 2$$
$$x = 5^2 = 25$$

25. Solving for x:
$$\log_x 0.01 = -2$$
$$x^{-2} = 0.01$$
$$x^2 = 100$$
$$x = 10$$

27. Graphing the equation:

29. Simplifying the logarithm:
$$\log_{27} 9 = x$$
$$27^x = 9$$
$$3^{3x} = 3^2$$
$$3x = 2$$
$$x = \frac{2}{3}$$

31. Expanding the logarithm: $\log_2 5x = \log_2 5 + \log_2 x$

33. Expanding the logarithm: $\log_a \dfrac{y^3 \sqrt{x}}{z} = \log_a y^3 + \log_a x^{1/2} - \log_a z = 3\log_a y + \dfrac{1}{2}\log_a x - \log_a z$

35. Writing as a single logarithm: $\log_2 x + \log_2 y = \log_2 xy$

37. Writing as a single logarithm: $2\log_a 5 - \dfrac{1}{2}\log_a 9 = \log_a 5^2 - \log_a 9^{1/2} = \log_a 25 - \log_a 3 = \log_a \dfrac{25}{3}$

39. Solving the equation:

$$\log_2 x + \log_2 4 = 3$$
$$\log_2 4x = 3$$
$$4x = 2^3$$
$$4x = 8$$
$$x = 2$$

41. Solving the equation:
$$\log_3 x + \log_3 (x-2) = 1$$
$$\log_3 (x^2 - 2x) = 1$$
$$x^2 - 2x = 3^1$$
$$x^2 - 2x - 3 = 0$$
$$(x-3)(x+1) = 0$$
$$x = 3, -1$$
The solution is 3 (-1 does not check).

43. Solving the equation:
$$\log_6 (x-1) + \log_6 x = 1$$
$$\log_6 (x^2 - x) = 1$$
$$x^2 - x = 6^1$$
$$x^2 - x - 6 = 0$$
$$(x-3)(x+2) = 0$$
$$x = 3, -2$$
The solution is 3 (-2 does not check).

45. Evaluating: $\log 346 \approx 2.5391$

47. Solving for x:
$$\log x = 3.9652$$
$$x = 10^{3.9652} \approx 9,230$$

49. Simplifying: $\ln e = \ln e^1 = 1$

51. Simplifying: $\ln e^2 = 2$

53. Finding the pH: $pH = -\log\left(7.9 \times 10^{-3}\right) \approx 2.1$

55. Finding $\left[H^+\right]$:
$$2.7 = -\log\left[H^+\right]$$
$$-2.7 = \log\left[H^+\right]$$
$$\left[H^+\right] = 10^{-2.7} \approx 2.0 \times 10^{-3}$$

57. Solving the equation:
$$4^x = 8$$
$$2^{2x} = 2^3$$
$$2x = 3$$
$$x = \frac{3}{2}$$

59. Using a calculator: $\log_{16} 8 = \dfrac{\ln 8}{\ln 16} = 0.75$

61. Using the compound interest formula:
$$5000\left(1 + 0.16\right)^t = 10000$$
$$1.16^t = 2$$
$$\ln 1.16^t = \ln 2$$
$$t \ln 1.16 = \ln 2$$
$$t = \frac{\ln 2}{\ln 1.16} \approx 4.67$$

It will take approximately 4.67 years for the amount to double.

Chapter 9 Cumulative Review

1. Simplifying: $-8 + 2\left[5 - 3(-2 - 3)\right] = -8 + 2\left[5 - 3(-5)\right] = -8 + 2(5 + 15) = -8 + 2(20) = -8 + 40 = 32$

3. Simplifying: $\dfrac{3}{4} \div \dfrac{1}{8} \cdot \dfrac{3}{2} = \dfrac{3}{4} \cdot \dfrac{8}{1} \cdot \dfrac{3}{2} = \dfrac{72}{8} = 9$

5. Simplifying: $3\sqrt{48} - 3\sqrt{75} + 2\sqrt{27} = 3\sqrt{16 \cdot 3} - 3\sqrt{25 \cdot 3} + 2\sqrt{9 \cdot 3} = 12\sqrt{3} - 15\sqrt{3} + 6\sqrt{3} = 3\sqrt{3}$

7. Simplifying: $\left[(6 + 2i) - (3 - 4i)\right] - (5 - i) = (6 + 2i - 3 + 4i) - (5 - i) = 3 + 6i - 5 + i = -2 + 7i$

9. Simplifying: $\log_6\left(\log_5 5\right) = \log_6 1 = 0$

11. Simplifying: $\dfrac{1 - \dfrac{1}{x-3}}{1 + \dfrac{1}{x-3}} = \dfrac{1 - \dfrac{1}{x-3}}{1 + \dfrac{1}{x-3}} \cdot \dfrac{x-3}{x-3} = \dfrac{x-3-1}{x-3+1} = \dfrac{x-4}{x-2}$

13. Reducing the fraction: $\dfrac{452}{791} = \dfrac{4 \cdot 113}{7 \cdot 113} = \dfrac{4}{7}$

15. Multiplying: $\left(3t^2 + \dfrac{1}{4}\right)\left(4t^2 - \dfrac{1}{3}\right) = 12t^4 + t^2 - t^2 - \dfrac{1}{12} = 12t^4 - \dfrac{1}{12}$

17. Using long division:

$$\begin{array}{r}
y+5 \\
y+2{\overline{\smash{\big)}\,y^2+7y+7}} \\
\underline{y^2+2y} \\
5y+7 \\
\underline{5y+10} \\
-3
\end{array}$$

The quotient is $y+5-\dfrac{3}{y+2}$.

19. Adding:

$$\frac{-3}{x^2-2x-8}+\frac{4}{x^2-16}=\frac{-3}{(x-4)(x+2)}\cdot\frac{x+4}{x+4}+\frac{4}{(x-4)(x+4)}\cdot\frac{x+2}{x+2}$$

$$=\frac{-3x-12}{(x+4)(x-4)(x+2)}+\frac{4x+8}{(x+4)(x-4)(x+2)}$$

$$=\frac{x-4}{(x+4)(x-4)(x+2)}$$

$$=\frac{1}{(x+4)(x+2)}$$

21. Solving the equation:

$$6-3(2x-4)=2$$
$$6-6x+12=2$$
$$-6x+18=2$$
$$-6x=-16$$
$$x=\frac{8}{3}$$

23. Solving the equation:

$$28x^2=3x+1$$
$$28x^2-3x-1=0$$
$$(7x+1)(4x-1)=0$$
$$x=-\frac{1}{7},\frac{1}{4}$$

25. Solving the equation:

$$\frac{1}{x+3}+\frac{1}{x-2}=1$$

$$(x+3)(x-2)\left(\frac{1}{x+3}+\frac{1}{x-2}\right)=(x+3)(x-2)\cdot1$$

$$x-2+x+3=x^2+x-6$$
$$2x+1=x^2+x-6$$
$$x^2-x-7=0$$

$$x=\frac{1\pm\sqrt{1+28}}{2}=\frac{1\pm\sqrt{29}}{2}$$

27. Solving the equation:

$$x-3\sqrt{x}+2=0$$
$$\left(\sqrt{x}-2\right)\left(\sqrt{x}-1\right)=0$$
$$\sqrt{x}=1,2$$
$$x=1,4$$

29. Solving the equation:

$$\log_3 x=3$$
$$x=3^3=27$$

31. Solving the equation:
$$\log_3 (x-3) - \log_3 (x+2) = 1$$
$$\log_3 \frac{x-3}{x+2} = 1$$
$$\frac{x-3}{x+2} = 3^1$$
$$x-3 = 3x+6$$
$$-2x = 9$$
$$x = -\frac{9}{2} \quad \text{(impossible)}$$

There is no solution.

33. Multiply the first equation by 2 and the second equation by 3:
$$-18x + 6y = -2$$
$$15x - 6y = -6$$

Adding yields:
$$-3x = -8$$
$$x = \frac{8}{3}$$

Substituting into the first equation:
$$-9\left(\frac{8}{3}\right) + 3y = -1$$
$$-24 + 3y = -1$$
$$3y = 23$$
$$y = \frac{23}{3}$$

The solution is $\left(\frac{8}{3}, \frac{23}{3}\right)$.

35. Multiply the second equation by -1 and add it to the first equation:
$$x + 2y = 4$$
$$-x - 2z = -1$$

Adding yields the equation $2y - 2z = 3$. So the system becomes:
$$2y - 2z = 3$$
$$y - 2z = 5$$

Multiplying the second equation by -1:
$$2y - 2z = 3$$
$$-y + 2z = -5$$

Adding yields $y = -2$. Substituting to find z:
$$2(-2) - 2z = 3$$
$$-4 - 2z = 3$$
$$-2z = 7$$
$$z = -\frac{7}{2}$$

Substituting to find x:
$$x + 2(-2) = 4$$
$$x - 4 = 4$$
$$x = 8$$

The solution is $\left(8, -2, -\frac{7}{2}\right)$.

37.   Graphing the line:

39.   Written in symbols: $7 - y = -x$

41.   Writing in scientific notation: $0.0000972 = 9.72 \times 10^{-5}$

43.   Simplifying: $\left(2.6 \times 10^{-6}\right)\left(1.4 \times 10^{19}\right) = 3.64 \times 10^{13}$

45.   Factoring completely: $8a^3 - 125 = (2a)^3 - (5)^3 = (2a - 5)\left(4a^2 + 10a + 25\right)$

47.   The domain is $\{-3, 2\}$ and the range is $\{-3, -1, 3\}$. This is not a function.

49.   The U.S. population is approximately: $6(50 \text{ million}) = 300 \text{ million}$

Chapter 9 Test

1.   Graphing the function:

2.   Graphing the function:

3.   Finding the inverse:
$$2y - 3 = x$$
$$2y = x + 3$$
$$y = \frac{x + 3}{2}$$

The inverse is $f^{-1}(x) = \dfrac{x+3}{2}$. Sketching the graphs:

4.   Graphing the function and its inverse (which is not a function):

5.   Solving for x:

$$\log_4 x = 3$$
$$x = 4^3 = 64$$

6.   Solving for x:

$$\log_x 5 = 2$$
$$x^2 = 5$$
$$x = \sqrt{5}$$

7.   Graphing the function:

8.   Graphing the function:

9. Evaluating the logarithm:

$$x = \log_8 4$$
$$8^x = 4$$
$$2^{3x} = 2^2$$
$$3x = 2$$
$$x = \frac{2}{3} \approx 0.67$$

10. Evaluating the logarithm: $\log_7 21 = \dfrac{\ln 21}{\ln 7} \approx 1.56$

11. Evaluating the logarithm: $\log 23,400 \approx 4.37$ 12. Evaluating the logarithm: $\log 0.0123 \approx -1.91$

13. Evaluating the logarithm: $\ln 46.2 \approx 3.83$ 14. Evaluating the logarithm: $\ln 0.0462 \approx -3.07$

15. Expanding the logarithm: $\log_2 \dfrac{6x^2}{y} = \log_2 6 + \log_2 x^2 - \log_2 y = \log_2 6 + 2\log_2 x - \log_2 y$

16. Expanding the logarithm: $\log \dfrac{\sqrt{x}}{y^4 \sqrt[5]{z}} = \log \dfrac{x^{1/2}}{y^4 z^{1/5}} = \log x^{1/2} - \log y^4 - \log z^{1/5} = \dfrac{1}{2}\log x - 4\log y - \dfrac{1}{5}\log z$

17. Writing as a single logarithm: $2\log_3 x - \dfrac{1}{2}\log_3 y = \log_3 x^2 - \log_3 y^{1/2} = \log_3 \dfrac{x^2}{\sqrt{y}}$

18. Writing as a single logarithm: $\dfrac{1}{3}\log x - \log y - 2\log z = \log x^{1/3} - \log y - \log z^2 = \log \dfrac{\sqrt[3]{x}}{yz^2}$

19. Solving for x:

$$\log x = 4.8476$$
$$x = 10^{4.8476} \approx 70,404.432$$

20. Solving for x:

$$\log x = -2.6478$$
$$x = 10^{-2.6478} \approx 0.002$$

21. Solving for x:

$$3^x = 5$$
$$\ln 3^x = \ln 5$$
$$x \ln 3 = \ln 5$$
$$x = \frac{\ln 5}{\ln 3} \approx 1.465$$

22. Solving for x:

$$4^{2x-1} = 8$$
$$2^{4x-2} = 2^3$$
$$4x - 2 = 3$$
$$4x = 5$$
$$x = \frac{5}{4} = 1.25$$

23. Solving for x:

$$\log_5 x - \log_5 3 = 1$$
$$\log_5 \frac{x}{3} = 1$$
$$\frac{x}{3} = 5^1$$
$$x = 15$$

24. Solving for x:

$$\log_2 x + \log_2 (x-7) = 3$$
$$\log_2 (x^2 - 7x) = 3$$
$$x^2 - 7x = 2^3$$
$$x^2 - 7x - 8 = 0$$
$$(x-8)(x+1) = 0$$
$$x = 8, -1$$

The solution is 8 (-1 does not check).

25. Finding the pH: $\text{pH} = -\log\left(6.6 \times 10^{-7}\right) \approx 6.18$

26. Using the compound interest formula: $A = 400\left(1 + \dfrac{0.10}{2}\right)^{2\cdot 5} = 400(1.05)^{10} \approx 651.56$

There will be \$651.56 in the account after 5 years.

27.   Using the compound interest formula:

$$600\left(1+\frac{0.08}{4}\right)^{4t}=1800$$

$$\left(1+\frac{0.08}{4}\right)^{4t}=3$$

$$\ln\left(1.02\right)^{4t}=\ln 3$$

$$4t\ln 1.02=\ln 3$$

$$t=\frac{\ln 3}{4\ln 1.02}\approx 13.87$$

It will take 13.87 years for the account to reach $1,800.

28.   Yes, this function is one-to-one.

29.   No, since both $x=1$ and $x=-1$ have been assigned to 0, the function is not one-to-one.

30.   Yes, this function is one-to-one.

31.   Yes, this function is one-to-one.

32.   Since this function does not pass the horizontal line test, it is not one-to-one.

33.   Since this function does not pass the horizontal line test, it is not one-to-one.

Chapter 10
Conic Sections

Getting Ready for Chapter 10

1. Evaluating: $\sqrt{4+25} = \sqrt{29}$

2. Evaluating: $\sqrt{16+4} = \sqrt{20} = 2\sqrt{5}$

3. Evaluating: $\sqrt{(-2)^2 + 3^2} = \sqrt{4+9} = \sqrt{13}$

4. Evaluating: $\sqrt{1^2 + (-5)^2} = \sqrt{1+25} = \sqrt{26}$

5. Solving the equation:
$$x^2 = 16$$
$$x = \pm\sqrt{16}$$
$$x = \pm 4$$

6. Solving the equation:
$$-y^2 = 25$$
$$y^2 = -25$$
$$y = \pm\sqrt{-25}$$
$$y = \pm 5i$$

7. Solving the equation:
$$0 = x^2 - 6x + 8$$
$$(x-4)(x-2) = 0$$
$$x = 2, 4$$

8. Solving the equation:
$$0 = y^2 + 2y - 8$$
$$(y+4)(y-2) = 0$$
$$y = -4, 2$$

9. Solving the equation:
$$2 = (x-3)^2 + 1$$
$$(x-3)^2 = 1$$
$$x - 3 = \pm 1$$
$$x - 3 = -1, 1$$
$$x = 2, 4$$

10. Solving the equation:
$$10 = (-1)^2 + (y+1)^2$$
$$10 = 1 + (y+1)^2$$
$$(y+1)^2 = 9$$
$$y + 1 = \pm 3$$
$$y + 1 = -3, 3$$
$$y = -4, 2$$

11. Completing the square: $x^2 + 6x + 9 = (x+3)^2$

12. Completing the square: $x^2 - 4x + 4 = (x-2)^2$

13. Completing the square: $y^2 + 2y + 1 = (y+1)^2$

14. Completing the square: $y^2 + 1y + \dfrac{1}{4} = \left(y + \dfrac{1}{2}\right)^2$

15. Factoring: $5y^2 + 16y + 12 = (5y+6)(y+2)$

16. Factoring: $3x^2 + 17x - 28 = (3x-4)(x+7)$

17. Dividing: $\dfrac{12x^2 + y^2}{36} = \dfrac{x^2}{3} + \dfrac{y^2}{36}$

18. Dividing: $\dfrac{x^2 + 4y^2}{16} = \dfrac{x^2}{16} + \dfrac{y^2}{4}$

19. Solving for y:
$$x + 3y = 3$$
$$3y = -x + 3$$
$$y = -\dfrac{1}{3}x + 1$$

20. Solving for y:
$$2x + y = 3$$
$$y = -2x + 3$$

329

10.1 The Circle

1. Using the distance formula: $d = \sqrt{(6-3)^2 + (3-7)^2} = \sqrt{9+16} = \sqrt{25} = 5$

3. Using the distance formula: $d = \sqrt{(5-0)^2 + (0-9)^2} = \sqrt{25+81} = \sqrt{106}$

5. Using the distance formula: $d = \sqrt{(-2-3)^2 + (1+5)^2} = \sqrt{25+36} = \sqrt{61}$

7. Using the distance formula: $d = \sqrt{(-10+1)^2 + (5+2)^2} = \sqrt{81+49} = \sqrt{130}$

9. Solving the equation:

$$\sqrt{(x-1)^2 + (2-5)^2} = \sqrt{13}$$
$$(x-1)^2 + 9 = 13$$
$$(x-1)^2 = 4$$
$$x-1 = \pm 2$$
$$x-1 = -2, 2$$
$$x = -1, 3$$

11. Solving the equation:

$$\sqrt{(7-8)^2 + (y-3)^2} = 1$$
$$(y-3)^2 + 1 = 1$$
$$(y-3)^2 = 0$$
$$y-3 = 0$$
$$y = 3$$

13. The equation is $(x-2)^2 + (y-3)^2 = 16$.

15. The equation is $(x-3)^2 + (y+2)^2 = 9$.

17. The equation is $(x+5)^2 + (y+1)^2 = 5$.

19. The equation is $x^2 + (y+5)^2 = 1$.

21. The equation is $x^2 + y^2 = 4$.

23. The center is $(0,0)$ and the radius is 2.

25. The center is $(1,3)$ and the radius is 5.

27. The center is $(-2,4)$ and the radius is $2\sqrt{2}$.

29. The center is $(-1,-1)$ and the radius is 1.

31. Completing the square:

$$x^2 + y^2 - 6y = 7$$
$$x^2 + \left(y^2 - 6y + 9\right) = 7 + 9$$
$$x^2 + \left(y - 3\right)^2 = 16$$

The center is $(0,3)$ and the radius is 4.

33. Completing the square:

$$x^2 + y^2 - 4x - 6y = -4$$
$$\left(x^2 - 4x + 4\right) + \left(y^2 - 6y + 9\right) = -4 + 4 + 9$$
$$\left(x - 2\right)^2 + \left(y - 3\right)^2 = 9$$

The center is $(2,3)$ and the radius is 3.

35. Completing the square:

$$x^2 + y^2 + 2x + y = \frac{11}{4}$$
$$\left(x^2 + 2x + 1\right) + \left(y^2 + y + \frac{1}{4}\right) = \frac{11}{4} + 1 + \frac{1}{4}$$
$$\left(x + 1\right)^2 + \left(y + \frac{1}{2}\right)^2 = 4$$

The center is $\left(-1, -\frac{1}{2}\right)$ and the radius is 2.

37. The equation is $\left(x - 3\right)^2 + \left(y - 4\right)^2 = 25$.

39. The equations are:

A: $\left(x - \frac{1}{2}\right)^2 + \left(y - 1\right)^2 = \frac{1}{4}$

B: $\left(x - 1\right)^2 + \left(y - 1\right)^2 = 1$

C: $\left(x - 2\right)^2 + \left(y - 1\right)^2 = 4$

41. The equations are:

A: $\left(x + 8\right)^2 + y^2 = 64$

B: $x^2 + y^2 = 64$

C: $\left(x - 8\right)^2 + y^2 = 64$

43. The equations are:

A: $(x+4)^2 + (y-4)^2 = 16$

B: $x^2 + y^2 = 16$

C: $(x-4)^2 + (y+4)^2 = 16$

45. The x-coordinate of the center is $x = 500$, the y-coordinate of the center is $12 + 120 = 132$, and the radius is 120. Thus the equation of the circle is $(x-500)^2 + (y-132)^2 = 120^2 = 14,400$.

47. Let $y = f(x)$. Switch x and y and solve for y:

$$3^y = x$$
$$y = \log_3 x$$

The inverse is $f^{-1}(x) = \log_3 x$.

49. Let $y = f(x)$. Switch x and y and solve for y:

$$2y + 3 = x$$
$$2y = x - 3$$
$$y = \frac{x-3}{2}$$

The inverse is $f^{-1}(x) = \frac{x-3}{2}$.

51. Let $y = f(x)$. Switch x and y and solve for y:

$$\frac{y+3}{5} = x$$
$$y + 3 = 5x$$
$$y = 5x - 3$$

The inverse is $f^{-1}(x) = 5x - 3$.

53. Solving the equation:

$$y^2 = 9$$
$$y = \pm\sqrt{9} = \pm 3$$

55. Solving the equation:

$$-y^2 = 4$$
$$y^2 = -4$$
$$y = \pm\sqrt{-4} = \pm 2i$$

57. Solving the equation:

$$\frac{-x^2}{9} = 1$$
$$x^2 = -9$$
$$x = \pm\sqrt{-9} = \pm 3i$$

59. Dividing: $\dfrac{4x^2 + 9y^2}{36} = \dfrac{4x^2}{36} + \dfrac{9y^2}{36} = \dfrac{x^2}{9} + \dfrac{y^2}{4}$

61. To find the x-intercept, let $y = 0$:

$$3x - 4(0) = 12$$
$$3x - 0 = 12$$
$$3x = 12$$
$$x = 4$$

To find the y-intercept, let $x = 0$:

$$3(0) - 4y = 12$$
$$0 - 4y = 12$$
$$-4y = 12$$
$$y = -3$$

63. Substituting $x = 3$:

$$\frac{3^2}{25} + \frac{y^2}{9} = 1$$
$$\frac{9}{25} + \frac{y^2}{9} = 1$$
$$\frac{y^2}{9} = \frac{16}{25}$$
$$y^2 = \frac{144}{25}$$
$$y = \pm\sqrt{\frac{144}{25}} = \pm\frac{12}{5} = \pm 2.4$$

65. The radius is 2, so the equation is $(x-2)^2 + (y-3)^2 = 4$.

67. The radius is 2, so the equation is $(x-2)^2 + (y-3)^2 = 4$.

69.   Completing the square:
$$x^2 + y^2 - 6x + 8y = 144$$
$$\left(x^2 - 6x + 9\right) + \left(y^2 + 8y + 16\right) = 144 + 9 + 16$$
$$(x-3)^2 + (y+4)^2 = 169$$

The center is $(3,-4)$, so the distance is: $d = \sqrt{(-3)^2 + 4^2} = \sqrt{9+16} = 5$

71.   Completing the square:
$$x^2 + y^2 - 6x - 8y = 144$$
$$\left(x^2 - 6x + 9\right) + \left(y^2 - 8y + 16\right) = 144 + 9 + 16$$
$$(x-3)^2 + (y-4)^2 = 169$$

The center is $(3,4)$, so the distance is: $d = \sqrt{3^2 + 4^2} = \sqrt{9+16} = 5$

73.   The equation $y = \sqrt{9 - x^2}$ corresponds to the top half of the circle, and the equation $y = -\sqrt{9 - x^2}$ corresponds to the bottom half of the circle.

10.2 Ellipses and Hyperbolas

1.   Graphing the ellipse:

3.   Graphing the ellipse:

5.   Graphing the ellipse:

7.   The standard form is $\dfrac{x^2}{25} + \dfrac{y^2}{4} = 1$. Graphing the ellipse:

9.   The standard form is $\dfrac{x^2}{16} + \dfrac{y^2}{2} = 1$. Graphing the ellipse:

11.   Graphing the hyperbola:

13.   Graphing the hyperbola:

15.   Graphing the hyperbola:

17.   Graphing the hyperbola:

19.   The standard form is $\dfrac{x^2}{4} - \dfrac{y^2}{1} = 1$. Graphing the hyperbola:

21.   The standard form is $\dfrac{y^2}{9} - \dfrac{x^2}{16} = 1$. Graphing the hyperbola:

23.   For the x-intercepts, set $y = 0$:

$$0.4x^2 = 3.6$$
$$x^2 = 9$$
$$x = \pm 3$$

For the y-intercepts, set $x = 0$:

$$0.9y^2 = 3.6$$
$$y^2 = 4$$
$$y = \pm 2$$

25.   For the x-intercepts, set $y = 0$:

$$\dfrac{x^2}{0.04} = 1$$
$$x^2 = 0.04$$
$$x = \pm 0.2$$

For the y-intercepts, set $x = 0$:

$$-\dfrac{y^2}{0.09} = 1$$
$$y^2 = -0.09$$

There are no y-intercepts.

27.   For the x-intercepts, set $y = 0$:

$$\dfrac{25x^2}{9} = 1$$
$$x^2 = \dfrac{9}{25}$$
$$x = \pm\dfrac{3}{5}$$

For the y-intercepts, set $x = 0$:

$$\dfrac{25y^2}{4} = 1$$
$$y^2 = \dfrac{4}{25}$$
$$y = \pm\dfrac{2}{5}$$

29.   Graphing the ellipse:

31.   Completing the square:

$$4x^2 + y^2 - 4y - 12 = 0$$
$$4x^2 + \left(y^2 - 4y + 4\right) = 12 + 4$$
$$4x^2 + (y - 2)^2 = 16$$
$$\frac{x^2}{4} + \frac{(y - 2)^2}{16} = 1$$

Graphing the ellipse:

33.   Completing the square:

$$x^2 + 9y^2 + 4x - 54y + 76 = 0$$
$$\left(x^2 + 4x + 4\right) + 9\left(y^2 - 6y + 9\right) = -76 + 4 + 81$$
$$(x + 2)^2 + 9(y - 3)^2 = 9$$
$$\frac{(x + 2)^2}{9} + \frac{(y - 3)^2}{1} = 1$$

Graphing the ellipse:

35.   Graphing the hyperbola:

37. Completing the square:
$$9y^2 - x^2 - 4x + 54y + 68 = 0$$
$$9(y^2 + 6y + 9) - (x^2 + 4x + 4) = -68 + 81 - 4$$
$$9(y + 3)^2 - (x + 2)^2 = 9$$
$$\frac{(y + 3)^2}{1} - \frac{(x + 2)^2}{9} = 1$$

Graphing the hyperbola:

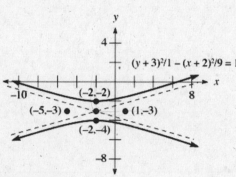

39. Completing the square:
$$4y^2 - 9x^2 - 16y + 72x - 164 = 0$$
$$4(y^2 - 4y + 4) - 9(x^2 - 8x + 16) = 164 + 16 - 144$$
$$4(y - 2)^2 - 9(x - 4)^2 = 36$$
$$\frac{(y - 2)^2}{9} - \frac{(x - 4)^2}{4} = 1$$

Graphing the hyperbola:

41. Substituting $x = 4$:
$$\frac{4^2}{25} + \frac{y^2}{9} = 1$$
$$\frac{y^2}{9} + \frac{16}{25} = 1$$
$$\frac{y^2}{9} = \frac{9}{25}$$
$$\frac{y}{3} = \pm \frac{3}{5}$$
$$y = \pm \frac{9}{5}$$

43. Substituting $x = 1.8$:
$$16(1.8)^2 + 9y^2 = 144$$
$$51.84 + 9y^2 = 144$$
$$9y^2 = 92.16$$
$$3y = \pm 9.6$$
$$y = \pm 3.2$$

45. The asymptotes are $y = \frac{3}{4}x$ and $y = -\frac{3}{4}x$.

47. The equation has the form $\frac{x^2}{16} + \frac{y^2}{b^2} = 1$. Substituting the point $\left(2, \sqrt{3}\right)$:
$$\frac{(2)^2}{16} + \frac{\left(\sqrt{3}\right)^2}{b^2} = 1$$
$$\frac{1}{4} + \frac{3}{b^2} = 1$$
$$\frac{3}{b^2} = \frac{3}{4}$$
$$b^2 = 4$$

The equation of the ellipse is $\frac{x^2}{16} + \frac{y^2}{4} = 1$.

49. The equation of the ellipse is $\dfrac{x^2}{20^2} + \dfrac{y^2}{10^2} = 1$. Substituting $y = 6$:

$$\frac{x^2}{20^2} + \frac{6^2}{10^2} = 1$$

$$\frac{x^2}{400} + \frac{9}{25} = 1$$

$$\frac{x^2}{400} = \frac{16}{25}$$

$$x^2 = 256$$

$$x = 16$$

The man can walk within 16 feet of the center.

51. Substituting $a = 4$ and $c = 3$:

$$4^2 = b^2 + 3^2$$

$$16 = b^2 + 9$$

$$b^2 = 7$$

$$b = \sqrt{7} \approx 2.65$$

The width should be approximately $2(2.65) = 5.3$ feet wide.

53. Evaluating: $f(4) = \dfrac{2}{4-2} = \dfrac{2}{2} = 1$

55. Evaluating: $f\big(g(0)\big) = f\left(\dfrac{2}{0+2}\right) = f(1) = \dfrac{2}{1-2} = \dfrac{2}{-1} = -2$

57. Simplifying: $f(x) + g(x) = \dfrac{2}{x-2} + \dfrac{2}{x+2} = \dfrac{2}{x-2} \cdot \dfrac{x+2}{x+2} + \dfrac{2}{x+2} \cdot \dfrac{x-2}{x-2} = \dfrac{2x+4+2x-4}{(x-2)(x+2)} = \dfrac{4x}{(x-2)(x+2)}$

59. Since $4^2 + 0^2 = 16$ and $0^2 + 5^2 = 25$, while $0^2 + 0^2 = 0$, only $(0,0)$ is a solution.

61. Multiplying: $(2y+4)^2 = (2y)^2 + 2(2y)(4) + 4^2 = 4y^2 + 16y + 16$

63. Solving for x:

$$x - 2y = 4$$

$$x = 2y + 4$$

65. Simplifying: $x^2 - 2\big(x^2 - 3\big) = x^2 - 2x^2 + 6 = -x^2 + 6$

67. Factoring: $5y^2 + 16y + 12 = (5y+6)(y+2)$

69. Solving the equation:

$$y^2 = 4$$

$$y = \pm\sqrt{4} = \pm 2$$

71. Solving the equation:

$$-x^2 + 6 = 2$$

$$-x^2 = -4$$

$$x^2 = 4$$

$$x = \pm\sqrt{4} = \pm 2$$

10.3 Second-Degree Inequalities and Nonlinear Systems

1. Graphing the inequality:

3. Graphing the inequality:

5. Graphing the inequality:

7. Graphing the inequality:

9. Solving the second equation for y yields $y = 3 - 2x$. Substituting into the first equation:

$$x^2 + (3 - 2x)^2 = 9$$
$$x^2 + 9 - 12x + 4x^2 = 9$$
$$5x^2 - 12x = 0$$
$$x(5x - 12) = 0$$
$$x = 0, \frac{12}{5}$$
$$y = 3, -\frac{9}{5}$$

The solutions are $(0, 3), \left(\frac{12}{5}, -\frac{9}{5} \right)$.

11. Solving the second equation for x yields $x = 8 - 2y$. Substituting into the first equation:

$$(8 - 2y)^2 + y^2 = 16$$
$$64 - 32y + 4y^2 + y^2 = 16$$
$$5y^2 - 32y + 48 = 0$$
$$(y - 4)(5y - 12) = 0$$
$$y = 4, \frac{12}{5}$$
$$x = 0, \frac{16}{5}$$

The solutions are $(0, 4), \left(\frac{16}{5}, \frac{12}{5} \right)$.

13.   Adding the two equations yields:

$$2x^2 = 50$$
$$x^2 = 25$$
$$x = -5, 5$$
$$y = 0$$

The solutions are $(-5,0), (5,0)$.

17.   Substituting into the first equation:

$$x^2 + \left(x^2 - 4\right)^2 = 16$$
$$x^2 + x^4 - 8x^2 + 16 = 16$$
$$x^4 - 7x^2 = 0$$
$$x^2\left(x^2 - 7\right) = 0$$
$$x = 0, -\sqrt{7}, \sqrt{7}$$
$$y = -4, 3, 3$$

The solutions are $\left(0,-4\right), \left(-\sqrt{7},3\right), \left(\sqrt{7},3\right)$.

21.   Substituting into the first equation:

$$-x + 1 = x^2 + 2x - 3$$
$$x^2 + 3x - 4 = 0$$
$$(x+4)(x-1) = 0$$
$$x = -4, 1$$
$$y = 5, 0$$

The solutions are $(-4,5), (1,0)$.

25.   Adding the two equations yields:

$$8x^2 = 72$$
$$x^2 = 9$$
$$x = \pm 3$$
$$y = 0$$

The solutions are $(-3,0), (3,0)$.

27.   Solving the first equation for x yields $x = y + 4$. Substituting into the second equation:

$$(y+4)^2 + y^2 = 16$$
$$y^2 + 8y + 16 + y^2 = 16$$
$$2y^2 + 8y = 0$$
$$2y(y+4) = 0$$
$$y = 0, -4$$
$$x = 4, 0$$

The solutions are $(0,-4), (4,0)$.

15.   Substituting into the first equation:

$$x^2 + \left(x^2 - 3\right)^2 = 9$$
$$x^2 + x^4 - 6x^2 + 9 = 9$$
$$x^4 - 5x^2 = 0$$
$$x^2\left(x^2 - 5\right) = 0$$
$$x = 0, -\sqrt{5}, \sqrt{5}$$
$$y = -3, 2, 2$$

The solutions are $\left(0,-3\right), \left(-\sqrt{5},2\right), \left(\sqrt{5},2\right)$.

19.   Substituting into the first equation:

$$3x + 2\left(x^2 - 5\right) = 10$$
$$3x + 2x^2 - 10 = 10$$
$$2x^2 + 3x - 20 = 0$$
$$(x+4)(2x-5) = 0$$
$$x = -4, \frac{5}{2}$$
$$y = 11, \frac{5}{4}$$

The solutions are $(-4,11), \left(\frac{5}{2}, \frac{5}{4}\right)$.

23.   Substituting into the first equation:

$$x - 5 = x^2 - 6x + 5$$
$$x^2 - 7x + 10 = 0$$
$$(x-2)(x-5) = 0$$
$$x = 2, 5$$
$$y = -3, 0$$

The solutions are $(2,-3), (5,0)$.

29.   a.   Subtracting the two equations yields:

$$(x+8)^2 - x^2 = 0$$
$$x^2 + 16x + 64 - x^2 = 0$$
$$16x = -64$$
$$x = -4$$

Substituting to find y:

$$(-4)^2 + y^2 = 64$$
$$y^2 + 16 = 64$$
$$y^2 = 48$$
$$y = \pm\sqrt{48} = \pm 4\sqrt{3}$$

The intersection points are $\left(-4, -4\sqrt{3}\right)$ and $\left(-4, 4\sqrt{3}\right)$.

b.   Subtracting the two equations yields:

$$x^2 - (x-8)^2 = 0$$
$$x^2 - x^2 + 16x - 64 = 0$$
$$16x = 64$$
$$x = 4$$

Substituting to find y:

$$4^2 + y^2 = 64$$
$$y^2 + 16 = 64$$
$$y^2 = 48$$
$$y = \pm\sqrt{48} = \pm 4\sqrt{3}$$

The intersection points are $\left(4, -4\sqrt{3}\right)$ and $\left(4, 4\sqrt{3}\right)$.

31.   Graphing the inequality:

33.   Graphing the inequality:

35.   There is no intersection.

37.   The system of inequalities is:

$$x^2 + y^2 < 16$$

$$y > 4 - \frac{1}{4}x^2$$

39.   The system of equations is:

$$x^2 + y^2 = 89$$

$$x^2 - y^2 = 39$$

Adding the two equations yields:

$$2x^2 = 128$$

$$x^2 = 64$$

$$x = \pm 8$$

$$y = \pm 5$$

The numbers are either 8 and 5, 8 and –5, –8 and 5, or –8 and –5.

41.   The system of equations is:

$$y = x^2 - 3$$

$$x + y = 9$$

Substituting into the second equation:

$$x + x^2 - 3 = 9$$

$$x^2 + x - 12 = 0$$

$$(x + 4)(x - 3) = 0$$

$$x = -4, 3$$

$$y = 13, 6$$

The numbers are either –4 and 13, or 3 and 6.

43.   Solving the equation:

$$g(x) = 0$$

$$\left(x + \frac{2}{5}\right)^2 = 0$$

$$x + \frac{2}{5} = 0$$

$$x = -\frac{2}{5}$$

45.   Solving the equation:

$$f(x) = g(x)$$

$$x^2 + 4x - 4 = 1$$

$$x^2 + 4x - 5 = 0$$

$$(x + 5)(x - 1) = 0$$

$$x = -5, 1$$

47.   Solving the equation:

$$f(x) = g(x)$$

$$x^2 + 4x - 4 = x - 6$$

$$x^2 + 3x + 2 = 0$$

$$(x + 2)(x + 1) = 0$$

$$x = -2, -1$$

Chapter 10 Review

1. Using the distance formula: $d = \sqrt{(-1-2)^2 + (5-6)^2} = \sqrt{9+1} = \sqrt{10}$

3. Using the distance formula: $d = \sqrt{(-4-0)^2 + (0-3)^2} = \sqrt{16+9} = \sqrt{25} = 5$

5. Solving the equation:

$$\sqrt{(x-2)^2 + (-1+4)^2} = 5$$
$$(x-2)^2 + 9 = 25$$
$$(x-2)^2 = 16$$
$$x-2 = \pm\sqrt{16}$$
$$x-2 = -4, 4$$
$$x = -2, 6$$

7. The equation is $x^2 + y^2 = 25$.

9. Finding the radius: $r = \sqrt{(-2-2)^2 + (3-0)^2} = \sqrt{16+9} = \sqrt{25} = 5$. The equation is $(x+2)^2 + (y-3)^2 = 25$.

11. The center is (0,0) and the radius is 2.

13. Completing the square:

$$x^2 + y^2 - 6x + 4y = -4$$
$$\left(x^2 - 6x + 9\right) + \left(y^2 + 4y + 4\right) = -4 + 9 + 4$$
$$(x-3)^2 + (y+2)^2 = 9$$

The center is (3,–2) and the radius is 3.

15.   Graphing the ellipse:

17.   Graphing the hyperbola:

19.   Graphing the inequality:

21.   Graphing the inequality:

23.   Graphing the solution set:

25.   Solving the second equation for y yields $y = 4 - 2x$. Substituting into the first equation:

$$x^2 + (4 - 2x)^2 = 16$$
$$x^2 + 16 - 16x + 4x^2 = 16$$
$$5x^2 - 16x = 0$$
$$x(5x - 16) = 0$$
$$x = 0, \frac{16}{5}$$
$$y = 4, -\frac{12}{5}$$

The solutions are $(0, 4), \left(\frac{16}{5}, -\frac{12}{5}\right)$.

27. Adding the two equations yields:

$$18x^2 = 72$$
$$x^2 = 4$$
$$x = \pm 2$$
$$y = 0$$

The solutions are $(-2, 0), (2, 0)$.

Chapter 10 Cumulative Review

1. Simplifying: $\dfrac{-9 - 6}{4 - 7} = \dfrac{-15}{-3} = 5$

3. Simplifying: $\dfrac{18a^7 b^{-4}}{36a^2 b^{-8}} = \dfrac{1}{2} a^{7-2} b^{-4+8} = \dfrac{1}{2} a^5 b^4 = \dfrac{a^5 b^4}{2}$

5. Simplifying: $\dfrac{12x^2 y^3 - 16x^2 y + 8xy^3}{4xy} = \dfrac{12x^2 y^3}{4xy} - \dfrac{16x^2 y}{4xy} + \dfrac{8xy^3}{4xy} = 3xy^2 - 4x + 2y^2$

7. Simplifying: $\log_4 16 = \log_4 4^2 = 2$

9. Factoring: $ab^3 + b^3 + 6a + 6 = b^3(a+1) + 6(a+1) = (a+1)(b^3 + 6)$

11. Solving the equation:

$$6 - 2(5x - 1) + 4x = 20$$
$$6 - 10x + 2 + 4x = 20$$
$$-6x + 8 = 20$$
$$-6x = 12$$
$$x = -2$$

13. Solving the equation:

$$(x+1)(x+2) = 12$$
$$x^2 + 3x + 2 = 12$$
$$x^2 + 3x - 10 = 0$$
$$(x+5)(x-2) = 0$$
$$x = -5, 2$$

15. Solving the equation:

$$t - 6 = \sqrt{t - 4}$$
$$(t-6)^2 = t - 4$$
$$t^2 - 12t + 36 = t - 4$$
$$t^2 - 13t + 40 = 0$$
$$(t-5)(t-8) = 0$$
$$t = 5, 8$$

The solution is 8 (5 does not check).

17. Solving the equation:

$$(4x - 3)^2 = -50$$
$$4x - 3 = \pm\sqrt{-50}$$
$$4x - 3 = \pm 5i\sqrt{2}$$
$$4x = 3 \pm 5i\sqrt{2}$$
$$x = \dfrac{3 \pm 5i\sqrt{2}}{4}$$

19. Solving the inequality:

$$|4x + 3| - 6 > 5$$
$$|4x + 3| > 11$$

$$4x + 3 > 11 \quad \text{or} \quad 4x + 3 < -11$$
$$4x > 8 \qquad\qquad 4x < -14$$
$$x > 2 \qquad\qquad x < -\dfrac{7}{2}$$

The solution set is $x < -\dfrac{7}{2}$ or $x > 2$. Graphing the solution set:

−7/2   2

21. Graphing the inequality:

23. Graphing the parabola:

25. Multiplying: $\dfrac{3y^2 - 3y}{3y - 12} \cdot \dfrac{y^2 - 2y - 8}{y^2 + 3y + 2} = \dfrac{3y(y-1)}{3(y-4)} \cdot \dfrac{(y-4)(y+2)}{(y+2)(y+1)} = \dfrac{3y(y-1)(y-4)(y+2)}{3(y-4)(y+2)(y+1)} = \dfrac{y(y-1)}{y+1} = \dfrac{y^2 - y}{y+1}$

27. Multiplying: $(2+3i)(1-4i) = 2 + 3i - 8i - 12i^2 = 2 - 5i + 12 = 14 - 5i$

29. Rationalizing the denominator: $\dfrac{3}{\sqrt{7}-\sqrt{3}} = \dfrac{3}{\sqrt{7}-\sqrt{3}} \cdot \dfrac{\sqrt{7}+\sqrt{3}}{\sqrt{7}+\sqrt{3}} = \dfrac{3(\sqrt{7}+\sqrt{3})}{7-3} = \dfrac{3\sqrt{7}+3\sqrt{3}}{4}$

31. Finding the inverse:

$$4y + 1 = x$$
$$4y = x - 1$$
$$y = \frac{1}{4}x - \frac{1}{4}$$

The inverse function is $f^{-1}(x) = \dfrac{1}{4}x - \dfrac{1}{4}$.

33. Evaluating: $x = 10^{3.9786} \approx 9{,}519.19$

35. Solving for x:

$$mx + 2 = nx - 3$$
$$mx - nx = -5$$
$$x(m-n) = -5$$
$$x = \frac{-5}{m-n}$$

37. Finding the slope: $m = \dfrac{-3-(-3)}{-1-2} = \dfrac{-3+3}{-3} = \dfrac{0}{-3} = 0$

39. Finding the value: $f(4) = -\dfrac{3}{2}(4) + 1 = -6 + 1 = -5$

41. Finding the equation:

$$(x-1)(3x-2) = 0$$
$$3x^2 - 3x - 2x + 2 = 0$$
$$3x^2 - 5x + 2 = 0$$

43. Finding the composition: $(f \circ g)(x) = f(4-x) = 3(4-x) - 7 = 12 - 3x - 7 = -3x + 5$

45. Computing: $-8 + (-7-4) = -8 - 11 = -19$

47. The variation equation is $y = Kx$. Substituting $y = 24$ and $x = 8$:

$$24 = K \cdot 8$$
$$K = 3$$

So $y = 3x$. Substituting $x = 2$: $y = 3 \cdot 2 = 6$

49.   Let x represent the gallons of 30% solution and y represent the gallons of 70% solution. The system of equations is:

$$x + y = 16$$
$$0.3x + 0.7y = 0.6(16)$$

Multiply the first equation by –0.3:

$$-0.3x - 0.3y = -4.8$$
$$0.3x + 0.7y = 9.6$$

Adding yields:

$$0.4y = 4.8$$
$$y = 12$$
$$x = 4$$

The mixture should contain 4 gal of 30% solution and 12 gal of 70% solution.

51.   Finding the amount: $0.62(150 \text{ people}) = 93 \text{ people}$

Chapter 10 Test

1.   Finding the distance: $d = \sqrt{(3+9)^2 + (-1-4)^2} = \sqrt{(12)^2 + (-5)^2} = \sqrt{144 + 25} = \sqrt{169} = 13$

2.   Finding the distance: $d = \sqrt{(0-4)^2 + (-4+6)^2} = \sqrt{(-4)^2 + (2)^2} = \sqrt{16 + 4} = \sqrt{20} = 2\sqrt{5}$

3.   Solving the equation:

$$\sqrt{(x+1)^2 + (2-4)^2} = \left(2\sqrt{5}\right)^2$$
$$(x+1)^2 + 4 = 20$$
$$(x+1)^2 = 16$$
$$x+1 = \pm\sqrt{16}$$
$$x+1 = -4, 4$$
$$x = -5, 3$$

4.   The equation is $(x+2)^2 + (y-4)^2 = 9$.

5.   Finding the radius: $r = \sqrt{(-3)^2 + (-4)^2} = \sqrt{9 + 16} = \sqrt{25} = 5$. The equation is $x^2 + y^2 = 25$.

6.   The center is $(0,-2)$ and the radius is $\sqrt{64} = 8$.

7.   Completing the square:

$$x^2 + y^2 - 10x + 6y = 5$$
$$\left(x^2 - 10x + 25\right) + \left(y^2 + 6y + 9\right) = 5 + 25 + 9$$
$$(x-5)^2 + (y+3)^2 = 39$$

The center is $(5,-3)$ and the radius is $\sqrt{39}$.

8.   This is a circle with center $(1,-2)$ and radius = 3. Graphing the circle:

9.   Graphing the ellipse:

10.   The standard form is $\dfrac{x^2}{4} - \dfrac{y^2}{16} = 1$. Graphing the hyperbola:

11.   Completing the square:

$$25x^2 + 150x + 4y^2 + 125 = 0$$
$$25\left(x^2 + 6x\right) + 4y^2 = -125$$
$$25\left(x^2 + 6x + 9\right) + 4y^2 = -125 + 225$$
$$25\left(x + 3\right)^2 + 4y^2 = 100$$
$$\frac{\left(x + 3\right)^2}{4} + \frac{y^2}{25} = 1$$

Graphing the ellipse:

12.   Graphing the inequality:

13.   Graphing the inequality:

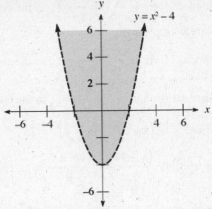

14.   Solving the second equation for y yields $y = 5 - 2x$. Substituting into the first equation:

$$x^2 + \left(5 - 2x\right)^2 = 25$$
$$x^2 + 25 - 20x + 4x^2 = 25$$
$$5x^2 - 20x = 0$$
$$5x\left(x - 4\right) = 0$$
$$x = 0, 4$$
$$y = 5, -3$$

The solutions are $(0,5)$, $(4,-3)$.

15.   Substituting into the first equation:

$$x^2 + \left(x^2 - 4\right)^2 = 16$$
$$x^2 + x^4 - 8x^2 + 16 = 16$$
$$x^4 - 7x^2 = 0$$
$$x^2\left(x^2 - 7\right) = 0$$
$$x = 0, \pm\sqrt{7}$$
$$y = -4, 3$$

The solutions are $\left(0,-4\right), \left(\sqrt{7},3\right), \left(-\sqrt{7},3\right)$.

16.   Graphing the solution set:

17.   The standard form is $\dfrac{x^2}{16} + \dfrac{y^2}{4} = 1$, so the correct graph is C.

18.   The correct graph is A.

19.   The standard form is $\dfrac{x^2}{16} - \dfrac{y^2}{4} = 1$, so the correct graph is D.

20.   The standard form is $\dfrac{x^2}{4} + \dfrac{y^2}{16} = 1$, so the correct graph is B.

21.   The equation is $(x+3)^2 + (y-4)^2 = 25$.

22.   The equation is $\dfrac{(x-4)^2}{25} + \dfrac{(y+2)^2}{9} = 1$.

23.   The equation is $\dfrac{y^2}{16} - \dfrac{x^2}{25} = 1$.

Chapter 11
Sequences and Series

Getting Ready for Chapter 11

1. Evaluating when $x = 1, 2, 3, 4$:

$$2(1) + 3 = 2 + 3 = 5$$
$$2(2) + 3 = 4 + 3 = 7$$
$$2(3) + 3 = 6 + 3 = 9$$
$$2(4) + 3 = 8 + 3 = 11$$

2. Evaluating when $x = 1, 2, 3, 4$:

$$4(1) - 1 = 4 - 1 = 3$$
$$4(2) - 1 = 8 - 1 = 7$$
$$4(3) - 1 = 12 - 1 = 11$$
$$4(4) - 1 = 16 - 1 = 15$$

3. Evaluating when $x = 1, 2, 3, 4$:

$$\frac{1}{1+3} = \frac{1}{4}$$
$$\frac{2}{2+3} = \frac{2}{5}$$
$$\frac{3}{3+3} = \frac{3}{6} = \frac{1}{2}$$
$$\frac{4}{4+3} = \frac{4}{7}$$

4. Evaluating when $x = 1, 2, 3, 4$:

$$\frac{1+1}{1+2} = \frac{2}{3}$$
$$\frac{2+1}{2+2} = \frac{3}{4}$$
$$\frac{3+1}{3+2} = \frac{4}{5}$$
$$\frac{4+1}{4+2} = \frac{5}{6}$$

5. Evaluating when $x = 1, 2, 3, 4$:

$$(-2)^1 = -2$$
$$(-2)^2 = 4$$
$$(-2)^3 = -8$$
$$(-2)^4 = 16$$

6. Evaluating when $x = 1, 2, 3, 4$:

$$(-3)^1 = -3$$
$$(-3)^2 = 9$$
$$(-3)^3 = -27$$
$$(-3)^4 = 81$$

7. Simplifying: $\dfrac{2\sqrt{2}}{2} = \sqrt{2}$

8. Simplifying: $\dfrac{2}{\sqrt{2}} \cdot \dfrac{\sqrt{2}}{\sqrt{2}} = \dfrac{2\sqrt{2}}{2} = \sqrt{2}$

9. Simplifying: $3 \cdot 3^{x-1} = 3^x$

10. Simplifying: $5 \cdot 5^{x-1} = 5^x$

11. Simplifying: $\dfrac{xy^7}{xy^3} = y^{7-3} = y^4$

12. Simplifying: $\dfrac{ab^5}{ab^3} = b^{5-3} = b^2$

13. Simplifying: $\dfrac{xy^{-6}}{xy^{-3}} = y^{-6+3} = y^{-3} = \dfrac{1}{y^3}$

14. Simplifying: $\dfrac{1}{b^9}$ is simplified

351

15.   Expanding: $(a+b)^2 = (a+b)(a+b) = a^2 + 2ab + b^2$

16.   Expanding: $(a-b)^2 = (a-b)(a-b) = a^2 - 2ab + b^2$

17.   Expanding: $(a+b)^3 = (a+b)(a+b)^2 = (a+b)(a^2 + 2ab + b^2) = a^3 + 3a^2b + 3ab^2 + b^3$

18.   Expanding: $(a-b)^3 = (a-b)(a-b)^2 = (a-b)(a^2 - 2ab + b^2) = a^3 - 3a^2b + 3ab^2 - b^3$

11.1 Sequences

1.   The first five terms are: $4, 7, 10, 13, 16$

3.   The first five terms are: $3, 7, 11, 15, 19$

5.   The first five terms are: $1, 2, 3, 4, 5$

7.   The first five terms are: $4, 7, 12, 19, 28$

9.   The first five terms are: $\dfrac{1}{4}, \dfrac{2}{5}, \dfrac{1}{2}, \dfrac{4}{7}, \dfrac{5}{8}$

11.   The first five terms are: $1, \dfrac{1}{4}, \dfrac{1}{9}, \dfrac{1}{16}, \dfrac{1}{25}$

13.   The first five terms are: $2, 4, 8, 16, 32$

15.   The first five terms are: $2, \dfrac{3}{2}, \dfrac{4}{3}, \dfrac{5}{4}, \dfrac{6}{5}$

17.   The first five terms are: $-2, 4, -8, 16, -32$

19.   The first five terms are: $3, 5, 3, 5, 3$

21.   The first five terms are: $1, -\dfrac{2}{3}, \dfrac{3}{5}, -\dfrac{4}{7}, \dfrac{5}{9}$

23.   The first five terms are: $\dfrac{1}{2}, 1, \dfrac{9}{8}, 1, \dfrac{25}{32}$

25.   The first five terms are: $3, -9, 27, -81, 243$

27.   The first five terms are: $1, 5, 13, 29, 61$

29.   The first five terms are: $2, 3, 5, 9, 17$

31.   The first five terms are: $5, 11, 29, 83, 245$

33.   The first five terms are: $4, 4, 4, 4, 4$

35.   The general term is: $a_n = 4n$

37.   The general term is: $a_n = n^2$

39.   The general term is: $a_n = 2^{n+1}$

41.   The general term is: $a_n = \dfrac{1}{2^{n+1}}$

43.   The general term is: $a_n = 3n + 2$

45.   The general term is: $a_n = -4n + 2$

47.   The general term is: $a_n = (-2)^{n-1}$

49.   The general term is: $a_n = \log_{n+1}(n+2)$

51.   **a.**   The sequence of salaries is: 28000, 29120, 30284.80, $31,496.19$, 32756.04

   b.   The general term is: $a_n = 28,000(1.04)^{n-1}$

53.   **a.**   The sequence of values is: 16 ft, 48 ft, 80 ft, 112 ft, 144 ft

   b.   The sum of the values is 400 feet.   **c.**   No, since the sum is less than 420 feet.

55.   **a.**   The sequence of angles is: $180°, 360°, 540°, 720°$ **b.**   Substituting $n = 20$: $a_{20} = 180°(20 - 2) = 3,240°$

   c.   The sum of the interior angles would be $0°$.

57.   Solving for x:

$$\log_9 x = \frac{3}{2}$$
$$x = 9^{3/2} = 27$$

59.   Simplifying: $\log_2 32 = \log_2 2^5 = 5$

61.   Simplifying: $\log_3 [\log_2 8] = \log_3 [\log_2 2^3] = \log_3 3 = 1$

63.   Simplifying: $-2 + 6 + 4 + 22 = 30$

65.   Simplifying: $-8 + 16 - 32 + 64 = 40$

67.   Simplifying: $(1 - 3) + (4 - 3) + (9 - 3) + (16 - 3) = -2 + 1 + 6 + 13 = 18$

69.   Simplifying: $-\dfrac{1}{3} + \dfrac{1}{9} - \dfrac{1}{27} + \dfrac{1}{81} = -\dfrac{27}{81} + \dfrac{9}{81} - \dfrac{3}{81} + \dfrac{1}{81} = -\dfrac{20}{81}$

71.   Simplifying: $\dfrac{1}{3} + \dfrac{1}{2} + \dfrac{3}{5} + \dfrac{2}{3} = \dfrac{10}{30} + \dfrac{15}{30} + \dfrac{18}{30} + \dfrac{20}{30} = \dfrac{63}{30} = \dfrac{21}{10}$

73.   Computing the values: $a_{100} \approx 2.7048, a_{1,000} \approx 2.7169, a_{10,000} \approx 2.7181, a_{100,000} \approx 2.7183$

75.   The first ten terms are: $1, 1, 2, 3, 5, 8, 13, 21, 34, 55$

11.2 Series

1. Expanding the sum: $\displaystyle\sum_{i=1}^{4}(2i+4)=6+8+10+12=36$

3. Expanding the sum: $\displaystyle\sum_{i=2}^{3}(i^2-1)=3+8=11$

5. Expanding the sum: $\displaystyle\sum_{i=1}^{4}(i^2-3)=-2+1+6+13=18$

7. Expanding the sum: $\displaystyle\sum_{i=1}^{4}\frac{i}{1+i}=\frac{1}{2}+\frac{2}{3}+\frac{3}{4}+\frac{4}{5}=\frac{30}{60}+\frac{40}{60}+\frac{45}{60}+\frac{48}{60}=\frac{163}{60}$

9. Expanding the sum: $\displaystyle\sum_{i=1}^{4}(-3)^i=-3+9-27+81=60$

11. Expanding the sum: $\displaystyle\sum_{i=3}^{6}(-2)^i=-8+16-32+64=40$

13. Expanding the sum: $\displaystyle\sum_{i=2}^{6}(-2)^i=4-8+16-32+64=44$

15. Expanding the sum: $\displaystyle\sum_{i=1}^{5}\left(-\frac{1}{2}\right)^i=-\frac{1}{2}+\frac{1}{4}-\frac{1}{8}+\frac{1}{16}-\frac{1}{32}=-\frac{16}{32}+\frac{8}{32}-\frac{4}{32}+\frac{2}{32}-\frac{1}{32}=-\frac{11}{32}$

17. Expanding the sum: $\displaystyle\sum_{i=2}^{5}\frac{i-1}{i+1}=\frac{1}{3}+\frac{1}{2}+\frac{3}{5}+\frac{2}{3}=1+\frac{5}{10}+\frac{6}{10}=\frac{21}{10}$

19. Expanding the sum: $\displaystyle\sum_{i=1}^{5}(x+i)=(x+1)+(x+2)+(x+3)+(x+4)+(x+5)=5x+15$

21. Expanding the sum: $\displaystyle\sum_{i=1}^{4}(x-2)^i=(x-2)+(x-2)^2+(x-2)^3+(x-2)^4$

23. Expanding the sum: $\displaystyle\sum_{i=1}^{5}\frac{x+i}{x-1}=\frac{x+1}{x-1}+\frac{x+2}{x-1}+\frac{x+3}{x-1}+\frac{x+4}{x-1}+\frac{x+5}{x-1}$

25. Expanding the sum: $\displaystyle\sum_{i=3}^{8}(x+i)^i=(x+3)^3+(x+4)^4+(x+5)^5+(x+6)^6+(x+7)^7+(x+8)^8$

27. Expanding the sum: $\displaystyle\sum_{i=3}^{6}(x-2i)^{i+3}=(x-6)^6+(x-8)^7+(x-10)^8+(x-12)^9$

29. Writing with summation notation: $\displaystyle 2+4+8+16=\sum_{i=1}^{4}2^i$

31. Writing with summation notation: $\displaystyle 4+8+16+32+64=\sum_{i=2}^{6}2^i$

33. Writing with summation notation: $\displaystyle 5+9+13+17+21=\sum_{i=1}^{5}(4i+1)$

35. Writing with summation notation: $\displaystyle -4+8-16+32=\sum_{i=2}^{5}-(-2)^i$

37. Writing with summation notation: $\displaystyle\frac{3}{4}+\frac{4}{5}+\frac{5}{6}+\frac{6}{7}+\frac{7}{8}=\sum_{i=3}^{7}\frac{i}{i+1}$

39. Writing with summation notation: $\displaystyle\frac{1}{3}+\frac{2}{5}+\frac{3}{7}+\frac{4}{9}=\sum_{i=1}^{4}\frac{i}{2i+1}$

41. Writing with summation notation: $\displaystyle(x-2)^6+(x-2)^7+(x-2)^8+(x-2)^9=\sum_{i=6}^{9}(x-2)^i$

43. Writing with summation notation: $\left(1+\dfrac{1}{x}\right)^2 + \left(1+\dfrac{2}{x}\right)^3 + \left(1+\dfrac{3}{x}\right)^4 + \left(1+\dfrac{4}{x}\right)^5 = \displaystyle\sum_{i=1}^{4}\left(1+\dfrac{i}{x}\right)^{i+1}$

45. Writing with summation notation: $\dfrac{x}{x+3} + \dfrac{x}{x+4} + \dfrac{x}{x+5} = \displaystyle\sum_{i=3}^{5}\dfrac{x}{x+i}$

47. Writing with summation notation: $x^2(x+2) + x^3(x+3) + x^4(x+4) = \displaystyle\sum_{i=2}^{4} x^i(x+i)$

49. **a.** Writing as a series: $\dfrac{1}{3} = 0.3 + 0.03 + 0.003 + 0.0003 + \ldots$

   b. Writing as a series: $\dfrac{2}{9} = 0.2 + 0.02 + 0.002 + 0.0002 + \ldots$

   c. Writing as a series: $\dfrac{3}{11} = 0.27 + 0.0027 + 0.000027 + \ldots$

51. The sequence of values he falls is: 16, 48, 80, 112, 144, 176, 208
During the seventh second he falls 208 feet, and the total he falls is 784 feet.

53. **a.** The series is $16 + 48 + 80 + 112 + 144$.   **b.** Writing in summation notation: $\displaystyle\sum_{i=1}^{5}(32i - 16)$

55. Expanding the logarithm: $\log_2 x^3 y = \log_2 x^3 + \log_2 y = 3\log_2 x + \log_2 y$

57. Expanding the logarithm: $\log_{10}\dfrac{\sqrt[3]{x}}{y^2} = \log_{10} x^{1/3} - \log_{10} y^2 = \dfrac{1}{3}\log_{10} x - 2\log_{10} y$

59. Writing as a single logarithm: $\log_{10} x - \log_{10} y^2 = \log_{10}\dfrac{x}{y^2}$

61. Writing as a single logarithm: $2\log_3 x - 3\log_3 y - 4\log_3 z = \log_3 x^2 - \log_3 y^3 - \log_3 z^4 = \log_3 \dfrac{x^2}{y^3 z^4}$

63. Solving the equation:

$$\log_4 x - \log_4 5 = 2$$
$$\log_4 \dfrac{x}{5} = 2$$
$$\dfrac{x}{5} = 4^2$$
$$\dfrac{x}{5} = 16$$
$$x = 80$$

The solution is 80.

65. Solving the equation:

$$\log_2 x + \log_2 (x-7) = 3$$
$$\log_2 (x^2 - 7x) = 3$$
$$x^2 - 7x = 2^3$$
$$x^2 - 7x = 8$$
$$x^2 - 7x - 8 = 0$$
$$(x+1)(x-8) = 0$$
$$x = -1, 8$$

The solution is 8 (–1 does not check).

67. Simplifying: $2 + 9(8) = 2 + 72 = 74$

69. Simplifying: $\dfrac{10}{2}\left(\dfrac{1}{2} + 5\right) = 5\left(\dfrac{1}{2} + \dfrac{10}{2}\right) = 5\left(\dfrac{11}{2}\right) = \dfrac{55}{2}$

71. Simplifying: $3 + (n-1)(2) = 3 + 2n - 2 = 2n + 1$

73. Solving the equation for x:

$$\sum_{i=1}^{4}(x - i) = 16$$
$$(x-1) + (x-2) + (x-3) + (x-4) = 16$$
$$4x - 10 = 16$$
$$4x = 26$$
$$x = \dfrac{13}{2}$$

11.3 Arithmetic Sequences

1. The sequence is arithmetic: $d = 1$

5. The sequence is arithmetic: $d = -5$

9. The sequence is arithmetic: $d = \dfrac{2}{3}$

3. The sequence is not arithmetic.

7. The sequence is not arithmetic.

11. Finding the general term: $a_n = 3 + (n-1) \cdot 4 = 3 + 4n - 4 = 4n - 1$. Therefore: $a_{24} = 4 \cdot 24 - 1 = 96 - 1 = 95$

13. Finding the required term: $a_{10} = 6 + (10 - 1) \cdot (-2) = 6 - 18 = -12$. Finding the sum: $S_{10} = \dfrac{10}{2}(6 - 12) = 5(-6) = -30$

15. Writing out the equations:
$$a_6 = a_1 + 5d \qquad a_{12} = a_1 + 11d$$
$$17 = a_1 + 5d \qquad 29 = a_1 + 11d$$
The system of equations is:
$$a_1 + 11d = 29$$
$$a_1 + 5d = 17$$
Subtracting yields:
$$6d = 12$$
$$d = 2$$
$$a_1 = 7$$
Finding the required term: $a_{30} = 7 + 29 \cdot 2 = 7 + 58 = 65$

17. Writing out the equations:
$$a_3 = a_1 + 2d \qquad a_8 = a_1 + 7d$$
$$16 = a_1 + 2d \qquad 26 = a_1 + 7d$$
The system of equations is:
$$a_1 + 7d = 26$$
$$a_1 + 2d = 16$$
Subtracting yields:
$$5d = 10$$
$$d = 2$$
$$a_1 = 12$$
Finding the required term: $a_{20} = 12 + 19 \cdot 2 = 12 + 38 = 50$. Finding the sum: $S_{20} = \dfrac{20}{2}(12 + 50) = 10 \cdot 62 = 620$

19. Finding the required term: $a_{20} = 3 + 19 \cdot 4 = 3 + 76 = 79$. Finding the sum: $S_{20} = \dfrac{20}{2}(3 + 79) = 10 \cdot 82 = 820$

21. Writing out the equations:
$$a_4 = a_1 + 3d \qquad a_{10} = a_1 + 9d$$
$$14 = a_1 + 3d \qquad 32 = a_1 + 9d$$
The system of equations is:
$$a_1 + 9d = 32$$
$$a_1 + 3d = 14$$
Subtracting yields:
$$6d = 18$$
$$d = 3$$
$$a_1 = 5$$
Finding the required term: $a_{40} = 5 + 39 \cdot 3 = 5 + 117 = 122$. Finding the sum: $S_{40} = \dfrac{40}{2}(5 + 122) = 20 \cdot 127 = 2540$

23.   Using the summation formula:

$$S_6 = \frac{6}{2}(a_1 + a_6)$$

$$-12 = 3(a_1 - 17)$$

$$a_1 - 17 = -4$$

$$a_1 = 13$$

Now find d:

$$a_6 = 13 + 5 \bullet d$$

$$-17 = 13 + 5d$$

$$5d = -30$$

$$d = -6$$

25.   Using $a_1 = 14$ and $d = -3$: $a_{85} = 14 + 84 \bullet (-3) = 14 - 252 = -238$

27.   Using the summation formula:

$$S_{20} = \frac{20}{2}(a_1 + a_{20})$$

$$80 = 10(-4 + a_{20})$$

$$-4 + a_{20} = 8$$

$$a_{20} = 12$$

Now finding d:

$$a_{20} = a_1 + 19d$$

$$12 = -4 + 19d$$

$$16 = 19d$$

$$d = \frac{16}{19}$$

Finding the required term: $a_{39} = -4 + 38\left(\dfrac{16}{19}\right) = -4 + 32 = 28$

29.   Using $a_1 = 5$ and $d = 4$: $a_{100} = a_1 + 99d = 5 + 99 \bullet 4 = 5 + 396 = 401$

Now finding the required sum: $S_{100} = \dfrac{100}{2}(5 + 401) = 50 \bullet 406 = 20,300$

31.   Using $a_1 = 12$ and $d = -5$: $a_{35} = a_1 + 34d = 12 + 34 \bullet (-5) = 12 - 170 = -158$

33.   Using $a_1 = \dfrac{1}{2}$ and $d = \dfrac{1}{2}$: $a_{10} = a_1 + 9d = \dfrac{1}{2} + 9 \bullet \dfrac{1}{2} = \dfrac{10}{2} = 5$. Finding the sum: $S_{10} = \dfrac{10}{2}\left(\dfrac{1}{2} + 5\right) = \dfrac{10}{2} \bullet \dfrac{11}{2} = \dfrac{55}{2}$

35.   a.   The first five terms are: $18,000, $14,700, $11,400, $8,100, $4,800
       b.   The common difference is –$3,300.
       c.   Constructing a line graph:

d. The value is approximately $9,750.

e. The recursive formula is: $a_0 = 18000; a_n = a_{n-1} - 3300$ for $n \geq 1$

37. a. The sequence of values is: 1500 ft, 1460 ft, 1420 ft, 1380 ft, 1340 ft, 1300 ft

b. It is arithmetic because the same amount is subtracted from each succeeding term.

c. The general term is: $a_n = 1500 + (n-1) \cdot (-40) = 1500 - 40n + 40 = 1540 - 40n$

39. a. The first 15 triangular numbers is: 1,3,6,10,15,21,28,36,45,55,66,78,91,105,120

b. The recursive formula is: $a_1 = 1; a_n = n + a_{n-1}$ for $n \geq 2$

c. It is not arithmetic because the same amount is not added to each term.

41. Finding the logarithm: $\log 576 \approx 2.7604$

43. Finding the logarithm: $\log 0.0576 \approx -1.2396$

45. Solving for x:
$$\log x = 2.6484$$
$$x = 10^{2.6484} \approx 445$$

47. Solving for x:
$$\log x = -7.3516$$
$$x = 10^{-7.3516} \approx 4.45 \times 10^{-8}$$

49. Simplifying: $\dfrac{1}{8}\left(\dfrac{1}{2}\right) = \dfrac{1}{16}$

51. Simplifying: $\dfrac{3\sqrt{3}}{3} = \sqrt{3}$

53. Simplifying: $2 \cdot 2^{n-1} = 2^{1+n-1} = 2^n$

55. Simplifying: $\dfrac{ar^6}{ar^3} = r^{6-3} = r^3$

57. Simplifying: $\dfrac{\frac{1}{5}}{1-\frac{1}{2}} = \dfrac{\frac{1}{5}}{1-\frac{1}{2}} \cdot \dfrac{10}{10} = \dfrac{2}{10-5} = \dfrac{2}{5}$

59. Simplifying: $\dfrac{-3\left[(-2)^8 - 1\right]}{-2-1} = \dfrac{-3(256-1)}{-2-1} = \dfrac{-3(255)}{-3} = 255$

61. Using the summation formulas:

$$S_7 = \frac{7}{2}(a_1 + a_7)$$
$$147 = \frac{7}{2}(a_1 + a_7)$$
$$a_1 + a_7 = 42$$
$$a_1 + (a_1 + 6d) = 42$$
$$2a_1 + 6d = 42$$
$$a_1 + 3d = 21$$

$$S_{13} = \frac{13}{2}(a_1 + a_{13})$$
$$429 = \frac{13}{2}(a_1 + a_{13})$$
$$a_1 + a_{13} = 66$$
$$a_1 + (a_1 + 12d) = 66$$
$$2a_1 + 12d = 66$$
$$a_1 + 6d = 33$$

So we have the system of equations:
$$a_1 + 6d = 33$$
$$a_1 + 3d = 21$$

Subtracting:
$$3d = 12$$
$$d = 4$$
$$a_1 = 33 - 6 \cdot 4 = 9$$

63. Solving the equation:
$$a_{15} - a_7 = -24$$
$$(a_1 + 14d) - (a_1 + 6d) = -24$$
$$a_1 + 14d - a_1 - 6d = -24$$
$$8d = -24$$
$$d = -3$$

358 Chapter 11 Sequences and Series

65. Using $a_1 = 1$ and $d = \sqrt{2}$, find the term: $a_{50} = a_1 + 49d = 1 + 49\sqrt{2}$

Finding the sum: $S_{50} = \dfrac{50}{2}(a_1 + a_{50}) = 25\left(1 + 1 + 49\sqrt{2}\right) = 50 + 1225\sqrt{2}$

11.4 Geometric Sequences

1. The sequence is geometric: $r = 5$

3. The sequence is geometric: $r = \dfrac{1}{3}$

5. The sequence is not geometric.

7. The sequence is geometric: $r = -2$

9. The sequence is not geometric.

11. Finding the general term: $a_n = 4 \cdot 3^{n-1}$

13. Finding the term: $a_6 = -2\left(-\dfrac{1}{2}\right)^{6-1} = -2\left(-\dfrac{1}{2}\right)^5 = -2\left(-\dfrac{1}{32}\right) = \dfrac{1}{16}$

15. Finding the term: $a_{20} = 3(-1)^{20-1} = 3(-1)^{19} = -3$

17. Finding the sum: $S_{10} = \dfrac{10\left(2^{10} - 1\right)}{2 - 1} = 10 \cdot 1023 = 10,230$

19. Finding the sum: $S_{20} = \dfrac{1\left((-1)^{20} - 1\right)}{-1 - 1} = \dfrac{1 \cdot 0}{-2} = 0$

21. Using $a_1 = \dfrac{1}{5}$ and $r = \dfrac{1}{2}$, the term is: $a_8 = \dfrac{1}{5} \cdot \left(\dfrac{1}{2}\right)^{8-1} = \dfrac{1}{5} \cdot \left(\dfrac{1}{2}\right)^7 = \dfrac{1}{5} \cdot \dfrac{1}{128} = \dfrac{1}{640}$

23. Using $a_1 = -\dfrac{1}{2}$ and $r = \dfrac{1}{2}$, the sum is: $S_5 = \dfrac{-\dfrac{1}{2}\left(\left(\dfrac{1}{2}\right)^5 - 1\right)}{\dfrac{1}{2} - 1} = \dfrac{-\dfrac{1}{2}\left(\dfrac{1}{32} - 1\right)}{-\dfrac{1}{2}} = \dfrac{1}{32} - 1 = -\dfrac{31}{32}$

25. Using $a_1 = \sqrt{2}$ and $r = \sqrt{2}$, the term is: $a_{10} = \sqrt{2}\left(\sqrt{2}\right)^9 = \left(\sqrt{2}\right)^{10} = 2^5 = 32$

The sum is: $S_{10} = \dfrac{\sqrt{2}\left(\left(\sqrt{2}\right)^{10} - 1\right)}{\sqrt{2} - 1} = \dfrac{\sqrt{2}(32 - 1)}{\sqrt{2} - 1} = \dfrac{31\sqrt{2}}{\sqrt{2} - 1} \cdot \dfrac{\sqrt{2} + 1}{\sqrt{2} + 1} = \dfrac{62 + 31\sqrt{2}}{2 - 1} = 62 + 31\sqrt{2}$

27. Using $a_1 = 100$ and $r = 0.1$, the term is: $a_6 = 100(0.1)^5 = 10^2\left(10^{-5}\right) = 10^{-3} = \dfrac{1}{1000}$

The sum is: $S_6 = \dfrac{100\left((0.1)^6 - 1\right)}{0.1 - 1} = \dfrac{100\left(10^{-6} - 1\right)}{-0.9} = \dfrac{-99.9999}{-0.9} = 111.111$

29. Since $a_4 \cdot r \cdot r = a_6$, we have the equation:

$a_4 r^2 = a_6$

$40r^2 = 160$

$r^2 = 4$

$r = \pm 2$

31. Since $a_1 = -3$ and $r = -2$, the values are:

$a_8 = -3(-2)^7 = -3(-128) = 384$

$S_8 = \dfrac{-3\left((-2)^8 - 1\right)}{-2 - 1} = \dfrac{-3(256 - 1)}{-3} = 255$

33. Since $a_7 \cdot r \cdot r \cdot r = a_{10}$, we have the equation:

$a_7 r^3 = a_{10}$

$13r^3 = 104$

$r^3 = 8$

$r = 2$

© 2012 Cengage Learning. All Rights Reserved. May not be scanned, copied or duplicated, or posted to a publicly accessible website, in whole or in part.

35. Using $a_1 = \dfrac{1}{2}$ and $r = \dfrac{1}{2}$ in the sum formula: $S = \dfrac{\dfrac{1}{2}}{1-\dfrac{1}{2}} = \dfrac{\dfrac{1}{2}}{\dfrac{1}{2}} = 1$

37. Using $a_1 = 4$ and $r = \dfrac{1}{2}$ in the sum formula: $S = \dfrac{4}{1-\dfrac{1}{2}} = \dfrac{4}{\dfrac{1}{2}} = 8$

39. Using $a_1 = 2$ and $r = \dfrac{1}{2}$ in the sum formula: $S = \dfrac{2}{1-\dfrac{1}{2}} = \dfrac{2}{\dfrac{1}{2}} = 4$

41. Using $a_1 = \dfrac{4}{3}$ and $r = -\dfrac{1}{2}$ in the sum formula: $S = \dfrac{\dfrac{4}{3}}{1+\dfrac{1}{2}} = \dfrac{\dfrac{4}{3}}{\dfrac{3}{2}} = \dfrac{4}{3} \cdot \dfrac{2}{3} = \dfrac{8}{9}$

43. Using $a_1 = \dfrac{2}{5}$ and $r = \dfrac{2}{5}$ in the sum formula: $S = \dfrac{\dfrac{2}{5}}{1-\dfrac{2}{5}} = \dfrac{\dfrac{2}{5}}{\dfrac{3}{5}} = \dfrac{2}{5} \cdot \dfrac{5}{3} = \dfrac{2}{3}$

45. Using $a_1 = \dfrac{3}{4}$ and $r = \dfrac{1}{3}$ in the sum formula: $S = \dfrac{\dfrac{3}{4}}{1-\dfrac{1}{3}} = \dfrac{\dfrac{3}{4}}{\dfrac{2}{3}} = \dfrac{3}{4} \cdot \dfrac{3}{2} = \dfrac{9}{8}$

47. Interpreting the decimal as an infinite sum with $a_1 = 0.4$ and $r = 0.1$: $S = \dfrac{0.4}{1-0.1} = \dfrac{0.4}{0.9} \cdot \dfrac{10}{10} = \dfrac{4}{9}$

49. Interpreting the decimal as an infinite sum with $a_1 = 0.27$ and $r = 0.01$: $S = \dfrac{0.27}{1-0.01} = \dfrac{0.27}{0.99} \cdot \dfrac{100}{100} = \dfrac{27}{99} = \dfrac{3}{11}$

51.   **a.** The first five terms are: \$450,000, \$315,000, \$220,500, \$154,350, \$108,045
   b. The common ratio is 0.7.
   c. Constructing a line graph:

   d. The value is approximately \$90,000.
   e. The recursive formula is: $a_0 = 450000; a_n = 0.7a_{n-1}$ for $n \geq 1$

**53.   a.   **Using $a_1 = \dfrac{1}{3}$ and $r = \dfrac{1}{3}$ in the sum formula: $S = \dfrac{\frac{1}{3}}{1 - \frac{1}{3}} = \dfrac{\frac{1}{3}}{\frac{2}{3}} = \dfrac{1}{2}$

**b.   **Finding the sum: $S_6 = \dfrac{\frac{1}{3}\left(\left(\frac{1}{3}\right)^6 - 1\right)}{\frac{1}{3} - 1} = \dfrac{\frac{1}{3}\left(\frac{1}{729} - 1\right)}{-\frac{2}{3}} = \dfrac{\frac{1}{3}\left(-\frac{728}{729}\right)}{-\frac{2}{3}} = -\dfrac{1}{2}\left(-\dfrac{728}{729}\right) = \dfrac{364}{729}$

**c.   **Finding the difference of these two answers: $S - S_6 = \dfrac{1}{2} - \dfrac{364}{729} = \dfrac{729}{1458} - \dfrac{728}{1458} = \dfrac{1}{1458}$

**55.   **There are two infinite series for these heights. For the amount the ball falls, use $a_1 = 20$ and $r = \dfrac{7}{8}$:

$$S_{\text{fall}} = \dfrac{20}{1 - \frac{7}{8}} = \dfrac{20}{\frac{1}{8}} = 160$$

For the amount the ball rises, use $a_1 = \dfrac{7}{8}(20) = \dfrac{35}{2}$ and $r = \dfrac{7}{8}$: $S_{\text{rise}} = \dfrac{\frac{35}{2}}{1 - \frac{7}{8}} = \dfrac{\frac{35}{2}}{\frac{1}{8}} = \dfrac{35}{2} \cdot 8 = 140$

The total distance traveled is then 160 feet + 140 feet = 300 feet

**57.   a.   **The general term is: $a_n = 15\left(\dfrac{4}{5}\right)^{n-1}$ **b.   **Finding the sum: $S = \dfrac{15}{1 - \frac{4}{5}} = \dfrac{15}{\frac{1}{5}} = 75$ feet

**59.   **Evaluating the logarithm: $\log_4 20 = \dfrac{\log 20}{\log 4} \approx 2.16$ **61.   **Evaluating the logarithm: $\ln 576 \approx 6.36$

**63.   **Solving for t:

$$A = 10e^{5t}$$
$$\dfrac{A}{10} = e^{5t}$$
$$5t = \ln\left(\dfrac{A}{10}\right)$$
$$t = \dfrac{1}{5}\ln\left(\dfrac{A}{10}\right)$$

**65.   **Expanding: $(x + y)^0 = 1$ **67.   **Expanding: $(x + y)^2 = (x + y)(x + y) = x^2 + 2xy + y^2$

**69.   **Simplifying: $\dfrac{6 \cdot 5 \cdot 4 \cdot 3 \cdot 2 \cdot 1}{(2 \cdot 1)(4 \cdot 3 \cdot 2 \cdot 1)} = \dfrac{6 \cdot 5}{2 \cdot 1} = \dfrac{30}{2} = 15$

**71.   **Interpreting the decimal as an infinite sum with $a_1 = 0.63$ and $r = 0.01$: $S = \dfrac{0.63}{1 - 0.01} = \dfrac{0.63}{0.99} = \dfrac{63}{99} = \dfrac{7}{11}$

**73.   **Finding the common ratio:

$$S = \dfrac{a_1}{1 - r}$$
$$6 = \dfrac{4}{1 - r}$$
$$6 - 6r = 4$$
$$-6r = -2$$
$$r = \dfrac{1}{3}$$

75. a. The areas are: stage 2: $\dfrac{3}{4}$; stage 3: $\dfrac{9}{16}$; stage 4: $\dfrac{27}{64}$

b. They form a geometric sequence.

c. The area is 0, since the sequence of areas is approaching 0.

d. The perimeters form an increasing sequence.

11.5 The Binomial Expansion

1. Using the binomial formula:

$$(x+2)^4 = \binom{4}{0}x^4 + \binom{4}{1}x^3(2) + \binom{4}{2}x^2(2)^2 + \binom{4}{3}x(2)^3 + \binom{4}{4}(2)^4$$
$$= x^4 + 4\cdot 2x^3 + 6\cdot 4x^2 + 4\cdot 8x + 16$$
$$= x^4 + 8x^3 + 24x^2 + 32x + 16$$

3. Using the binomial formula:

$$(x+y)^6 = \binom{6}{0}x^6 + \binom{6}{1}x^5 y + \binom{6}{2}x^4 y^2 + \binom{6}{3}x^3 y^3 + \binom{6}{4}x^2 y^4 + \binom{6}{5}xy^5 + \binom{6}{6}y^6$$
$$= x^6 + 6x^5 y + 15x^4 y^2 + 20x^3 y^3 + 15x^2 y^4 + 6xy^5 + y^6$$

5. Using the binomial formula:

$$(2x+1)^5 = \binom{5}{0}(2x)^5 + \binom{5}{1}(2x)^4(1) + \binom{5}{2}(2x)^3(1)^2 + \binom{5}{3}(2x)^2(1)^3 + \binom{5}{4}(2x)(1)^4 + \binom{5}{5}(1)^5$$
$$= 32x^5 + 5\cdot 16x^4 + 10\cdot 8x^3 + 10\cdot 4x^2 + 5\cdot 2x + 1$$
$$= 32x^5 + 80x^4 + 80x^3 + 40x^2 + 10x + 1$$

7. Using the binomial formula:

$$(x-2y)^5 = \binom{5}{0}x^5 + \binom{5}{1}x^4(-2y) + \binom{5}{2}x^3(-2y)^2 + \binom{5}{3}x^2(-2y)^3 + \binom{5}{4}x(-2y)^4 + \binom{5}{5}(-2y)^5$$
$$= x^5 - 5\cdot 2x^4 y + 10\cdot 4x^3 y^2 - 10\cdot 8x^2 y^3 + 5\cdot 16xy^4 - 32y^5$$
$$= x^5 - 10x^4 y + 40x^3 y^2 - 80x^2 y^3 + 80xy^4 - 32y^5$$

9. Using the binomial formula:

$$(3x-2)^4 = \binom{4}{0}(3x)^4 + \binom{4}{1}(3x)^3(-2) + \binom{4}{2}(3x)^2(-2)^2 + \binom{4}{3}(3x)(-2)^3 + \binom{4}{4}(-2)^4$$
$$= 81x^4 - 4\cdot 54x^3 + 6\cdot 36x^2 - 4\cdot 24x + 16$$
$$= 81x^4 - 216x^3 + 216x^2 - 96x + 16$$

11. Using the binomial formula:

$$(4x-3y)^3 = \binom{3}{0}(4x)^3 + \binom{3}{1}(4x)^2(-3y) + \binom{3}{2}(4x)(-3y)^2 + \binom{3}{3}(-3y)^3$$
$$= 64x^3 - 3\cdot 48x^2 y + 3\cdot 36xy^2 - 27y^3$$
$$= 64x^3 - 144x^2 y + 108xy^2 - 27y^3$$

13. Using the binomial formula:

$$(x^2+2)^4 = \binom{4}{0}(x^2)^4 + \binom{4}{1}(x^2)^3(2) + \binom{4}{2}(x^2)^2(2)^2 + \binom{4}{3}(x^2)(2)^3 + \binom{4}{4}(2)^4$$
$$= x^8 + 4\cdot 2x^6 + 6\cdot 4x^4 + 4\cdot 8x^2 + 16$$
$$= x^8 + 8x^6 + 24x^4 + 32x^2 + 16$$

15.   Using the binomial formula:

$$\left(x^2 + y^2\right)^3 = \binom{3}{0}\left(x^2\right)^3 + \binom{3}{1}\left(x^2\right)^2\left(y^2\right) + \binom{3}{2}\left(x^2\right)\left(y^2\right)^2 + \binom{3}{3}\left(y^2\right)^3 = x^6 + 3x^4 y^2 + 3x^2 y^4 + y^6$$

17.   Using the binomial formula:

$$(2x + 3y)^4 = \binom{4}{0}(2x)^4 + \binom{4}{1}(2x)^3(3y) + \binom{4}{2}(2x)^2(3y)^2 + \binom{4}{3}(2x)(3y)^3 + \binom{4}{4}(3y)^4$$

$$= 16x^4 + 4 \bullet 24x^3 y + 6 \bullet 36x^2 y^2 + 4 \bullet 54xy^3 + 81y^4$$

$$= 16x^4 + 96x^3 y + 216x^2 y^2 + 216xy^3 + 81y^4$$

19.   Using the binomial formula:

$$\left(\frac{x}{2} + \frac{y}{3}\right)^3 = \binom{3}{0}\left(\frac{x}{2}\right)^3 + \binom{3}{1}\left(\frac{x}{2}\right)^2\left(\frac{y}{3}\right) + \binom{3}{2}\left(\frac{x}{2}\right)\left(\frac{y}{3}\right)^2 + \binom{3}{3}\left(\frac{y}{3}\right)^3$$

$$= \frac{x^3}{8} + 3 \bullet \frac{x^2 y}{12} + 3 \bullet \frac{xy^2}{18} + \frac{y^3}{27}$$

$$= \frac{x^3}{8} + \frac{x^2 y}{4} + \frac{xy^2}{6} + \frac{y^3}{27}$$

21.   Using the binomial formula:

$$\left(\frac{x}{2} - 4\right)^3 = \binom{3}{0}\left(\frac{x}{2}\right)^3 + \binom{3}{1}\left(\frac{x}{2}\right)^2(-4) + \binom{3}{2}\left(\frac{x}{2}\right)(-4)^2 + \binom{3}{3}(-4)^3$$

$$= \frac{x^3}{8} - 3 \bullet x^2 + 3 \bullet 8x - 64$$

$$= \frac{x^3}{8} - 3x^2 + 24x - 64$$

23.   Using the binomial formula:

$$\left(\frac{x}{3} + \frac{y}{2}\right)^4 = \binom{4}{0}\left(\frac{x}{3}\right)^4 + \binom{4}{1}\left(\frac{x}{3}\right)^3\left(\frac{y}{2}\right) + \binom{4}{2}\left(\frac{x}{3}\right)^2\left(\frac{y}{2}\right)^2 + \binom{4}{3}\left(\frac{x}{3}\right)\left(\frac{y}{2}\right)^3 + \binom{4}{4}\left(\frac{y}{2}\right)^4$$

$$= \frac{x^4}{81} + 4 \bullet \frac{x^3 y}{54} + 6 \bullet \frac{x^2 y^2}{36} + 4 \bullet \frac{xy^3}{24} + \frac{y^4}{16}$$

$$= \frac{x^4}{81} + \frac{2x^3 y}{27} + \frac{x^2 y^2}{6} + \frac{xy^3}{6} + \frac{y^4}{16}$$

25.   Writing the first four terms:

$$\binom{9}{0}x^9 + \binom{9}{1}x^8(2) + \binom{9}{2}x^7(2)^2 + \binom{9}{3}x^6(2)^3$$

$$= x^9 + 9 \bullet 2x^8 + 36 \bullet 4x^7 + 84 \bullet 8x^6$$

$$= x^9 + 18x^8 + 144x^7 + 672x^6$$

27.   Writing the first four terms:

$$\binom{10}{0}x^{10} + \binom{10}{1}x^9(-y) + \binom{10}{2}x^8(-y)^2 + \binom{10}{3}x^7(-y)^3 = x^{10} - 10x^9 y + 45x^8 y^2 - 120x^7 y^3$$

29.   Writing the first four terms:

$$\binom{25}{0}x^{25} + \binom{25}{1}x^{24}(3) + \binom{25}{2}x^{23}(3)^2 + \binom{25}{3}x^{22}(3)^3$$

$$= x^{25} + 25 \bullet 3x^{24} + 300 \bullet 9x^{23} + 2300 \bullet 27x^{22}$$

$$= x^{25} + 75x^{24} + 2700x^{23} + 62100x^{22}$$

31. Writing the first four terms:

$$\binom{60}{0}x^{60} + \binom{60}{1}x^{59}(-2) + \binom{60}{2}x^{58}(-2)^2 + \binom{60}{3}x^{57}(-2)^3$$

$$= x^{60} - 60 \cdot 2x^{59} + 1770 \cdot 4x^{58} - 34220 \cdot 8x^{57}$$

$$= x^{60} - 120x^{59} + 7080x^{58} - 273760x^{57}$$

33. Writing the first four terms:

$$\binom{18}{0}x^{18} + \binom{18}{1}x^{17}(-y) + \binom{18}{2}x^{16}(-y)^2 + \binom{18}{3}x^{15}(-y)^3 = x^{18} - 18x^{17}y + 153x^{16}y^2 - 816x^{15}y^3$$

35. Writing the first three terms: $\binom{15}{0}x^{15} + \binom{15}{1}x^{14}(1) + \binom{15}{2}x^{13}(1)^2 = x^{15} + 15x^{14} + 105x^{13}$

37. Writing the first three terms: $\binom{12}{0}x^{12} + \binom{12}{1}x^{11}(-y) + \binom{12}{2}x^{10}(-y)^2 = x^{12} - 12x^{11}y + 66x^{10}y^2$

39. Writing the first three terms:

$$\binom{20}{0}x^{20} + \binom{20}{1}x^{19}(2) + \binom{20}{2}x^{18}(2)^2 = x^{20} + 20 \cdot 2x^{19} + 190 \cdot 4x^{18} = x^{20} + 40x^{19} + 760x^{18}$$

41. Writing the first two terms: $\binom{100}{0}x^{100} + \binom{100}{1}x^{99}(2) = x^{100} + 100 \cdot 2x^{99} = x^{100} + 200x^{99}$

43. Writing the first two terms: $\binom{50}{0}x^{50} + \binom{50}{1}x^{49}y = x^{50} + 50x^{49}y$

45. Finding the required term: $\binom{12}{8}(2x)^4(3y)^8 = 495 \cdot 2^4 \cdot 3^8 x^4 y^8 = 51,963,120x^4y^8$

47. Finding the required term: $\binom{10}{4}x^6(-2)^4 = 210 \cdot 16x^6 = 3,360x^6$

49. Finding the required term: $\binom{12}{5}x^7(-2)^5 = -792 \cdot 32x^7 = -25,344x^7$

51. Finding the required term: $\binom{25}{2}x^{23}(-3y)^2 = 300 \cdot 9x^{23}y^2 = 2,700x^{23}y^2$

53. Finding the required term: $\binom{20}{11}(2x)^9(5y)^{11} = \dfrac{20!}{11!9!}(2x)^9(5y)^{11}$

55. Writing the first three terms:

$$\binom{10}{0}(x^2y)^{10} + \binom{10}{1}(x^2y)^9(-3) + \binom{10}{2}(x^2y)^8(-3)^2$$

$$= x^{20}y^{10} - 10 \cdot 3x^{18}y^9 + 45 \cdot 9x^{16}y^8$$

$$= x^{20}y^{10} - 30x^{18}y^9 + 405x^{16}y^8$$

57. Finding the third term: $\binom{7}{2}\left(\dfrac{1}{2}\right)^5\left(\dfrac{1}{2}\right)^2 = 21 \cdot \dfrac{1}{128} = \dfrac{21}{128}$

59. Solving the equation:

$$5^x = 7$$

$$\log 5^x = \log 7$$

$$x \log 5 = \log 7$$

$$x = \frac{\log 7}{\log 5} \approx 1.21$$

61. Solving the equation:

$$8^{2x+1} = 16$$

$$2^{6x+3} = 2^4$$

$$6x + 3 = 4$$

$$6x = 1$$

$$x = \frac{1}{6} \approx 0.17$$

63. Using the compound interest formula:

$$400\left(1+\frac{0.10}{4}\right)^{4t} = 800$$

$$(1.025)^{4t} = 2$$

$$\ln(1.025)^{4t} = \ln 2$$

$$4t\ln 1.025 = \ln 2$$

$$t = \frac{\ln 2}{4\ln 1.025} \approx 7.02$$

It will take 7.02 years.

65. They both equal 56.

67. They both equal 125,970.

69. Smaller triangles (Serpinski triangles) begin to emerge.

Chapter 11 Review

1. Writing the first four terms: 7,9,11,13

3. Writing the first four terms: 0,3,8,15

5. Writing the first four terms: 4,16,64,256

7. The general term is: $a_n = 3n - 1$

9. The general term is: $a_n = n^4$

11. The general term is: $a_n = \left(\frac{1}{2}\right)^n = 2^{-n}$

13. Expanding the sum: $\sum_{i=1}^{4}(2i+3) = 5+7+9+11 = 32$

15. Expanding the sum: $\sum_{i=2}^{3}\frac{i^2}{i+2} = 1 + \frac{9}{5} = \frac{14}{5}$

17. Expanding the sum: $\sum_{i=3}^{5}(4i+i^2) = 21+32+45 = 98$

19. Writing in summation notation: $\sum_{i=1}^{4} 3i$

21. Writing in summation notation: $\sum_{i=1}^{5}(2i+3)$

23. Writing in summation notation: $\sum_{i=1}^{4}\frac{1}{i+2}$

25. Writing in summation notation: $\sum_{i=1}^{3}(x-2i)$

27. The sequence is geometric.

29. The sequence is arithmetic.

31. The sequence is geometric.

33. The sequence is arithmetic.

35. Finding the general term: $a_n = 2+(n-1)3 = 2+3n-3 = 3n-1$. Now finding a_{20}: $a_{20} = 3(20)-1 = 59$

37. Finding a_{10}: $a_{10} = -2+9\cdot4 = 34$. Now finding S_{10}: $S_{10} = \frac{10}{2}(-2+34) = 5\cdot32 = 160$

39. First write the equations:

$$a_5 = a_1 + 4d \qquad a_8 = a_1 + 7d$$
$$21 = a_1 + 4d \qquad 33 = a_1 + 7d$$

We have the system of equations:

$$a_1 + 7d = 33$$
$$a_1 + 4d = 21$$

Subtracting yields:

$$3d = 12$$
$$d = 4$$
$$a_1 = 33 - 28 = 5$$

Now finding a_{10}: $a_{10} = 5 + 9\cdot4 = 41$

41. First write the equations:

$$a_4 = a_1 + 3d \qquad a_8 = a_1 + 7d$$
$$-10 = a_1 + 3d \qquad -18 = a_1 + 7d$$

We have the system of equations:

$$a_1 + 7d = -18$$
$$a_1 + 3d = -10$$

Subtracting yields:

$$4d = -8$$
$$d = -2$$
$$a_1 = -18 + 14 = -4$$

Now finding a_{20}: $a_{20} = -4 + 19 \cdot (-2) = -42$. Finding S_{20}: $S_{20} = \frac{20}{2}(-4 - 42) = 10 \cdot (-46) = -460$

43. Using $a_1 = 100$ and $d = -5$: $a_{40} = 100 + 39 \cdot (-5) = 100 - 195 = -95$

45. The general term is: $a_n = 5(-2)^{n-1}$. Now finding a_{16}: $a_{16} = 5(-2)^{15} = -163,840$

47. Finding the sum: $S = \dfrac{-2}{1 - \dfrac{1}{3}} = \dfrac{-2}{\dfrac{2}{3}} = -3$

49. Since $a_3 \cdot r = a_4$, we have:

$$12r = 24$$
$$r = 2$$

Finding the first term:

$$a_3 = a_1 r^2$$
$$12 = a_1 \cdot 2^2$$
$$12 = 4a_1$$
$$a_1 = 3$$

Now finding a_6: $a_6 = a_1 r^5 = 3 \cdot 2^5 = 96$

51. Using the binomial formula:

$$(x - 2)^4 = \binom{4}{0}x^4 + \binom{4}{1}x^3(-2) + \binom{4}{2}x^2(-2)^2 + \binom{4}{3}x(-2)^3 + \binom{4}{4}(-2)^4$$
$$= x^4 - 4 \cdot 2x^3 + 6 \cdot 4x^2 - 4 \cdot 8x + 16$$
$$= x^4 - 8x^3 + 24x^2 - 32x + 16$$

53. Using the binomial formula:

$$(3x + 2y)^3 = \binom{3}{0}(3x)^3 + \binom{3}{1}(3x)^2(2y) + \binom{3}{2}(3x)(2y)^2 + \binom{3}{3}(2y)^3$$
$$= 27x^3 + 3 \cdot 18x^2 y + 3 \cdot 12xy^2 + 8y^3$$
$$= 27x^3 + 54x^2 y + 36xy^2 + 8y^3$$

55. Using the binomial formula:

$$\left(\frac{x}{2} + 3\right)^4 = \binom{4}{0}\left(\frac{x}{2}\right)^4 + \binom{4}{1}\left(\frac{x}{2}\right)^3(3) + \binom{4}{2}\left(\frac{x}{2}\right)^2(3)^2 + \binom{4}{3}\left(\frac{x}{2}\right)(3)^3 + \binom{4}{4}(3)^4$$
$$= \frac{1}{16}x^4 + 4 \cdot \frac{3}{8}x^3 + 6 \cdot \frac{9}{4}x^2 + 4 \cdot \frac{27}{2}x + 81$$
$$= \frac{1}{16}x^4 + \frac{3}{2}x^3 + \frac{27}{2}x^2 + 54x + 81$$

57.   Writing the first three terms:

$$\binom{10}{0}x^{10} + \binom{10}{1}x^9(3y) + \binom{10}{2}x^8(3y)^2 = x^{10} + 10 \cdot 3x^9 y + 45 \cdot 9x^8 y^2 = x^{10} + 30x^9 y + 405x^8 y^2$$

59.   Writing the first three terms:

$$\binom{11}{0}x^{11} + \binom{11}{1}x^{10}y + \binom{11}{2}x^9 y^2 = x^{11} + 11x^{10}y + 55x^9 y^2$$

61.   Writing the first two terms: $\binom{16}{0}x^{16} + \binom{16}{1}x^{15}(-2y) = x^{16} - 16 \cdot 2x^{15}y = x^{16} - 32x^{15}y$

63.   Writing the first two terms: $\binom{50}{0}x^{50} + \binom{50}{1}x^{49}(-1) = x^{50} - 50x^{49}$

65.   Finding the sixth term: $\binom{10}{5}x^5(-3)^5 = 252 \cdot (-243)x^5 = -61,236x^5$

Chapter 11 Cumulative Review

1.   Simplifying: $|-2| = 2$

3.   Simplifying: $2^3 + 3(2 + 20 \div 4) = 8 + 3(2 + 5) = 8 + 3(7) = 8 + 21 = 29$

5.   Simplifying: $-5(2x + 3) + 8x = -10x - 15 + 8x = -2x - 15$

7.   Simplifying: $(3y + 2)^2 - (3y - 2)^2 = 9y^2 + 12y + 4 - 9y^2 + 12y - 4 = 24y$

9.   Simplifying: $\dfrac{\frac{1}{5} - \frac{1}{4}}{\frac{1}{2} + \frac{3}{4}} = \dfrac{\frac{1}{5} - \frac{1}{4}}{\frac{1}{2} + \frac{3}{4}} \cdot \dfrac{20}{20} = \dfrac{4 - 5}{10 + 15} = -\dfrac{1}{25}$

11.   Simplifying: $x^{2/3} \cdot x^{1/5} = x^{2/3+1/5} = x^{10/15+3/15} = x^{13/15}$

13.   Simplifying: $\sqrt{48x^5 y^3} = \sqrt{16x^4 y^2 \cdot 3xy} = 4x^2 y\sqrt{3xy}$

15.   Solving the equation:

$$5y - 2 = -3y + 6$$
$$8y = 8$$
$$y = 1$$

17.   Solving the equation:

$$|3x - 1| - 2 = 6$$
$$|3x - 1| = 8$$
$$3x - 1 = -8, 8$$
$$3x = -7, 9$$
$$x = -\frac{7}{3}, 3$$

19.   Solving the equation:

$$2x^2 = 5x + 3$$
$$2x^2 - 5x - 3 = 0$$
$$(2x + 1)(x - 3) = 0$$
$$x = -\frac{1}{2}, 3$$

21.   Solving the equation:

$$\frac{2}{x+1} = \frac{4}{5}$$
$$4x + 4 = 10$$
$$4x = -6$$
$$x = -\frac{3}{2}$$

23.   Solving the equation:

$$4x^2 + 6x = -5$$
$$4x^2 + 6x + 5 = 0$$
$$x = \frac{-6 \pm \sqrt{36 - 80}}{8} = \frac{-6 \pm \sqrt{-44}}{8} = \frac{-6 \pm 2i\sqrt{11}}{8} = \frac{-3 \pm i\sqrt{11}}{4}$$

25. Solving the equation:

$$\log_2 x = 3$$
$$x = 2^3 = 8$$

27. Solving the equation:

$$8^{x+3} = 4$$
$$2^{3x+9} = 2^2$$
$$3x + 9 = 2$$
$$3x = -7$$
$$x = -\frac{7}{3}$$

29. Substituting into the first equation:

$$4x + 2(-3x + 1) = 4$$
$$4x - 6x + 2 = 4$$
$$-2x = 2$$
$$x = -1$$
$$y = 4$$

The solution is $(-1, 4)$.

31. Form the augmented matrix: $\begin{bmatrix} -1 & 2 & -3 & | & -4 \\ 2 & -2 & -3 & | & -1 \\ -1 & 2 & 2 & | & -3 \end{bmatrix}$

Multiply the first row by -1: $\begin{bmatrix} 1 & -2 & 3 & | & 4 \\ 2 & -2 & -3 & | & -1 \\ -1 & 2 & 2 & | & -3 \end{bmatrix}$

Add -2 times row 1 to row 2 and 1 times row 1 to row 3: $\begin{bmatrix} 1 & -2 & 3 & | & 4 \\ 0 & 2 & -9 & | & -9 \\ 0 & 0 & 5 & | & 1 \end{bmatrix}$

Divide row 2 by 2 and row 3 by 5: $\begin{bmatrix} 1 & -2 & 3 & | & 4 \\ 0 & 1 & -\frac{9}{2} & | & -\frac{9}{2} \\ 0 & 0 & 1 & | & \frac{1}{5} \end{bmatrix}$

So $z = \frac{1}{5}$. Substituting to find y:

$$y - \frac{9}{2}\left(\frac{1}{5}\right) = -\frac{9}{2}$$
$$y - \frac{9}{10} = -\frac{9}{2}$$
$$y = -\frac{18}{5}$$

Substituting to find x:

$$x - 2\left(-\frac{18}{5}\right) + 3\left(\frac{1}{5}\right) = 4$$

$$x + \frac{39}{5} = 4$$

$$x = -\frac{19}{5}$$

The solution is $\left(-\frac{19}{5}, -\frac{18}{5}, \frac{1}{5}\right)$.

33. Multiplying: $\left(x^{1/5} + 3\right)\left(x^{1/5} - 3\right) = \left(x^{1/5}\right)^2 - (3)^2 = x^{2/5} - 9$

35. Graphing the line:

37. Graphing the curve:

39. Graphing the curve:

41. Completing the square:

$$x^2 + 4x + y^2 - 6y = 12$$

$$\left(x^2 + 4x + 4\right) + \left(y^2 - 6y + 9\right) = 12 + 4 + 9$$

$$\left(x + 2\right)^2 + \left(y - 3\right)^2 = 25$$

Graphing the circle:

$(x+2)^2 + (y-3)^2 = 25$

43. The sequence is arithmetic, and the general term is $a_n = 11n - 8$.

45. First find the slope: $m = \dfrac{-5+1}{-3+6} = -\dfrac{4}{3}$. Using the point-slope formula:

$$y + 5 = -\frac{4}{3}(x + 3)$$

$$y + 5 = -\frac{4}{3}x - 4$$

$$y = -\frac{4}{3}x - 9$$

47. Using the distance formula: $d = \sqrt{\left(4 + 3\right)^2 + \left(5 - 1\right)^2} = \sqrt{49 + 16} = \sqrt{65}$

49. Let $s = 20$ and solve for t:

$$48t - 16t^2 = 20$$

$$16t^2 - 48t + 20 = 0$$

$$4t^2 - 12t + 5 = 0$$

$$\left(2t - 1\right)\left(2t - 5\right) = 0$$

$$t = \frac{1}{2}, \frac{5}{2}$$

The object will be 20 feet above the ground after $\dfrac{1}{2}$ second and $\dfrac{5}{2}$ seconds.

51. Expanding the sum: $\displaystyle\sum_{i=1}^{5}\left(2i - 1\right) = 1 + 3 + 5 + 7 + 9 = 25$

53. Writing in scientific notation: $164.1 \times 10^6 = 1.641 \times 10^8$ computers

Chapter 11 Test

1. The first five terms are: $-2, 1, 4, 7, 10$

2. The first five terms are: $2, 5, 10, 17, 26$

3. The first five terms are: $2, 16, 54, 128, 250$

4. The first five terms are: $2, \dfrac{3}{4}, \dfrac{4}{9}, \dfrac{5}{16}, \dfrac{6}{25}$

5. The first five terms are: $3, 7, 11, 15, 19$

6. The first five terms are: $4, -8, 16, -32, 64$

7. Writing the general term: $a_n = 4n + 2$

8. Writing the general term: $a_n = 2^{n-1}$

9. Writing the general term: $a_n = \left(\dfrac{1}{2}\right)^n = 2^{-n}$

10. Writing the general term: $a_n = (-3)^n$

11. Expanding the sum: $\displaystyle\sum_{i=1}^{5}(5i + 3) = 8 + 13 + 18 + 23 + 28 = 90$

12. Expanding the sum: $\displaystyle\sum_{i=3}^{5}\left(2^{i-1} - 1\right) = 2^2 + 2^3 + 2^4 = 4 + 8 + 16 = 28$

13. Expanding the sum: $\displaystyle\sum_{i=2}^{6}\left(i^2 + 2i\right) = 8 + 15 + 24 + 35 + 48 = 130$

14. Expanding the sum: $\displaystyle\sum_{i=1}^{3}(x + 2i) = (x + 2) + (x + 4) + (x + 6) = 3x + 12$

15. Writing with summation notation: $4 + 7 + 10 + 13 + 16 = \displaystyle\sum_{i=1}^{5}(3i + 1)$

16. Writing with summation notation: $\dfrac{2}{4} + \dfrac{3}{5} + \dfrac{4}{6} + \dfrac{5}{7} = \displaystyle\sum_{i=1}^{4}\dfrac{i+1}{i+3}$

17. Writing with summation notation: $(x + 4)^5 + (x + 4)^6 + (x + 4)^7 + (x + 4)^8 = \displaystyle\sum_{i=5}^{8}(x + 4)^i$

18. Writing with summation notation: $\dfrac{x}{x-1} + \dfrac{x^2}{x-2} + \dfrac{x^3}{x-3} = \displaystyle\sum_{i=1}^{3}\dfrac{x^i}{x-i}$

19. The sequence is geometric with $r = -5$.

20. This sequence is neither arithmetic nor geometric.

21. This sequence is arithmetic with $d = -\dfrac{1}{2}$.

22. This sequence is arithmetic with $d = 5$.

23. First write the equations:

$$a_5 = a_1 + 4d \qquad a_9 = a_1 + 8d$$
$$11 = a_1 + 4d \qquad 19 = a_1 + 8d$$

We have the system of equations:

$$a_1 + 8d = 19$$
$$a_1 + 4d = 11$$

Subtracting yields:

$$4d = 8$$
$$d = 2$$
$$a_1 = 19 - 8 \cdot 2 = 3$$

24. Since $a_3 \cdot r \cdot r = a_5$, we have:

$$18 \cdot r^2 = 162$$
$$r^2 = 9$$
$$r = \pm 3$$

Since $a_2 \cdot r = a_3$, $a_2 = \pm 6$.

25. Using $a_1 = 5$ and $d = 6$: $a_{10} = 5 + 9 \cdot 6 = 59$. Now finding the sum: $S_{10} = \dfrac{10}{2}(5 + 59) = 5 \cdot 64 = 320$

26. Using $a_1 = 25$ and $d = -5$: $a_{10} = 25 + 9 \cdot (-5) = -20$. Now finding the sum: $S_{10} = \dfrac{10}{2}(25 - 20) = 5 \cdot 5 = 25$

27. Using $a_1 = 3$ and $r = 2$: $S_{50} = \dfrac{3(2^{50} - 1)}{2 - 1} = 3(2^{50} - 1)$

28. Using $a_1 = \dfrac{1}{2}$ and $r = \dfrac{1}{3}$: $S = \dfrac{\frac{1}{2}}{1 - \frac{1}{3}} = \dfrac{\frac{1}{2}}{\frac{2}{3}} = \dfrac{1}{2} \cdot \dfrac{3}{2} = \dfrac{3}{4}$

29. Using the binomial formula:
$$(x - 3)^4 = \binom{4}{0}x^4 + \binom{4}{1}x^3(-3) + \binom{4}{2}x^2(-3)^2 + \binom{4}{3}x(-3)^3 + \binom{4}{4}(-3)^4$$
$$= x^4 - 4 \cdot 3x^3 + 6 \cdot 9x^2 - 4 \cdot 27x + 81$$
$$= x^4 - 12x^3 + 54x^2 - 108x + 81$$

30. Using the binomial formula:
$$(2x - 1)^5 = \binom{5}{0}(2x)^5 + \binom{5}{1}(2x)^4(-1) + \binom{5}{2}(2x)^3(-1)^2 + \binom{5}{3}(2x)^2(-1)^3 + \binom{5}{4}(2x)(-1)^4 + \binom{5}{5}(-1)^5$$
$$= 32x^5 - 5 \cdot 16x^4 + 10 \cdot 8x^3 - 10 \cdot 4x^2 + 5 \cdot 2x - 1$$
$$= 32x^5 - 80x^4 + 80x^3 - 40x^2 + 10x - 1$$

31. Finding the first three terms: $\binom{20}{0}x^{20} + \binom{20}{1}x^{19}(-1) + \binom{20}{2}x^{18}(-1)^2 = x^{20} - 20x^{19} + 190x^{18}$

32. Finding the sixth term: $\binom{8}{5}(2x)^3(-3y)^5 = -56 \cdot 1944x^3y^5 = -108,864x^3y^5$

33. Constructing Pascal's triangle:
```
                1
               1 1
              1 2 1
             1 3 3 1
            1 4 6 4 1
           1 5 10 10 5 1
          1 6 15 20 15 6 1
         1 7 21 35 35 21 7 1
        1 8 28 56 70 56 28 8 1
```

Appendices

Appendix A Introduction to Determinants

1. Evaluating the determinant: $\begin{vmatrix} 1 & 0 \\ 2 & 3 \end{vmatrix} = 1 \cdot 3 - 0 \cdot 2 = 3 - 0 = 3$

3. Evaluating the determinant: $\begin{vmatrix} 2 & 1 \\ 3 & 4 \end{vmatrix} = 2 \cdot 4 - 1 \cdot 3 = 8 - 3 = 5$

5. Evaluating the determinant: $\begin{vmatrix} 0 & 1 \\ 1 & 0 \end{vmatrix} = 0 \cdot 0 - 1 \cdot 1 = 0 - 1 = -1$

7. Evaluating the determinant: $\begin{vmatrix} -3 & 2 \\ 6 & -4 \end{vmatrix} = (-3) \cdot (-4) - 6 \cdot 2 = 12 - 12 = 0$

9. Solving the equation:
$$\begin{vmatrix} 2x & 1 \\ x & 3 \end{vmatrix} = 10$$
$$6x - x = 10$$
$$5x = 10$$
$$x = 2$$

11. Solving the equation:
$$\begin{vmatrix} 1 & 2x \\ 2 & -3x \end{vmatrix} = 21$$
$$-3x - 4x = 21$$
$$-7x = 21$$
$$x = -3$$

13. Solving the equation:
$$\begin{vmatrix} 2x & -4 \\ x & 2 \end{vmatrix} = -16$$
$$2x(2) + 4x = -16$$
$$4x + 4x = -16$$
$$8x = -16$$
$$x = -2$$

15. Solving the equation:
$$\begin{vmatrix} 11x & -7x \\ 3 & -2 \end{vmatrix} = 3$$
$$11x(-2) + 7x(3) = 3$$
$$-22x + 21x = 3$$
$$-x = 3$$
$$x = -3$$

17. Duplicating the first two columns:
$$\begin{vmatrix} 1 & 2 & 0 \\ 0 & 2 & 1 \\ 1 & 1 & 1 \end{vmatrix} \begin{matrix} 1 & 2 \\ 0 & 2 \\ 1 & 1 \end{matrix} = 1 \cdot 2 \cdot 1 + 2 \cdot 1 \cdot 1 + 0 \cdot 0 \cdot 1 - 1 \cdot 2 \cdot 0 - 1 \cdot 1 \cdot 1 - 1 \cdot 0 \cdot 2 = 2 + 2 + 0 - 0 - 1 - 0 = 3$$

19. Duplicating the first two columns:
$$\begin{vmatrix} 1 & 2 & 3 \\ 3 & 2 & 1 \\ 1 & 1 & 1 \end{vmatrix} \begin{matrix} 1 & 2 \\ 3 & 2 \\ 1 & 1 \end{matrix} = 1 \cdot 2 \cdot 1 + 2 \cdot 1 \cdot 1 + 3 \cdot 3 \cdot 1 - 1 \cdot 2 \cdot 3 - 1 \cdot 1 \cdot 1 - 1 \cdot 3 \cdot 2 = 2 + 2 + 9 - 6 - 1 - 6 = 0$$

21. Expanding across the first row:
$$\begin{vmatrix} 0 & 1 & 2 \\ 1 & 0 & 1 \\ -1 & 2 & 0 \end{vmatrix} = 0\begin{vmatrix} 0 & 1 \\ 2 & 0 \end{vmatrix} - 1\begin{vmatrix} 1 & 1 \\ -1 & 0 \end{vmatrix} + 2\begin{vmatrix} 1 & 0 \\ -1 & 2 \end{vmatrix} = 0(0-2) - 1(0+1) + 2(2-0) = 0 - 1 + 4 = 3$$

23. Expanding across the first row:
$$\begin{vmatrix} 3 & 0 & 2 \\ 0 & -1 & -1 \\ 4 & 0 & 0 \end{vmatrix} = 3\begin{vmatrix} -1 & -1 \\ 0 & 0 \end{vmatrix} - 0\begin{vmatrix} 0 & -1 \\ 4 & 0 \end{vmatrix} + 2\begin{vmatrix} 0 & -1 \\ 4 & 0 \end{vmatrix} = 3(0-0) - 0(0+4) + 2(0+4) = 0 - 0 + 8 = 8$$

25. Expanding across the first row: $\begin{vmatrix} 2 & -1 & 0 \\ 1 & 0 & -2 \\ 0 & 1 & 2 \end{vmatrix} = 2\begin{vmatrix} 0 & -2 \\ 1 & 2 \end{vmatrix} + 1\begin{vmatrix} 1 & -2 \\ 0 & 2 \end{vmatrix} + 0\begin{vmatrix} 1 & 0 \\ 0 & 1 \end{vmatrix} = 2(0+2) + 1(2-0) + 0 = 4 + 2 = 6$

27. Expanding across the first row:
$$\begin{vmatrix} 1 & 3 & 7 \\ -2 & 6 & 4 \\ 3 & 7 & -1 \end{vmatrix} = 1\begin{vmatrix} 6 & 4 \\ 7 & -1 \end{vmatrix} - 3\begin{vmatrix} -2 & 4 \\ 3 & -1 \end{vmatrix} + 7\begin{vmatrix} -2 & 6 \\ 3 & 7 \end{vmatrix} = 1(-6-28) - 3(2-12) + 7(-14-18) = -34 + 30 - 224 = -228$$

Appendix B Cramer's Rule

1. First find the determinants:
$$D = \begin{vmatrix} 2 & -3 \\ 4 & -2 \end{vmatrix} = -4 + 12 = 8$$
$$D_x = \begin{vmatrix} 3 & -3 \\ 10 & -2 \end{vmatrix} = -6 + 30 = 24$$
$$D_y = \begin{vmatrix} 2 & 3 \\ 4 & 10 \end{vmatrix} = 20 - 12 = 8$$

   Now use Cramer's rule:
$$x = \frac{D_x}{D} = \frac{24}{8} = 3 \qquad y = \frac{D_y}{D} = \frac{8}{8} = 1$$
   The solution is (3,1).

3. First find the determinants:
$$D = \begin{vmatrix} 5 & -2 \\ -10 & 4 \end{vmatrix} = 20 - 20 = 0$$
$$D_x = \begin{vmatrix} 4 & -2 \\ 1 & 4 \end{vmatrix} = 16 + 2 = 18$$
$$D_y = \begin{vmatrix} 5 & 4 \\ -10 & 1 \end{vmatrix} = 5 + 40 = 45$$

   Since $D = 0$ and other determinants are nonzero, there is no solution, or $\varnothing$.

5. First find the determinants:

$$D = \begin{vmatrix} 4 & -7 \\ 5 & 2 \end{vmatrix} = 8 + 35 = 43$$

$$D_x = \begin{vmatrix} 3 & -7 \\ -3 & 2 \end{vmatrix} = 6 - 21 = -15$$

$$D_y = \begin{vmatrix} 4 & 3 \\ 5 & -3 \end{vmatrix} = -12 - 15 = -27$$

Now use Cramer's rule:

$$x = \frac{D_x}{D} = -\frac{15}{43} \qquad y = \frac{D_y}{D} = -\frac{27}{43}$$

The solution is $\left(-\frac{15}{43}, -\frac{27}{53} \right)$.

7. First find the determinants:

$$D = \begin{vmatrix} 9 & -8 \\ 2 & 3 \end{vmatrix} = 27 + 16 = 43$$

$$D_x = \begin{vmatrix} 4 & -8 \\ 6 & 3 \end{vmatrix} = 12 + 48 = 60$$

$$D_y = \begin{vmatrix} 9 & 4 \\ 2 & 6 \end{vmatrix} = 54 - 8 = 46$$

Now use Cramer's rule:

$$x = \frac{D_x}{D} = \frac{60}{43} \qquad y = \frac{D_y}{D} = \frac{46}{43}$$

The solution is $\left(\frac{60}{43}, \frac{46}{43} \right)$.

9. First find the determinants:

$$D = \begin{vmatrix} 1 & 1 & 1 \\ 1 & -1 & -1 \\ 2 & 2 & -1 \end{vmatrix} = 1 \begin{vmatrix} -1 & -1 \\ 2 & -1 \end{vmatrix} - 1 \begin{vmatrix} 1 & -1 \\ 2 & -1 \end{vmatrix} + 1 \begin{vmatrix} 1 & -1 \\ 2 & 2 \end{vmatrix} = 3 - 1 + 4 = 6$$

$$D_x = \begin{vmatrix} 4 & 1 & 1 \\ 2 & -1 & -1 \\ 2 & 2 & -1 \end{vmatrix} = 4 \begin{vmatrix} -1 & -1 \\ 2 & -1 \end{vmatrix} - 1 \begin{vmatrix} 2 & -1 \\ 2 & -1 \end{vmatrix} + 1 \begin{vmatrix} 2 & -1 \\ 2 & 2 \end{vmatrix} - 12 - 0 + 6 = 18$$

$$D_y = \begin{vmatrix} 1 & 4 & 1 \\ 1 & 2 & -1 \\ 2 & 2 & -1 \end{vmatrix} = 1 \begin{vmatrix} 2 & -1 \\ 2 & -1 \end{vmatrix} - 4 \begin{vmatrix} 1 & -1 \\ 2 & -1 \end{vmatrix} + 1 \begin{vmatrix} 1 & 2 \\ 2 & 2 \end{vmatrix} = 0 - 4 - 2 = -6$$

$$D_z = \begin{vmatrix} 1 & 1 & 4 \\ 1 & -1 & 2 \\ 2 & 2 & 2 \end{vmatrix} = 1 \begin{vmatrix} -1 & 2 \\ 2 & 2 \end{vmatrix} - 1 \begin{vmatrix} 1 & 2 \\ 2 & 2 \end{vmatrix} + 4 \begin{vmatrix} 1 & -1 \\ 2 & 2 \end{vmatrix} = -6 + 2 + 16 = 12$$

Now use Cramer's rule:

$$x = \frac{D_x}{D} = \frac{18}{6} = 3 \qquad y = \frac{D_y}{D} = \frac{-6}{6} = -1 \qquad z = \frac{D_z}{D} = \frac{12}{6} = 2$$

The solution is $(3, -1, 2)$.

11. First find the determinants:

$$D = \begin{vmatrix} 1 & 1 & -1 \\ -1 & 1 & 1 \\ 1 & 1 & 1 \end{vmatrix} = 1 \begin{vmatrix} 1 & 1 \\ 1 & 1 \end{vmatrix} - 1 \begin{vmatrix} -1 & 1 \\ 1 & 1 \end{vmatrix} - 1 \begin{vmatrix} -1 & 1 \\ 1 & 1 \end{vmatrix} = 0 + 2 + 2 = 4$$

$$D_x = \begin{vmatrix} 2 & 1 & -1 \\ 3 & 1 & 1 \\ 4 & 1 & 1 \end{vmatrix} = 2 \begin{vmatrix} 1 & 1 \\ 1 & 1 \end{vmatrix} - 1 \begin{vmatrix} 3 & 1 \\ 4 & 1 \end{vmatrix} - 1 \begin{vmatrix} 3 & 1 \\ 4 & 1 \end{vmatrix} = 0 + 1 + 1 = 2$$

$$D_y = \begin{vmatrix} 1 & 2 & -1 \\ -1 & 3 & 1 \\ 1 & 4 & 1 \end{vmatrix} = 1 \begin{vmatrix} 3 & 1 \\ 4 & 1 \end{vmatrix} - 2 \begin{vmatrix} -1 & 1 \\ 1 & 1 \end{vmatrix} - 1 \begin{vmatrix} -1 & 3 \\ 1 & 4 \end{vmatrix} = -1 + 4 + 7 = 10$$

$$D_z = \begin{vmatrix} 1 & 1 & 2 \\ -1 & 1 & 3 \\ 1 & 1 & 4 \end{vmatrix} = 1 \begin{vmatrix} 1 & 3 \\ 1 & 4 \end{vmatrix} - 1 \begin{vmatrix} -1 & 3 \\ 1 & 4 \end{vmatrix} + 2 \begin{vmatrix} -1 & 1 \\ 1 & 1 \end{vmatrix} = 1 + 7 - 4 = 4$$

Now use Cramer's rule:

$$x = \frac{D_x}{D} = \frac{2}{4} = \frac{1}{2} \qquad y = \frac{D_y}{D} = \frac{10}{4} = \frac{5}{2} \qquad z = \frac{D_z}{D} = \frac{4}{4} = 1$$

The solution is $\left(\dfrac{1}{2}, \dfrac{5}{2}, 1\right)$.

13. First find the determinants:

$$D = \begin{vmatrix} 3 & -1 & 2 \\ 6 & -2 & 4 \\ 1 & -5 & 2 \end{vmatrix} = 3\begin{vmatrix} -2 & 4 \\ -5 & 2 \end{vmatrix} + 1\begin{vmatrix} 6 & 4 \\ 1 & 2 \end{vmatrix} + 2\begin{vmatrix} 6 & -2 \\ 1 & -5 \end{vmatrix} = 48 + 8 - 56 = 0$$

$$D_x = \begin{vmatrix} 4 & -1 & 2 \\ 8 & -2 & 4 \\ 1 & -5 & 2 \end{vmatrix} = 4\begin{vmatrix} -2 & 4 \\ -5 & 2 \end{vmatrix} + 1\begin{vmatrix} 8 & 4 \\ 1 & 2 \end{vmatrix} + 2\begin{vmatrix} 8 & -2 \\ 1 & -5 \end{vmatrix} = 64 + 12 - 76 = 0$$

$$D_y = \begin{vmatrix} 3 & 4 & 2 \\ 6 & 8 & 4 \\ 1 & 1 & 2 \end{vmatrix} = 3\begin{vmatrix} 8 & 4 \\ 1 & 2 \end{vmatrix} - 4\begin{vmatrix} 6 & 4 \\ 1 & 2 \end{vmatrix} + 2\begin{vmatrix} 6 & 8 \\ 1 & 1 \end{vmatrix} = 36 - 32 - 4 = 0$$

$$D_z = \begin{vmatrix} 3 & -1 & 4 \\ 6 & -2 & 8 \\ 1 & -5 & 1 \end{vmatrix} = 3\begin{vmatrix} -2 & 8 \\ -5 & 1 \end{vmatrix} + 1\begin{vmatrix} 6 & 8 \\ 1 & 1 \end{vmatrix} + 4\begin{vmatrix} 6 & -2 \\ 1 & -5 \end{vmatrix} = 114 - 2 - 112 = 0$$

Since $D = 0$ and the other determinants are also 0, there is no unique solution (dependent).

15. First find the determinants:

$$D = \begin{vmatrix} 2 & -1 & 3 \\ 1 & -5 & -2 \\ -4 & -2 & 1 \end{vmatrix} = 2\begin{vmatrix} -5 & -2 \\ -2 & 1 \end{vmatrix} + 1\begin{vmatrix} 1 & -2 \\ -4 & 1 \end{vmatrix} + 3\begin{vmatrix} 1 & -5 \\ -4 & -2 \end{vmatrix} = -18 - 7 - 66 = -91$$

$$D_x = \begin{vmatrix} 4 & -1 & 3 \\ 1 & -5 & -2 \\ 3 & -2 & 1 \end{vmatrix} = 4\begin{vmatrix} -5 & -2 \\ -2 & 1 \end{vmatrix} + 1\begin{vmatrix} 1 & -2 \\ 3 & 1 \end{vmatrix} + 3\begin{vmatrix} 1 & -5 \\ 3 & -2 \end{vmatrix} = -36 + 7 + 39 = 10$$

$$D_y = \begin{vmatrix} 2 & 4 & 3 \\ 1 & 1 & -2 \\ -4 & 3 & 1 \end{vmatrix} = 2\begin{vmatrix} 1 & -2 \\ 3 & 1 \end{vmatrix} - 4\begin{vmatrix} 1 & -2 \\ -4 & 1 \end{vmatrix} + 3\begin{vmatrix} 1 & 1 \\ -4 & 3 \end{vmatrix} = 14 + 28 + 21 = 63$$

$$D_z = \begin{vmatrix} 2 & -1 & 4 \\ 1 & -5 & 1 \\ -4 & -2 & 3 \end{vmatrix} = 2\begin{vmatrix} -5 & 1 \\ -2 & 3 \end{vmatrix} + 1\begin{vmatrix} 1 & 1 \\ -4 & 3 \end{vmatrix} + 4\begin{vmatrix} 1 & -5 \\ -4 & -2 \end{vmatrix} = -26 + 7 - 88 = -107$$

Now use Cramer's rule:

$$x = \frac{D_x}{D} = -\frac{10}{91} \qquad y = \frac{D_y}{D} = -\frac{63}{91} = -\frac{9}{13} \qquad z = \frac{D_z}{D} = \frac{-107}{-91} = \frac{107}{91}$$

The solution is $\left(-\dfrac{10}{91}, -\dfrac{9}{13}, \dfrac{107}{91}\right)$.

17. First find the determinants:

$$D = \begin{vmatrix} -1 & -7 & 0 \\ 1 & 0 & 3 \\ 0 & 2 & 1 \end{vmatrix} = -1 \begin{vmatrix} 0 & 3 \\ 2 & 1 \end{vmatrix} + 7 \begin{vmatrix} 1 & 3 \\ 0 & 1 \end{vmatrix} + 0 \begin{vmatrix} 1 & 0 \\ 0 & 2 \end{vmatrix} = 6 + 7 + 0 = 13$$

$$D_x = \begin{vmatrix} 1 & -7 & 0 \\ 11 & 0 & 3 \\ 0 & 2 & 1 \end{vmatrix} = 1 \begin{vmatrix} 0 & 3 \\ 2 & 1 \end{vmatrix} + 7 \begin{vmatrix} 11 & 3 \\ 0 & 1 \end{vmatrix} + 0 \begin{vmatrix} 11 & 0 \\ 0 & 2 \end{vmatrix} = -6 + 77 + 0 = 71$$

$$D_y = \begin{vmatrix} -1 & 1 & 0 \\ 1 & 11 & 3 \\ 0 & 0 & 1 \end{vmatrix} = -1 \begin{vmatrix} 11 & 3 \\ 0 & 1 \end{vmatrix} - 1 \begin{vmatrix} 1 & 3 \\ 0 & 1 \end{vmatrix} + 0 \begin{vmatrix} 1 & 11 \\ 0 & 0 \end{vmatrix} = -11 - 1 + 0 = -12$$

$$D_z = \begin{vmatrix} -1 & -7 & 1 \\ 1 & 0 & 11 \\ 0 & 2 & 0 \end{vmatrix} = -1 \begin{vmatrix} 0 & 11 \\ 2 & 0 \end{vmatrix} + 7 \begin{vmatrix} 1 & 11 \\ 0 & 0 \end{vmatrix} + 1 \begin{vmatrix} 1 & 0 \\ 0 & 2 \end{vmatrix} = 22 + 0 + 2 = 24$$

Now use Cramer's rule:

$$x = \frac{D_x}{D} = \frac{71}{13} \qquad y = \frac{D_y}{D} = -\frac{12}{13} \qquad z = \frac{D_z}{D} = \frac{24}{13}$$

The solution is $\left(\dfrac{71}{13}, -\dfrac{12}{13}, \dfrac{24}{13} \right)$.

19. First find the determinants:

$$D = \begin{vmatrix} 1 & -1 & 0 \\ 3 & 0 & 1 \\ 0 & 1 & -2 \end{vmatrix} = 1 \begin{vmatrix} 0 & 1 \\ 1 & -2 \end{vmatrix} + 1 \begin{vmatrix} 3 & 1 \\ 0 & -2 \end{vmatrix} + 0 \begin{vmatrix} 3 & 0 \\ 0 & 1 \end{vmatrix} = -1 - 6 + 0 = -7$$

$$D_x = \begin{vmatrix} 2 & -1 & 0 \\ 11 & 0 & 1 \\ -3 & 1 & -2 \end{vmatrix} = 2 \begin{vmatrix} 0 & 1 \\ 1 & -2 \end{vmatrix} + 1 \begin{vmatrix} 11 & 1 \\ -3 & -2 \end{vmatrix} + 0 \begin{vmatrix} 11 & 0 \\ -3 & 1 \end{vmatrix} = -2 - 19 + 0 = -21$$

$$D_y = \begin{vmatrix} 1 & 2 & 0 \\ 3 & 11 & 1 \\ 0 & -3 & -2 \end{vmatrix} = 1 \begin{vmatrix} 11 & 1 \\ -3 & -2 \end{vmatrix} - 2 \begin{vmatrix} 3 & 1 \\ 0 & -2 \end{vmatrix} + 0 \begin{vmatrix} 3 & 11 \\ 0 & -3 \end{vmatrix} = -19 + 12 + 0 = -7$$

$$D_z = \begin{vmatrix} 1 & -1 & 2 \\ 3 & 0 & 11 \\ 0 & 1 & -3 \end{vmatrix} = 1 \begin{vmatrix} 0 & 11 \\ 1 & -3 \end{vmatrix} + 1 \begin{vmatrix} 3 & 11 \\ 0 & -3 \end{vmatrix} + 2 \begin{vmatrix} 3 & 0 \\ 0 & 1 \end{vmatrix} = -11 - 9 + 6 = -14$$

Now use Cramer's rule:

$$x = \frac{D_x}{D} = \frac{-21}{-7} = 3 \qquad y = \frac{D_y}{D} = \frac{-7}{-7} = 1 \qquad z = \frac{D_z}{D} = \frac{-14}{-7} = 2$$

The solution is $(3,1,2)$.

Appendix C Synthetic Division

1. Using synthetic division:

$$-2\overline{)\begin{array}{ccc} 1 & -5 & 6 \\ & -2 & 14 \\ \hline 1 & -7 & 20 \end{array}}$$

The quotient is $x - 7 + \dfrac{20}{x+2}$.

3. Using synthetic division:

$$1\overline{)\begin{array}{ccc} 3 & -4 & 1 \\ & 3 & -1 \\ \hline 3 & -1 & 0 \end{array}}$$

The quotient is $3x - 1$.

5. Using synthetic division:

$$2\overline{)\begin{array}{cccc} 1 & 2 & 3 & 4 \\ & 2 & 8 & 22 \\ \hline 1 & 4 & 11 & 26 \end{array}}$$

The quotient is $x^2 + 4x + 11 + \dfrac{26}{x-2}$.

7. Using synthetic division:

$$3\overline{)\begin{array}{cccc} 3 & -1 & 2 & 5 \\ & 9 & 24 & 78 \\ \hline 3 & 8 & 26 & 83 \end{array}}$$

The quotient is $3x^2 + 8x + 26 + \dfrac{83}{x-3}$.

9. Using synthetic division:

$$1\overline{)\begin{array}{cccc} 2 & 0 & 1 & -3 \\ & 2 & 2 & 3 \\ \hline 2 & 2 & 3 & 0 \end{array}}$$

The quotient is $2x^2 + 2x + 3$.

11. Using synthetic division:

$$-4\overline{)\begin{array}{ccccc} 1 & 0 & 2 & 0 & 1 \\ & -4 & 16 & -72 & 288 \\ \hline 1 & -4 & 18 & -72 & 289 \end{array}}$$

The quotient is $x^3 - 4x^2 + 18x - 72 + \dfrac{289}{x+4}$.

13. Using synthetic division:

$$2\overline{)\begin{array}{cccccc} 1 & -2 & 1 & -3 & -1 & 1 \\ & 2 & 0 & 2 & -2 & -6 \\ \hline 1 & 0 & 1 & -1 & -3 & -5 \end{array}}$$

The quotient is $x^4 + x^2 - x - 3 - \dfrac{5}{x-2}$.

15. Using synthetic division:

$$1\overline{)\begin{array}{ccc} 1 & 1 & 1 \\ & 1 & 2 \\ \hline 1 & 2 & 3 \end{array}}$$

The quotient is $x + 2 + \dfrac{3}{x-1}$.

17. Using synthetic division:

$$-1\overline{)\begin{array}{ccccc} 1 & 0 & 0 & 0 & -1 \\ & -1 & 1 & -1 & 1 \\ \hline 1 & -1 & 1 & -1 & 0 \end{array}}$$

The quotient is $x^3 - x^2 + x - 1$.

19. Using synthetic division:

$$1\overline{)\begin{array}{cccc} 1 & 0 & 0 & -1 \\ & 1 & 1 & 1 \\ \hline 1 & 1 & 1 & 0 \end{array}}$$

The quotient is $x^2 + x + 1$.

Appendix D Conditional Statements

1. The hypothesis is "you argue for your limitations", the conclusion is "they are yours".
3. The hypothesis is "x is an even number", the conclusion is "x is divisible by 2".
5. The hypothesis is "a triangle is equilateral", the conclusion is "all of its angles are equal".
7. The hypothesis is "$x + 5 = -2$", the conclusion is "$x = -7$".
9. The converse is "If $a^2 = 64$, then $a = 8$", the inverse is "If $a \neq 8$, then $a^2 \neq 64$", and the contrapositive is "If $a^2 \neq 64$, then $a \neq 8$".

11. The converse is "If $a = b$, then $\dfrac{a}{b} = 1$", the inverse is "If $\dfrac{a}{b} \neq 1$, then $a \neq b$", and the

    contrapositive is "If $a \neq b$, then $\dfrac{a}{b} \neq 1$".

13. The converse is "If it is a rectangle, then it is a square", the inverse is "If it is not a square, then it is not a rectangle", and the contrapositive is "If it is not a rectangle, then it is not a square".
15. The converse is "If good is not enough, then better is possible", the inverse is "If better is not possible, then good is enough", and the contrapositive is "If good is enough, then better is not possible".
17. If E, then F.
19. If it is misery, then it loves company.   21. If the wheel is squeaky, then it gets the grease.
23. c   25. a
27. c   29. b